December 2013

Dear Patrick,

mastering the VUCA-challenges in the new ChemCo-world requires new ways of working. Aligning volatility with agile response and variability with balanced utilization indicates the paradigm change in SCM.

Josef Packowski

LEAN Supply Chain Planning

The New Supply Chain Management Paradigm for Process Industries to Master Today's VUCA World

LEAN Supply Chain Planning

The New Supply Chain Management Paradigm for Process Industries to Master Today's VUCA World

Josef Packowski

CRC Press
Taylor & Francis Group
Boca Raton   London   New York

CRC Press is an imprint of the
Taylor & Francis Group, an **informa** business

A PRODUCTIVITY PRESS BOOK

CRC Press
Taylor & Francis Group
6000 Broken Sound Parkway NW, Suite 300
Boca Raton, FL 33487-2742

© 2014 by Taylor & Francis Group, LLC
CRC Press is an imprint of Taylor & Francis Group, an Informa business

No claim to original U.S. Government works

Printed on acid-free paper by CPI Group (UK) Ltd, Croydon, CR0 4YY
Version Date: 20130830

International Standard Book Number-13: 978-1-4822-0533-6 (Hardback)

Library of Congress Cataloging-in-Publication Data

Packowski, Josef.
    LEAN supply chain planning : the new supply chain management paradigm for process industries to master today's vuca world / Josef Packowski.
        pages cm
    Includes bibliographical references and index.
    ISBN 978-1-4822-0533-6 (hardcover)
    1. Business logistics. 2. Lean manufacturing. I. Title.

HD38.5.P33 2014
658.701--dc23   2013031214

Visit the Taylor & Francis Web site at
http://www.taylorandfrancis.com

and the CRC Press Web site at
http://www.crcpress.com

Thanks to all our customers for their

confidence and ongoing trust in working with us

as we strive to motivate them and encourage them to adopt

new ways to solve new challenges in business.

Contents

PART II   How to Design and Build LEAN SCM

PART IV How Your Industry Peers Gained Benefits by LEAN SCM

Introduction: What the Book Is All About

Today, many global supply chains in process industries are neither equipped nor orchestrated to cope effectively with the new VUCA world we are facing. VUCA—volatility, uncertainty, complexity, and ambiguity—is an acronym that originated in the military back in the late 1990s and was quickly adapted to the business environment. It describes precisely the conditions of increasing variability and uncertainty of demand, and the complexity and ambiguity of product portfolios and supply chain networks in which companies operate today.

Facing the threat of increasing VUCA challenges, manufacturers are left grasping for what it means to build a superior supply chain management (SCM) organization that is capable of managing these challenges effectively. Which enablers for agility are required to manage future VUCA dynamics? Those in global network structures (the network footprint) or others in the extended supplier relationship configuration (contract manufacturers, service providers, or suppliers)? Which aspects of today's operational and organizational lean initiatives are delivering tangible cost and efficiency results? How can supply chain organizations sustain reliable supply in an era of ever-widening virtualization of supply networks and increasing exposure to global risk? Finally, where can supply chain managers turn for the answers to these questions?

In response to these challenges, CAMELOT Consulting Group has worked jointly with leading research institutes and key global industry players to come up with a *"New Supply Chain Planning Paradigm"* to face the VUCA challenges in SCM in a new way. The paradigm change in orchestrating supply chains is best explained by laying out a new approach to managing variability, uncertainty, and complexity in today's planning processes and systems.

A few pioneering supply chain organizations in the process industry have already embraced the new way of coordinating and synchronizing their global networks. The reports and industry cases included in this book (see Figure 0.1).

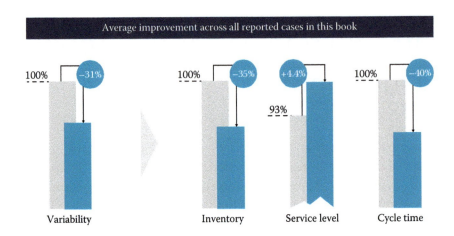

FIGURE 0.1
A step change in variability management improves key supply chain metrics.

Before we move on to present this new supply chain planning approach, we want you to clearly understand the need for a paradigm shift first. In process industries, today's usual supply chain planning practices aim to determine manufacturing decisions up to 12 months prior to delivering actual products to the customers. To do so, planners reach out to their sales and marketing colleagues and ask them for forecasts—preferably as detailed and accurate as possible at the SKU (stock-keeping unit) level. It is obvious that the supply chain performance resulting from such a forecast-based SCM approach is directly linked to the quality of sales forecasts. Therefore, it is understandable that all excellence initiatives in the past have started inevitably by attempting to improve on forecast accuracy, establishing the forecast myth that all activities could be perfectly planned and which still dominates corporate practices. However, ask yourself if we do not all experience difficulty in determining our own personal futures 12 months out, even regarding the subjects we ought to know most about. How then can we expect our sales organizations to know what the future holds for our products in volatile marketplaces at this detailed level of granularity?

So the real issue in SCM is not about improving the accuracy of the sales forecast and reducing the amount of uncertainty in the future, it is rather about *eliminating the need for certainty* in operational planning. We have therefore anchored our LEAN SCM Planning approach in freeing supply chain planners from the need for certainty, ushering in a paradigm change for most planning practices.

A major change that accompanies our *LEAN SCM Planning paradigm* is the management of demand variability. In traditional planning concepts, this is solved in a one-sided way, through planning and scheduling of manufacturing capacities only. This is because in today's supply chain practices, and in the ERP or APS systems that support them, safety stock levels are used as fixed planning parameters and not touched from a planning perspective to buffer variability. This has negative consequences for operational performance and the way in which companies react to demand fluctuations in planning. In this way, the traditional planning approaches represent a conceptual dead-end for today's variability management problems.

Within the new LEAN SCM Planning paradigm, we are *mastering variability with a two-sided approach.* We manage the demand variability in supply chain planning now on both sides, on manufacturing capacities and in inventories. To be more precise, the safety stock elements in all SKU-based inventories are now actively used in planning runs, as they have been designed for, to level replenishment signals and keep market noise out of manufacturing to the extent possible. To make this happen, we have developed a disciplined approach to the *dynamic adaptation of inventory target levels* to changing conditions along the supply chain. This allows SCM to keep a key component of demand variability—demand peaks—out of manufacturing, smoothing capacity utilization, and spending less time resolving production planning and schedule problems. This might sound intuitive, but represents a paradigm shift in the operation of today's planning processes and systems.

The conceptual foundation for managing variability and leveling capacity utilization in local manufacturing sites is the *cyclic scheduling* with "product wheels." Industry experts such as Ian F. Glenday, Peter L. King, and Raymond C. Floyd have already been able to connect the general lean (manufacturing) concepts, and the underlying elements of simplicity, flow and pull, with physical restrictions that are typical in process industries. These concepts have already been influential in many process manufacturing organizations. We have built on these experiences but needed to go further to apply product wheels in a high-product-mix and high-volatility environment—which we named *"Breathing"* and *"High-Mix" Rhythm Wheels.* They are built around optimal product sequences and cycle times. But the most valuable conceptual advancement we have incorporated is our approach to manage variability with two control parameters: the cycle time boundaries. With these new conceptual elements, we are providing appropriate flexibility in manufacturing to enable companies to manage increasing market volatility, and we also hold the key for smoothing variability and volatility propagation upstream along the supply chain in our hands.

The LEAN SCM Planning concepts we present here have been worked out in light of and for the purpose of *end-to-end supply chain synchronization*. So the central question is how to manage multi-echelon synchronization along supply chains in process industries, with typically long lead times starting, for example, with chemical conversion processes and moving downstream to shorter physical bulk production and packaging processes? In particular, how should supply chain organizations apply cyclic planning at manufacturing sites while aiming for real consumption-based pull replenishment?

In response, we have formalized a *"global takt" for synchronization* and achieving end-to-end flow. In a stable supply chain environment, this might seem easy, but not in situations characterized by high demand volatility and high product mixes in manufacturing portfolios. We have to make the Rhythm Wheel approach more flexible, to "breathe" in sync with cycle times, but in a well-structured, disciplined way, within the defined variability control parameters. The key is to "funnel" variability with the Rhythm Wheel cycle time boundaries along the supply chain and in this way actively counteracting the infamous bullwhip effect and achieve a step change in supply chain performance.

With traditional supply chain concepts, the line between planning parameterization (configuration) and the planning run (execution) is blurred. In contrast to this classical planning approach, in LEAN SCM Planning, we have sliced the given planning complexity precisely. We slice the planning task horizontally into global tactical *pre-parameterization* (conditioning) and local planning run areas. Having done so, we have devised a *new LEAN SCM Planning Framework* to better cope with global synchronization needs.

While working with industry pioneers on this new supply chain planning approach, we were confronted almost immediately with additional questions when we stepped into the first implementations:

- How should the organizational model be adapted to the significant change in supply chain planning?
- What are the new roles and responsibilities required in the global supply chain community?

- Which factors should be aligned in corporate performance management to the new planning principles?
- What system gaps can be closed without discarding prior IT (information technology) investments?
- How can this new planning paradigm be implemented to achieve a step change in performance?

To answer these questions, we have consolidated all our conceptual research results and organizational project experience in this book, developed new IT add-on solutions to complement the existing SCM systems for implementation, and given a name to the holistic transformation approach—*LEAN SCM*. This new planning paradigm answers the VUCA challenges in process industries and overcomes the insufficiencies of traditional planning approaches. To highlight the distinction between lean (in small letters)—with its focus on manufacturing objectives—and LEAN—with its focus on end-to-end supply chain synchronization—we coined the all-capitalized term "LEAN" (see Figure 0.2).

Our implementation experience shows that there are three major obstacles to managing a *LEAN SCM transformation* program. First, a company's executive leadership must understand that this is not a single-project initiative, but rather a journey—in other words, sticking to LEAN SCM once the journey has started is crucial for success. Introducing the new paradigm of integrated supply chain planning and variability management requires a *new SCM operating model* with clear end-to-end accountabilities. This will make end-to-end integration possible between, for example, global inventory and local asset management. It is a new way of coordinating and synchronizing operations and throughput in a multi-step value chain. Top management support, training (and incentives) for all stakeholders, and strong commitment to the paradigm change are the preconditions for successful transformation. But bear in mind that you are aiming for nothing less than a step change in supply chain performance.

Second, aligned *performance management* is a critical success factor in the LEAN SCM transformation. The new conceptual elements and the new planning processes require new process performance indicators, such as Rhythm Wheel cycle time attainment and cycle time variation, to be monitored carefully. Therefore, an effectively adapted and well-designed performance management system is fundamental. But this typically does not imply the need to reinvent current performance management systems.

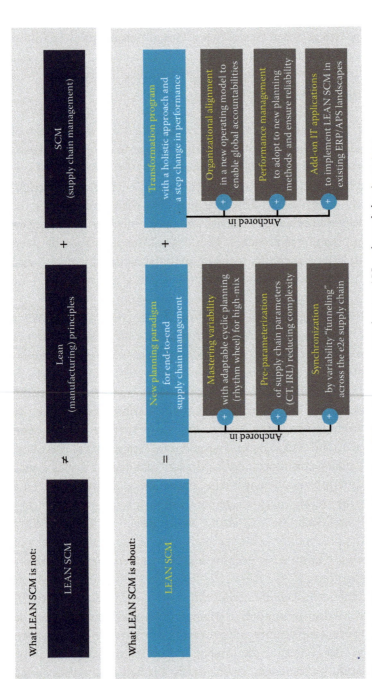

CT = cycle time, IRL = inventory replenishment level, ERP = enterprise resource planning, APS = advanced planning systems.

FIGURE 0.2

What lean SCM and LEAN SCM are about.

We will provide a set of meaningful metrics on the basis of which to generate improved supply chain performance through LEAN SCM. Finally, we depict a pragmatic way of creating the right accountabilities within performance management and show you how to anchor it in your planning organization.

Third, technology is instrumental in helping LEAN SCM create sustainable results. Many lean improvement initiatives depend on few individuals and manual techniques—and if those individuals change positions, much of the planning knowledge, enthusiasm, and leadership are lost. In this light, IT applications are even more critical to capture and standardize processes sustainably in a global end-to-end transformation. These additional *IT technologies* are also supposed to institutionalize LEAN SCM Planning. Applications such as the "Rhythm Wheel Designer" or the "Dynamic Target Stock Planner" provide interlocks with concepts such as cyclic planning and balanced variability management in supply chain organizations, ensuring that common LEAN SCM Planning techniques and best practices have staying power in your SCM organization.

You are holding the results of our LEAN SCM work in your hands right now: it is a holistic practitioner's guide to mastering variability, uncertainty, complexity, and ambiguity in process industry supply chains. It also includes detailed concept descriptions and process explanations. To make it even more practical and valuable for your own reflection, we have enriched all topics with relevant industry cases. We believe that the performance improvements achieved through LEAN SCM initiatives are best described by your industry pioneers themselves. You can therefore also find in this book accounts of how your peers have already lived the LEAN SCM paradigm, used the relevant instruments successfully, and gained:

- Improved customer service and increased supply chain agility through reduced cycle times for Rhythm Wheel-managed products.
- Significant improvements in overall equipment effectiveness (OEE) through leveled and takted material flows that are synchronized to customer demand.
- Significant reductions in working capital through actionable supply chain analytics on variability and risk allocation of stocks across the end-to-end supply network.

I am certain you will enjoy the same outstanding results along your company's supply chain by reading this book and adopting LEAN SCM—because now you are targeting nothing less than a quantum leap in your operations and supply chain performance.

Dr. Josef Packowski
Mannheim, Germany

Reader's Guide

This book will guide your company in undertaking the paradigm shift from traditional planning to LEAN SCM. After learning about and implementing LEAN SCM concepts, your company's supply chain will be able to meet the VUCA challenges of today's global marketplace. You will learn how industry leaders changed their approach to tackle the challenges of variability and uncertainty in supply chain management and how they achieved better customer service at lower cost through LEAN SCM.

In this reader's guide, we provide an overview on

- The contents of this book
- A short summary of each chapter
- A guide to who should read what

What This Book Contains

In this book, we offer you a practitioner's guide to approaching and preparing for all levels of supply chain planning, from the strategic dimension to the daily operational level in light of VUCA challenges. It includes *comprehensive how-to-do-schemes* for flexible Rhythm Wheel planning and *when-to-apply guidelines*. Furthermore, the book contains the argumentation, concepts, and tools related to dynamic inventory target stock-setting and variability buffering across the full range of planning horizons. We describe variability control mechanisms for inventory and capacity that will allow your company to implement all the key elements of the new supply chain synchronization approach.

Designed as a pragmatic and practicable LEAN SCM Planning approach, the concepts we present in this book will enable your company to relieve the real pain points along supply chains. As such, we will show you how your company can achieve an end-to-end LEAN SCM transformation, incorporating all key enablers: organization and stakeholders, accountabilities and performance management, data harmonization, and IT systems.

The book is structured into four parts (see Figure 0.3). In the first part of the book, we focus on the current challenges in the process industries, the insufficiencies of traditional planning approaches, and the ways in which LEAN SCM overcomes these challenges. In the second part, we outline

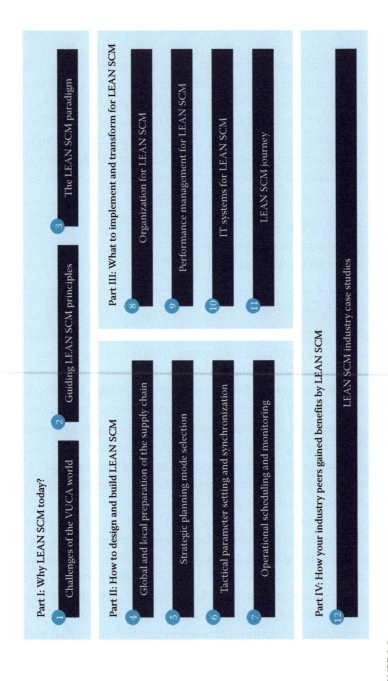

Part I: Why LEAN SCM today?

1 — Challenges of the VUCA world

2 — Guiding LEAN SCM principles

3 — The LEAN SCM paradigm

Part II: How to design and build LEAN SCM

4 — Global and local preparation of the supply chain

5 — Strategic planning mode selection

6 — Tactical parameter setting and synchronization

7 — Operational scheduling and monitoring

Part III: What to implement and transform for LEAN SCM

8 — Organization for LEAN SCM

9 — Performance management for LEAN SCM

10 — IT systems for LEAN SCM

11 — LEAN SCM journey

Part IV: How your industry peers gained benefits by LEAN SCM

12 — LEAN SCM industry case studies

FIGURE 0.3

The book is structured into four parts.

how your company should prepare its supply chain for LEAN SCM from a global and a local perspective. We then explain what planning modes you should choose for your company's supply chains from a strategic perspective, how you can synchronize supply chain operations, and how operational planning and scheduling is conducted with LEAN SCM. In the third part, we draw out the implications of LEAN SCM for your organization, describe its impact on performance management, and explain how IT systems support the entire LEAN SCM process. Furthermore, we present valuable recommendations for your own "LEAN journey." In the fourth part, we conclude the book with selected, useful case studies of thought leader companies that have recently implemented LEAN SCM methods.

Overall, this book contains 12 chapters, which are all readable on their own. Each chapter is summarized to highlight the key takeaways, as are the most important sections within a chapter. Also, the book contains a wealth of useful graphics, as we believe they facilitate the understanding of the contents and serve as quick links when you want to locate key points. Furthermore, we have included a number of text boxes, which contain additional and valuable information with interesting insights from companies of the process industries.

Brief Chapter Overview

Over the course of 12 chapters, we explain the need for a paradigm shift in supply chain planning, the way LEAN SCM and the associated concepts work, and what needs to be changed within your company to successfully implement LEAN SCM.

PART I: WHY LEAN SCM TODAY?

1 Supply Chain Management in Process Industries

In this chapter, we discuss successful supply chain planning as the backbone of modern SCM. As evidence shows, traditional supply chain planning cannot effectively manage the new challenges of the VUCA world, which is an acronym for today's new business reality imposing so much pressure on supply chains. You will see why traditional planning fails and that the naïve transfer of lean methods to process industries cannot solve these challenges.

2 Guiding Principles of LEAN SCM Planning: Facing VUCA Challenges

LEAN SCM principles are introduced to guide your company along its journey to the paradigm change in SCM in process industries. You will learn about LEAN demand, supply, and synchronization principles and what is fundamentally different between LEAN SCM and traditional planning approaches.

3 Fundamentals of LEAN SCM Planning: A Paradigm Shift in Planning

In this chapter, we show how LEAN SCM ensures resilience along the supply chain against the VUCA world, how LEAN SCM works fundamentally, and what benefits you can expect regarding customer service improvements and greater cost efficiency.

PART II: HOW TO DESIGN AND BUILD LEAN SCM

4 Prepare Your Supply Chain for LEAN SCM

This chapter describes what your company should do before implementing LEAN SCM. You will see how to align the supply chain strategy with overall business goals, how to prepare the supply chain from a global perspective, and how to prepare local shop floor operations and material flows for LEAN SCM.

5 Strategic LEAN Supply Chain Planning Configuration

Once your company's supply chain is prepared for the transition to LEAN SCM, you need to analyze and select appropriate planning modes according to the specific needs of your supply chain. This approach has proven successful in the past for distinguishing between production and replenishment modes. In this chapter, you will learn how to select the best-suited production and replenishment modes for your company's supply chains.

6 Tactical LEAN Supply Chain Planning Parameterization

After having chosen the right production and replenishment modes, you will learn in this chapter how to parameterize them, as we describe how to build and configure Rhythm Wheels and how to right-size stocks along the supply chain. Furthermore, you will learn how to synchronize your company's supply chain from an end-to-end perspective.

7 Operational LEAN Supply Chain Planning Execution

After having set the parameters for the supply chain, the next step is planning execution. In this chapter, you will learn how Rhythm Wheels are applied and executed in day-to-day business with dynamic and variable demand, and how the entire planning process is monitored and continuously improved.

PART III: WHAT TO IMPLEMENT AND TRANSFORM FOR LEAN SCM

8 Building an Organization for LEAN SCM

This chapter provides clear guidelines for establishing an effective LEAN supply chain organization. We investigate what needs to be changed "below" and "above" ground in your organization to successfully implement LEAN SCM and how to best manage the "transition."

9 Performance Management for LEAN SCM

In this chapter, you will learn how planning and decision making for LEAN SCM can be supported by performance management, which includes important performance metrics that are unique to LEAN SCM. The chapter also covers the most important aspects you should consider for successfully running a LEAN supply chain in your company.

10 Planning System Landscape for LEAN SCM

To fully seize the benefits of the concepts and processes introduced in this book, IT assumes an essential role. In this chapter, you will find out what needs to change in your company's IT system landscape to sustainably support LEAN end-to-end supply chain planning.

11 The LEAN SCM Journey

Through the first 10 chapters of the book, you will have come to understand all the key LEAN SCM concepts, and the impact of LEAN SCM on your company's processes, the organization, performance management, and IT systems. In this chapter, you will find out how and where you should start your company's journey toward LEAN SCM in order to enjoy its benefits.

PART IV: HOW YOUR INDUSTRY PEERS GAINED BENEFITS BY LEAN SCM

12 LEAN SCM Industry Case Studies

In this chapter, we present key industry insights into how thought leader companies in process industries have moved toward LEAN SCM, as we share the experiences they have had with the new concepts.

Who Should Read What?

In general, anyone interested in or representing your company's supply chain organization should read the entire book to fully understand LEAN SCM. However, we know that time is scarce in the business world, so we have added this section of guidelines that indicate who should read what in the book. Regardless of the role, however, everyone should read the first three chapters, since they explain why a paradigm shift in supply chain planning is required and outline the differences between LEAN SCM and traditional planning and scheduling approaches.

Senior Supply Chain and Operations Managers

If you are a senior supply chain and operations manager, it is important to have a thorough understanding of LEAN SCM and the concepts behind it, in order to set priorities and allocate valuable resources within your company. Senior supply chain and operations managers should therefore especially read:

- Chapter 1—to understand what challenges will arise in process industries and why traditional ways of planning fail to manage these effectively.

- Chapter 2—to figure out the major differences between LEAN SCM and traditional planning approaches.
- Chapter 3—to thoroughly comprehend LEAN SCM and its benefits.
- Chapter 8—to understand the implications of LEAN SCM for the supply chain organization and the corresponding shifts in roles and responsibilities.
- Chapter 11—to know how to set out on the "LEAN SCM journey."
- Chapter 12—to learn how thought leaders in the process industry have successfully raised their companies' supply chain performance with LEAN concepts.

IT Experts for SCM

Next to understanding the rationale and necessity for adopting LEAN SCM that we make clear in Chapters 1 through 3, those in IT who are responsible for SCM should definitely read:

- Chapter 5—to understand LEAN concepts and recognize the new business requirements of LEAN SCM.
- Chapter 10—to understand the impact of LEAN SCM on the current IT planning system landscape.
- Chapter 11—to comprehend the importance of IT support as enabler for the "LEAN journey."
- Chapter 12—to see what impact LEAN SCM has on SCM performance.

Tactical and Operational Planners

Planners on the global and local levels ultimately apply LEAN concepts and methods in everyday planning. They should therefore especially focus on

- Chapter 4—to understand which global and local supply chain conditions to address before undertaking a LEAN SCM implementation.
- Chapter 5—to comprehend the production and replenishment concepts behind LEAN SCM and to select those that are most appropriate for their companies.
- Chapter 6—to optimally set global and local supply chain parameters and synchronize them on an end-to-end basis.
- Chapter 7—to understand how LEAN concepts work in everyday business planning and learn how to interpret LEAN KPIs for continuous improvement.

- Chapter 10—to learn which IT systems support global and local parameter setting for LEAN SCM.

Lean Manufacturing Experts

Experts in lean manufacturing who are already familiar with some LEAN SCM concepts can enhance their knowledge especially by reading:

- Chapter 4—to complement the end-to-end perspective on LEAN SCM.
- Chapter 5—to reflect on and learn about the latest LEAN SCM concepts.
- Chapter 6—to enhance parameter-setting skills for production processes.
- Chapter 7—to understand the operational mechanisms associated with LEAN SCM concepts.
- Chapter 12—to learn about the LEAN tools that industry thought leader companies have implemented.

Business Consultants

To support their clients' efforts to successfully implement LEAN SCM, business consultants should especially read:

- Chapter 4—to ensure that a supply chain is prepared for LEAN SCM prior to implementation.
- Chapter 5—to be up-to-date on developments related to LEAN SCM concepts and to give appropriate recommendations regarding their selection.
- Chapter 6—to guarantee optimal and thorough parameter setting for clients.
- Chapter 7—to enable clients to improve continuously by themselves—without consultants.
- Chapter 9—to anchor LEAN SCM in performance management for staying power.
- Chapter 10—to create sustainable IT solutions for clients.

About the Author and the Motivation for This Book

Dr. Josef Packowski is co-founder and CEO of the CAMELOT Consulting Group, an international organization of leading specialists focused on value chain management in core industries comprising chemical, pharmaceutical, and consumer goods manufacturers. He received his doctoral degree in business and IT from Saarland University, and in addition to his professional work he is today a lecturer on advanced planning systems and supply chain management at the University of Mannheim, one of the leading business schools in Germany. He now actively supports the University in the form of an additional CAMELOT-endowed professorship for SCM, helping to establish a think-tank for modern SCM studies.

Dr. Packowski is a respected industry consultant with over 25 years of experience and a visionary leader in operations management and strategy in process industries. During this time, he has worked for several of CAMELOT's most prominent clients and global industry leaders such as Astellas, AstraZeneca, Bayer, BASF, DSM, Henkel, Lyondell Basell, Merck, Novartis, Roche, Sabic, and others.

However, he has dedicated most of his professional life to advising these companies on how to operate and work more effectively in SCM. Indeed, from the beginning of his academic and professional career, Dr. Packowski has consistently maintained scholarly and professional interest in production and supply chain planning. Starting in the late 1980s, he was privileged to support manufacturing teams in achieving "Class A" certification in MRP II and Sales & Operations Planning, in one of the first initiatives of that kind in the United Kingdom. This was followed by an intensive engagement in an EU-funded research program focusing on the development of *ERP-based "Production Planning for the Process Industry,"* resulting in the SAP R/3 PP-PI application offered by SAP AG in the early 1990s.

Dr. Packowski's subsequent PhD work, applying theoretical operations research methods in planning for process manufacturers, was more about the new "Advanced Planning" concepts that seemed to offer solutions for coping more effectively with typical planning issues in process industries, issues that ERP platforms had failed to solve and which he was experiencing in practice during his parallel consultancy work. Consequently, this brought him in the late 1990s into implementation projects and a close collaboration within his own CAMELOT consulting organization with Tom Baker, a pioneer in the area of *Advanced Planning and Scheduling (APS)* technology. Tom's mission statement, "Bring us the planning problem, we will solve it," was not only very motivating for him, but also raised the bar high. He experienced this challenge in the beginning of the year 2000, when they began implementing the new SAP SCM/APO planning applications with CAMELOT teams for the very first time in pharmaceutical and chemical companies. Today they can look back on more than 100 successful implementation and transformation projects in the industry around the globe.

During his professional career, Dr. Packowski had the opportunity to apply two major technology-driven SCM paradigms for their respective times: the MRP/ERP-based and the SCM/APS-based planning paradigms. In contrast to other business-related technology, the business-driven SCM paradigms have been changing much more frequently in all those years. Depending on the economic ups or downs of the time and the accompanying business objectives, he has had to emphasize either the *lean* or the *agile* SCM paradigm. Today, all these concepts and associated tools are well known and applied in the industry.

The motivation for this work was triggered by the need of a Big PharmaCo to achieve a step change in supply chain performance. That meant in the first place undertaking a multi-echelon synchronization of *lean* takt and *lean*-controlled process manufacturers, with the objective of achieving greater flexibility. Furthermore, the company wanted greater reliability despite increasing demand variability, and it needed to do this without jeopardizing its achievements in cost improvement.

After the first assessments, it became obvious that this focus on SCM had evolved into an area of high interest across the entire industry, with several practitioners' reports and guidelines already available. However, early reviews indicated that the bulk of the relevant publications concentrated on providing an isolated manufacturing view. Also, the first academic literature review revealed that this was a research area that had been given little attention. At that stage, Dr. Packowski became excited at the prospect of engaging with this *new supply chain planning domain* as he leveraged his more than 25 years of professional experience.

For further information, please visit the homepage of CAMELOT Management Consultants AG (www.camelot-mc.com) or write an e-mail to office@camelot-mc.com.

Acknowledgments

The concepts and methods presented in this book have been researched and developed during the past 4 years with great support from leading universities, process industry champions, and supply chain experts at CAMELOT Management Consultants. Although acknowledging the assistance of everyone who contributed to making this happen would be impossible, I would like to acknowledge those individuals and organizations that have provided invaluable suggestions and encouragement throughout this project.

First, I would like to thank *Professor Dr. Thonemann* from the University of Cologne. We kicked off a research program aiming to formalize the multi-echelon supply chain synchronization challenge for process industry networks that apply the concepts of Product Wheels or Rhythm Wheels at local sites. It would have taken me even longer than the last 4 years to complete this book without the dedicated involvement of Diploma and PhD students who contributed challenging ideas regarding various aspects of the new supply chain planning paradigm described herein. Therefore, it is my great pleasure to thank all those academic mentors and project supporters: *Professor Dr. Fleischmann* from the University of Mannheim, *Professor Dr. Briskorn* from the University of Siegen, and finally *Julian Amey* from the University Warwick, UK offered great support and valuable analytical evaluations.

I also wish to acknowledge several courageous industry leaders for their frank discussions and early support when we began transforming the LEAN SCM insights and new ways of working into supply chain practice. In particular, I am grateful to Andy Evans, head of Global Supply Chain Planning at *AstraZeneca*, *Dr. Thomas Proell*, SCM head, and *Tom van Laar*, TechOPS head, at *Novartis Pharma* for their challenging questions and further encouragement. Special thanks go also to *Dr. Robert Blackburn*, president, Information Services and Supply Chain, BASF, who we supported through the global restructuring of his operations and organization. The industry cases collected so far have been the best reality checks for us when transforming intellectual rigor into practicable solutions.

It is a central tenet of SCM that collaboration leads to better insights and results. The same is true for writing, and I gratefully acknowledge the many coworkers who helped to shape this text. My special thanks go to Michail Heinmann, Ernesto Knein, Philipp Streuber, Dr. David Francas, Michael Hamann, Marco Klein, Melanie Lenhardt, Anna Fitzer, and the CAMELOT Innovation Team, who conscientiously and industriously labored in contributing research and conceptual development and helped with the formulation of the new LEAN SCM Planning paradigm. Furthermore, I am grateful for discussions and reality checks with my Camelot IT Lab colleagues Tobias Heckmann, Christoph Habla, and Steffen Joswig, who transferred and implemented the relevant concepts into a unique LEAN SCM Planning Suite, first based on SAP SCM and later on the SAP HANA platform. Without the great support of all the above-mentioned individuals, the concepts, methods, and tools of LEAN SCM would not have risen to the level of maturity they enjoy today.

Finally, no acknowledgment would be complete without extending my sincere and heartfelt thanks to my great children and wonderful partner for their continued patience and understanding, encouragement, and continued love.

Dr. Josef Packowski
Mannheim, Germany

Part I

Why LEAN SCM Today?

1

Supply Chain Management in Process Industries

Process industries represent a key driver of global value creation. Industry segments such as chemicals, pharmaceuticals, food and beverages, and consumer goods comprise more than 50% of industrial production in the United States and Europe. Their products guarantee nutrition and health for the individual consumer and form the basis of virtually all products in our daily lives. However, a recent issue of *Supply Chain Management Review* clearly showed that companies in process industries are experiencing dramatic challenges in fulfilling their business targets and downward pressure on margins due to the unprecedented complexity and variability of today's global economy (see Table 1.1). The need for visibility and optimization across all elements of a supply chain has never been greater.

In the pharmaceutical industry, the rise of emerging markets and increasing price pressure are forcing companies to review established operating models. The dramatic shift in markets toward the so-called "pharmerging" countries adds to increasing pressure to redesign global value chains. Furthermore, the established pharmaceutical companies are threatened by patent expiry and generic competition, which could lead to sales losses of as much as 100 billion USD over a period of 2009–2015.

The chemical sector faces similar dynamics on the market side: saturation in traditional markets and the shift to the rapidly growing and dynamic BRIC (Brazil, Russia, India, and China) and SMIT (South Korea, Mexico, Indonesia, and Turkey) countries. On the supply side, dramatic changes in feedstock sources might alter the rules of the game in the entire industry sector. The shale gas boom in the United States led to a huge decline in feedstock prices and energy costs in a very short time and is thus widely regarded as a potential "game changer" in chemicals and

TABLE 1.1

Trends in Supply Chain Financial Ratios in Process Industries (Data Basis: Annual Reports (2000–2011))

| Industry | Average Operating Margin | Average Changes in Financial Ratios Over the Period 2000–2011 | | | |
		Operating Margin (%)	SG&A Margin (%)	Return On Assets (%)	Revenue Per Employee (K$)
Chemical	0.09	−1	1	2	16
Consumer goods	0.16	−2	0	2	29
Food	0.15	−1	1	−2	29
Pharmaceutical	0.24	−4	−1	−6	46

energy-intensive industries, rapidly and substantially changing global investment patterns and global network footprints.

Volatility has always been the key challenge for fast-moving consumer goods manufacturers. Today, however, many industry experts see many markets in this industry sector already in an era of hyper-competition, with product life cycles measured in a few months and sales promotions happening almost every day. Products are more and more tailored to the individual customer, adding a new level of complexity to supply chains, which therefore needs to be more effective and agile than ever before.

The bottom-line result of these changes is intense pressure on supply chain management (SCM) across all process industries: Growth has been slowing, inventories have been climbing, and costs have been escalating, leading to negative trends in operating margins and other key supply chain measures, as summarized in Table 1.1. In the face of these changes and ongoing pressure from customers to deliver outstanding service, it would appear that only a reinvention of best practices along value chains will make it possible for companies of the process industries to meet the expectations of internal and external stakeholders as well as financial markets.

The foundations for successfully orchestrating a global value chain are effective supply chain planning and reliable coordination of customer demand fulfillment. The performance of planning and coordination in SCM is directly impacting top-line results, costs, and capital. Yet we see reasonable and increasing doubt that the old recipes for supply chain success no longer suffice to keep up with the pace of change experienced in business reality today. Obviously, alternatives must be found to approach supply chain planning more effectively. In this book, we show that this requires not merely minor modifications of business processes, but a change in the entire planning paradigm!

1.1 SUPPLY CHAIN MANAGEMENT MUST MASTER THE VUCA WORLD

The acronym VUCA—volatility, uncertainty, complexity, and ambiguity—accurately describes the conditions under which companies and SCM operate in process industries today. It summarizes the key pressures felt today in SCM. The term VUCA originated in the military in the late 1990s, but was quickly adapted for use in business environments and now stands for strategies designed to cope with increasing volatility, unavoidable changes, and all manner of unpredictable issues that may arise—anything from a change in consumer taste to the onslaught of a recession.

1.1.1 Supply Chain Management Orchestrates Global Functions and Networks

Supply chains in process industries encompass production facilities, distribution centers, and suppliers around the entire globe and connect those entities to global markets. SCM is tasked with integrating all organizational units along value chains and coordinating material, information, and financial flows to fulfill customer demand. The objective of SCM is to maximize customer satisfaction and ensure the most efficient use of required resources, including distribution capacity, inventory, and labor. SCM subsumes all activities related to the design, planning, execution, and monitoring of material procurement, production, and distribution activities along end-to-end value chains, including managing the required information flow. From a functional point of view, all operations-related departments are involved and must collaborate: procurement, manufacturing, quality control and assurance, SCM, planning, customer service, warehousing, and logistics.

As most companies are organized, these departments operate as separate functional units. Thus SCM, in its role as an enabler of end-to-end process interactions along the supply chain, needs to ensure that the required collaboration is achieved. Consequently, SCM requires harmonized and globally interlinked processes. This task also demands appropriate IT (information technology) solutions for advanced planning to ensure the necessary data management and the creation of global transparency.

Business leaders today are well aware of the fundamental role of SCM in managing corporate value chains in a competitive environment and ensuring customer satisfaction at minimum cost and with minimum investment of working capital. However, the majority of supply chain improvement projects struggle to achieve their performance and pay-back targets. To bring these functions back on track, SCM must address the root causes of the VUCA world, which also means that new solutions must be found as the old ones have obviously failed to master volatility, uncertainty, complexity, and ambiguity effectively.

1.1.2 Key Pain Points in Supply Chain Organizations Today

Given the market conditions we have described above, it is not surprising that rising concern about the VUCA world is reflected in what supply chain organizations today regard as their key pain points. According to a recent survey published in *Supply Chain Management Review* in 2013 (addressing the top 10 pain points as shown in Figure 1.1), supply chain leaders are becoming increasingly concerned over the growing lack of supply chain visibility, demand volatility, and supply chain complexity. These three key pain points are directly rooted in the VUCA world.

The lack of visibility in today's ever-widening supply networks is challenging many companies with a growing ambiguity within global planning organizations, substantially reducing the overall supply chain performance. The impact of weak visibility is further amplified by the strong demand volatility and growing supply chain complexity that are sensed by the majority of supply chain managers. Many supply chain managers thus fear a

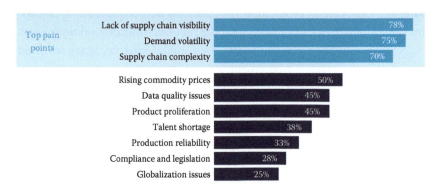

FIGURE 1.1
Top 10 pain points according to supply chain managers.

growing loss of control and deteriorating performance in their global supply chain operations.

In contrast, formerly key issues such as the impact of globalization or compliance with international regulations seem to have been resolved by most supply chain organizations as shown by the results of the survey. Today, it is indeed the VUCA world that is regarded as the most important challenge; consequently, its impact and the resulting pain points are on the top of most supply chain managers' agendas today.

1.1.3 Why Leadership Is Concerned about the Impact of Volatility

As we all individually experience every day, the pace of change in the global business environment has been increasing dramatically over the past several years. Of all the factors responsible for the VUCA world, variability and volatility are challenging supply chain performance most. Volatility is everywhere and ever-increasing: on the demand side, we face major changes in global demographics, shorter life cycles, and more products, but simultaneously with relatively lower order volumes per product. On the supply side, there are major changes in global feedstock availability, supply disruptions, and price volatility.

Characteristics that are unique to process industries only add to the challenge. Typically, production takes place in continuous processes or in significant batch sizes. As the manufacturing process is not based on small individual units, the corresponding manufacturing assets are considerably less flexible compared with those deployed in discrete manufacturing. Therefore, changeover effort and product sequencing are important factors in supply chain and production planning. By-products and even material waste may occur during the conversion of raw materials into final products. Production yield depends on process conditions as well as on campaign sizes and raw material quality. All those specific factors need to be taken into account by SCM when planning and coordinating operations across global manufacturing networks.

The consequences of the lack of concepts, tools, and capabilities with which to manage volatility in SCM are severe. Looking at inventories in many chemical and pharmaceutical companies shows that some stock levels in the supply chain amount to as much as 60% of annual demand, and there are in many cases up to 20 weeks' worth of finished goods stocks. Consequently, the amount of working capital tied up in those supply chains is by far beyond what management and financial shareholders expect.

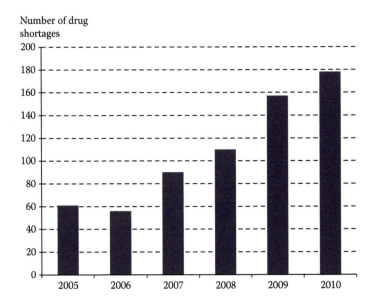

FIGURE 1.2

The number of drug shortages in the United States is steadily increasing.

But, more working capital tied in inventory and increasing operating costs are not the only consequences of weak response to volatility. In today's fast-paced business environment in process industries, there is an increasing gap between supply lead times and customer expectations of order lead times. Even worse, product shortages might lead to lost sales and reduce long-term revenues. Moreover, stock-outs not only lead to additional costs, lost revenues, and customer dissatisfaction but also cost lives when, for example, dealing with life-saving drugs in the pharmaceutical industry. However, the growing number of drug shortages reported in the United States shows all too clearly that keeping the right stock availability is becoming a more challenging task in the industry (see Figure 1.2). The reported steady increase in drug shortages over the last 5 years can be regarded as one of the main issues resulting from insufficient volatility management.

1.2 SUPPLY CHAIN PLANNING IN THE VUCA WORLD TODAY

Given the dynamics of the VUCA world, SCM is more demanding than ever in process industries. The increasing number of stock-out situations and continuously rising inventory levels that many companies observe are just two indicators of the growing issues in supply chains that are involved in keeping track of those dynamics in the business environment.

Supply chain planning is crucial for efficiently deploying resources and coordinating all activities along globally dispersed value chains; planning is thus the backbone of SCM. However, despite significant investments made in demand excellence programs or the introduction of the most sophisticated advanced planning techniques, supply chains struggle to manage volatility on both the demand and the supply side. One of the key reasons for this is that there are inherent flaws in the design of global planning approaches that prevent companies from achieving the targeted supply chain performance.

1.2.1 Planning and Control as the Backbone of Supply Chain Management

The backbone of efficient SCM has always been effective planning. Without proper planning, a company risks sacrificing cost efficiency as well as losing customers due to poor service. When properly executed, supply chain planning ensures that all processes along the supply chain are smoothly orchestrated and that the company can match supply and demand on a daily basis. In this vein, supply chain planning ensures competitive inventory levels as well as low costs of goods sold by using a company's resources and assets in the best possible way.

To manage this task, supply chain planning ensures that all customer demands and market needs are taken into account when making replenishment, production, and supply decisions. By effectively balancing supply and demand, planning ensures cost-efficiency and high market responsiveness in line with business objectives and targeted customer service.

In SCM, the planning task is typically hierarchically organized according to a range of time horizons to reduce planning complexity. In addition to long-term strategic planning that is conducted for the next 2–10 years in alignment with a company's overall business strategy, tactical supply chain planning addresses mid-term planning needs for the next 4–36 months. Tactical planning is evaluated at an aggregated level and delivers the basis for sales & operations planning (S&OP). In addition, short-term planning is maintained at the SKU (stock-keeping unit) level, covering, for example, the next 0–12 weeks, which is the basis for fine scheduling of production and order fulfillment.

At all these levels, planners must strike a delicate balance between over-engineering their plans—including complex "black box" methodologies that few stakeholders understand and even fewer buy into—and over-simplified approaches that rely entirely on individual experience and vague rules of thumb.

1.2.1.1 Forecasting and Demand Planning

Since a supply chain should ultimately be driven by customer demand, the planning starts with available and planned customer orders. Long lead times in production force every supply chain manager to consider demand forecasts. The process of forecasting future customer demand is crucial to various aspects of the supply chain—from next month's production schedule to yearly reviewed supply plans with major contractors to market estimates for identifying capacity requirements in the coming years.

The time horizon of forecast and demand plans must exceed the overall production lead time. The key challenge for demand planning is that longer planning horizons require a greater share of demand to be forecasted; and, typically, the longer the forecast horizon the lower the accuracy of the obtained forecasts.

Companies typically differentiate between long-term forecasts that are inputs to strategic planning and operational forecasts that drive production planning and scheduling for the coming weeks and months. To estimate future demand, many companies employ statistical forecasting methods whose results are reviewed by the sales force and enriched by market intelligence. The final gross sales demand is then balanced against available inventories to derive the net replenishment demand for the production supply.

1.2.1.2 Supply Planning and Production Scheduling

The goal of supply planning is ensuring customer satisfaction in terms of trustworthy order promising, delivery reliability, and responsiveness at the lowest possible cost. However, this means not only responding as quickly as possible to customer requests but also being flexible enough to manage customer request changes. From a financial perspective, supply planning has to minimize the associated purchasing, manufacturing, and distribution costs, including the "costs of change."

Sales forecasting and net replenishment demand at production sites define the required supply from a company's own production facilities and external parties such as suppliers or contract manufacturers. With respect to production and supply lead times, supply planning must first find the most efficient way to fulfill demand forecasts over the medium-term planning interval.

Starting with replenishment and production planning for finished goods, supply chain planning ensures that the right amounts of upstream intermediate materials and capacities are available with regard to lead times, while avoiding excess inventories along the supply chain. In addition to coordinating material flows between sites, effective planning is not possible without identifying optimal production sequences and batch sizes. In process industries, long production processes and complex changeover operations imply that a significant amount of inventory as well as capacity (production costs) is directly determined by production sequences and the corresponding lot sizes. Effective scheduling ensures that optimal decisions are taken for each asset at the sites and globally coordinated schedules contribute to an agile and cost-effective supply chain.

1.2.1.3 Supply and Demand Synchronization

Limited availability of raw materials and capacity constraints may lead to shortfalls against requirements. If bottlenecks occur, the supply chain has to respond promptly to avoid stock outs. In collaboration with sales and production, the supply chain organization must resolve such demand and supply imbalances. Potential solutions to unresolved issues are escalated to S&OP meetings for final decision making.

In most companies' experience, the very first step of supply chain planning, forecasting, and demand planning is actually the Achilles heel of their entire supply chain. Although all commonly used approaches for supply and production planning require very reliable forecasts even at the product level, forecast accuracy remains unsatisfactory in virtually all supply chains.

1.2.2 The VUCA World Poses New Challenges to Supply Chain Planning

The systems under which individual businesses and supply chains operate today are vast and complex—interconnected to the point of confusion and uncertainty. The stable and predictable times illustrated by the straight line in Figure 1.3 are history in today's VUCA world for virtually all players in process industries. Simple linear root-cause analytics have obviously

FIGURE 1.3
The days of stable and predictable SCM are over.

become less applicable. Therefore, it is necessary for supply chain managers to begin thinking in new ways and exploring new solutions. As there is no predictability or way to plan for *every* event that may arise, it becomes necessary to accept plans under uncertainty and conjoin with preparations to respond to *any* supply chain issue that may arise.

Your company must now become aware of and learn how to operate in a VUCA environment. This requires, however, adopting a novel and innovative perspective, broader organizational understanding, new supply chain concepts, and more innovative skills and tools than those that are required in a more stable environment. To assess the new needs for planning and daily supply chain orchestration, it is necessary first to develop a better understanding of the major factors behind the new VUCA world. What makes planning today so different and the planner's life so difficult?

1.2.2.1 Variability and Volatility Are on the Rise

Variability in all forms is the natural antagonist of supply chain efficiency. As evidenced in a wide range of surveys, supply chain managers continue to rank variability as one of the top challenges to achieving their goals and objectives. But what is variability? Variability in SCM is anchored in the difference between a company's market expectations and the actual customer requirements that need to be fulfilled every day. It has become axiomatic in recent years that the statistical spread between future demand expectations and actual customer orders is widening. As a consequence, supply chain managers have to anticipate, plan for, and react to a widening array of demand and supply scenarios (see Figure 1.4).

Although the terms "variability" and "volatility" are often used interchangeably, one should be aware of the difference between them. Jay Forrester, business innovator and MIT scholar, introduced this distinction in the 1960s. He explained that a company must plan for variability through buffering in inventory and capacity and at the same time respond

Manufacturer   Customer

FIGURE 1.4
Variability and volatility are on the rise.

to volatility through flexible processes and decisions in execution. To differentiate between variability and volatility, we can also use order lead time in execution: volatility applies within the time horizon of the customer order and variability applies to time horizons that precede the order lead time. SCM must therefore "plan for variability" across the entire planning horizon and prepare operations to "respond to volatility" as the plans are executed. This clear distinction between variability and volatility will be essential for understanding the fundamental concepts introduced later with the new supply chain planning paradigm.

Variability implies introducing "the unexpected" and is therefore a natural antagonist to planning. Variability is not only increasing on the market and customer side due to the globalization of value chains. Furthermore, the physical distance involved in global sourcing concepts and bottlenecks in logistics make raw material supply more unreliable as well. This leads also to greater short-term volatility in intermediate and finished goods supplies. Increasing quality and regulatory requirements may lead to further yield variation, especially in the chemical industry. However, by far the biggest driver of variability and volatility is the ever-changing customer behavior. Today, customers request more customized products and higher service levels, despite their need for shorter lead times and decreasing order quantities. This makes detailed "planning for variability," with demand forecasting at the product level, almost impossible. Instead, it is causing inferior operational performance and frustration for sales and supply chain organizations.

1.2.2.2 Uncertainty Keeps You Guessing

Conditions in the supply chain world are changing so rapidly and in so many unexpected ways that it is often overwhelming the ability to cope and understand what's going on. The accelerating rate of change in the business environment creates uncertainty, often also accompanied by a lack of clarity, which hinders management's ability to conceptualize the threats and

Manufacturer   Customer

FIGURE 1.5
Uncertainty keeps you guessing.

challenges that supply chain organizations face. As uncertainty is increasing, companies need to find better ways to face and address it in SCM.

Uncertainty leads, in turn, to missing trust. Every planner will start to "guesstimate" his or her own future demand, typically creating even more uncertainty for upstream supply chain operations (see Figure 1.5). To be prepared for what might come next, planners often use historical data, extrapolate them into the future, and then often add their own additional uncertainty on top before they pass their plans upstream to the next stage along the supply chain. This creates additional waves of uncertainty that are propagated through the value chain by interlinked planning systems.

Additionally, when the supply chain environment is changing unpredictably, a company simply cannot rely on statistics to form the so-called adaptive expectation regarding customer demand. Even if we rely on historical information and extrapolating that into the future, the prediction will undoubtedly be inaccurate. Relying too heavily on historical data in environments characterized by fundamental uncertainty might lead to the wrong assumption that yesterday's solution to a seemingly similar situation is appropriate today. What is needed here is a novel way of thinking.

Being flexible in an uncertain environment is crucial. Detailed plans are great, but as they say in the military, the plan never survives first contact with the enemy. Fighting the plan and not adapting to a changed situation can get an organization into severe trouble. This might sound obvious or easy, but when planners believe they have a really good schedule, it is hard for them not to be wedded to it, especially when they have been involved in its creation. A good plan and surrounding planning process should incorporate flexibility and options for adaptation.

1.2.2.3 Complexity Becomes Overwhelming and Synchronization Challenging

Growth strategies and global operations provide access to new customer groups and markets. However, the resulting trends toward global

manufacturing and distribution networks, outsourcing to best-cost countries, and the increasing number of toll manufacturers create significant
complexity for supply chains. Given the additional trend toward product
customization, most companies offer a broader product portfolio, but this
often results in more individual customer orders and smaller order quantities. These changes together create significant turbulence in production
schedules, asset and resource utilization, supply chain synchronization,
and inventory management. Unfortunately, very few companies apply the
analytical rigor needed to fully understand the tradeoffs between beneficial and wasteful complexity. As shown by a recent study from 2012
by Schey and Roesgen, complexity has become simply overwhelming for
most companies (as illustrated in Figure 1.6).

The greatest SCM challenge in working with all stakeholders in global
networks is getting them coordinated and in sync, especially considering
the increasing market pace with which companies must cope. Globally
scattered material and information flows typically result in a complex
strategic and tactical supply chain planning design. The increasing number of products and the shortening of product life cycles increase planning
complexity still further. By way of analogy to an appropriate solution,
consider the "complexity approach" of the African pygmy people when
they hunt a large elephant. They jointly "slice the elephant" and carry it
individually on their shoulders to their village. The key to making supply chain planning work is in this sense the common visibility (seeing
the whole elephant), collaborative information sharing, and differentiated
supply chain planning approaches (slicing the elephant) for the various
market segments with their individualized customer needs.

However, even with sophisticated planning solutions in place, the accumulation of lead times in globally operating value chains, with product replenishment lead times of 12 months or more, makes accurate demand forecasting
virtually impossible. In this case, upstream production stages are operating
as disconnected islands on the basis of gut feeling and past experience.

Manufacturer   Customer

FIGURE 1.6
Complexity becomes overwhelming.

1.2.2.4 Ambiguity Leads to Confusion and Inefficiency

The global span of value chains makes it more and more necessary that organizations work in a timely, interlinked way and that all supply chain planners from widely dispersed regions work collaboratively with their assets and processes aligned. In supply chain planning, ambiguity is the confusion arising from weakly harmonized processes, missing data, and poor definition of organizational interfaces (see Figure 1.7). For instance, although safety stocks play a fundamental role in any supply chain, there is so much unclarity and inconsistency regarding the definition and the right use of them. Ambiguity embodies the difficulty—and sometimes the seeming impossibility—in SCM of solving complex problems and making clear decisions.

In many companies, the poor standardization of planning concepts and activities in combination with divergent functional priorities prevents SCM from delivering the targeted results. The underlying ambiguity has two significant causes: conflicts and inefficiency. First, the inability of leadership to provide clear direction and synchronize supply chain activities results in individual misreads, poor decisions, or even no decisions. Second, there is increasing frustration among supply chain planners who work hard, but with unclear and impermanent directions that do not add up to satisfying results or comprehensive success.

Along with volatility, uncertainty, and complexity, it is ambiguity that ultimately prevents supply chain organizations from delivering the results that are demanded by internal and external stakeholders; consequently, supply chain costs, inventories, and service are not what they should (or could) be. But is traditional planning capable of solving these issues and mastering the challenges of the VUCA world?

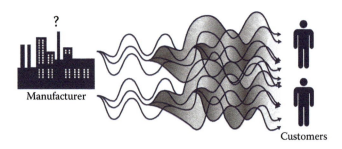

FIGURE 1.7
Ambiguity leads to confusion.

1.2.3 Today's Supply Chain Planning Approaches and Their Limitations

Over the past three decades, global supply chain planning has become more and more demanding, with each decade bringing additional challenges, finally resulting in the challenges of today's VUCA world. What was initially true, and remained mostly true even in the early 1990s, was that the planning task was largely a local one, such that the individual plant and asset could be planned separately considering solely the products dedicated to it. Today, however, production processes are globally connected, they process a myriad of distinct products, and they need to operate successfully under conditions of high demand and supply variability.

The evolution of supply chain planning systems has always responded to emerging new challenges. Since the advent of computers and the Internet, the implementation of new business concepts for planning has been intertwined with the use of IT; in some cases, it was the availability of new technologies that led to major breakthroughs in planning and SCM. Three concepts—material requirement planning (MRP), enterprise resource planning (ERP), and advanced planning systems (APS)—resulted in major changes in planning approaches; below, we briefly outline how these three work and highlight why they struggle to deliver acceptable results when operating in a VUCA world.

1.2.3.1 From MRP to ERP

In many companies, supply chain planning typically centers on the concept of MRP, which became popular in the 1960s as a solution for addressing a growing number of products and production steps. Based on demand for finished goods, MRP supports the calculation of required production volumes and precursor materials. As it grew in popularity, MRP also grew in scope, and evolved in the 1980s into manufacturing resource planning (MRP II), which combined MRP with master scheduling, rough-cut capacity planning, capacity requirements planning (CRP), and other functions.

With the development of client/server IT architecture, it became feasible to integrate virtually all of a corporation's business applications with a common database. This technological advancement led to the development of ERP, offering integration of internal and external information across an entire organization, and integrating all MRP/MRP II

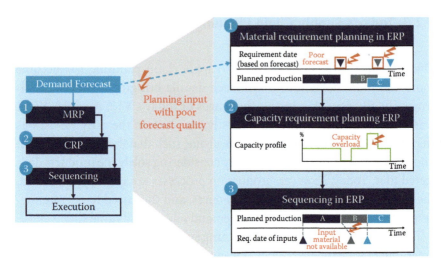

FIGURE 1.8
ERP-based planning and its limitations.

functionalities in one common platform; today, ERP systems form, at least in the execution of orders, the backbone of virtually all supply chain planning organizations.

As shown in Figure 1.8, traditional supply chain and production planning using MRP II is based primarily on three successive steps: MRP, CRP, and the sequencing of orders; all of these steps are typically integrated into an ERP system. The ultimate starting point for planning is always the demand forecast, which triggers all subsequent steps and activities; it is therefore the quality and accuracy of forecast data that determines the value and outcome of supply chain planning—at least when common planning "recipes" are used.

In the first step—material requirement planning—the required material quantities are calculated based on detailed product-level forecasts under consideration of current inventory levels and lot sizes. The result is a material requirement plan that covers expected demand. However, as capacity requirements are not considered in this step, the resulting production plan may lead to capacity overload. Therefore, in the second step—capacity requirement planning—the impact of the material requirement plan on available capacity is evaluated. If the available capacity is not sufficient, measures need to be taken to solve the identified capacity shortages. The production planner then tries to resolve capacity issues by, for example, adjusting shift models or postponing product delivery. In the third step, the planner sequences the required production volumes into a production schedule with a defined production start and end date. Remaining gaps between demand and production capacity and related issues are resolved in S&OP meetings, after which the plan is finally adjusted and ready for execution by the production unit.

1.2.3.2 Advanced Planning Systems and Supply Chain Management

At the end of the 1990s, in the face of globalized manufacturing and delivery processes, SCM as a corporate function rose to prominence. In parallel with the growing number of SCM departments and functions across companies, APS technology became an important cornerstone of most supply chain initiatives. The combination of SCM business concepts and APS as a technology platform provided companies with the means to enable globally integrated planning processes, encompassing multiple sites and countries.

Modern APS solutions essentially stick to the same principles as MRP II, but are designed to cope with complex supply networks across plants and regions. They are capable of integrating all material flows of intermediates between production plants. In contrast to locally and site-oriented planning in ERP systems, APS provides additional functionalities for global visibility and planning. Equipped with modern in-memory database technology and enabling advanced mathematical optimization methods, APS promised to solve complex planning problems in global value chains.

In contrast to previous ERP-based approaches, APS technology allows the planning and optimization of material and capacity volumes, including sequence scheduling, in one step. As shown in Figure 1.9, the determination of material and capacity requirements as well as production sequencing can be conducted simultaneously. This resolves a number of issues that arose for MRP II methodology within ERP systems. Infeasible supply chain plans can be avoided because shortages of material or capacity can be considered; the included sequencing capabilities with dynamic changeover time ensure a more realistic production model for process manufacturers, which in the end enables customer lead time expectations to be met more efficiently. Furthermore, APS technology provides important features such as the linking of all customer demand or stock-replenishment demand with the related multi-level production orders, a key enabler of global transparency. Finally, the used in-memory computing capabilities in APS systems offer integrated mathematical functionalities for automated optimization.

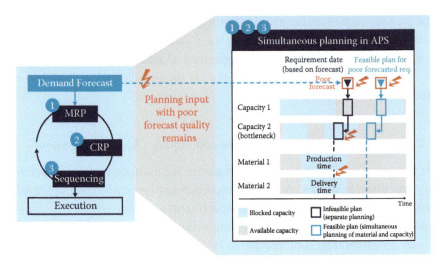

FIGURE 1.9
APS-based planning and its limitations.

However, one important issue has not been resolved by APS: dependency on sales forecasts. It should be noted here as well that APS requires highly accurate product-level forecasts for a planning horizon of up to several months. However, as forecast quality deteriorates substantially in the VUCA world, a lack of good quality inputs directly results in poor planning outcomes provided by any APS solution.

1.2.3.3 Drawbacks of Dependency on Forecasts and Ineffective Use of Inventories

Although APS and its predecessors such as MRP and ERP delivered substantial benefits to many companies, they all have an Achilles heel: the strong dependency on accurate input for planning in the form of demand forecasts. If the input does not have the required quality, planning faces multifold issues regarding costs and service.

And as the painful experiences of many companies show, forecast accuracy is often not sufficient. There are many possible reasons for that, from failing to integrate sales organizations into the planning process to the unpredictability of business in today's VUCA world. Consequently, many companies still labor incessantly to improve forecast accuracy. However, although some improvements might be achieved from time to time, most such initiatives do not solve the fundamental problem as they will never result in a level of forecast accuracy that is required by either ERP or APS solutions. Thus, we have to find a way to reduce the need for detailed forecasts in supply chain planning.

In all planning systems today every forecasted demand signal on the tactical planning horizon is netted during planning runs against a fixed-parameter inventory target, which includes a significant amount of safety stocks that is kept fixed as well. However, this substantial amount of stock is not really used by planning systems to balance the demand variations that occur so frequently. As a consequence, traditional planning systems actively plan for the so-called dead stock and have thus to be considered as a root cause of working capital levels that rise above market expectations.

Overall, all these issues place a huge question mark behind traditional planning approaches in either ERP or APS systems. And, the bad news here is that such issues cannot be fixed by adjusting a few settings in planning systems or introducing an additional supply chain role in your organization. What we have is a fundamental problem that needs to be solved by taking a new direction in supply chain planning.

1.3 WHY WE NEED A PARADIGM SHIFT IN SUPPLY CHAIN PLANNING NOW

All the challenges companies face in SCM today and the resulting pressure on costs, inventories, and service have triggered an intense debate about how to back out of the dead end of today's planning practices. In almost all companies, there are experts favoring either improving traditional planning approaches by investing in even more sophisticated systems and forecast improvement initiatives or going for lean approaches and extending them along the entire supply chain.

One of the primary objectives of both lean and traditional supply chain planning is that of achieving a stable and reliable *flow* of material through the networks and a timely *flow* of information along the supply chain. More than ever, a decisive competitive advantage can be achieved by companies with a balanced flow from and to their customers. The better the *balanced flow*, the better the customer service level and the deployment of working capital.

Manufacturing in process industries aims typically to achieve efficiency by producing output in large manufacturing campaigns through multiple batches. Large campaigns indeed promise at first glance to minimize production costs. However, large campaigns limit opportunities to introduce flow and come at the expense of higher inventories and lower responsiveness to customer demand; large campaigns can therefore also reduce the overall performance.

Several formal disciplines, such as value stream mapping and S&OP, are designed to create balanced flows along the supply chain and to tie production to sales forecasts or customer demand. However, these approaches need to be extended to successfully address the increasing variability and volatility that all companies face in the VUCA world. The same generally holds true for both traditional planning and conventional lean approaches. To master the quest for balanced flows successfully, it is less promising to go for one of the two directions alone. Instead, a new planning paradigm is required that combines the best of both worlds.

1.3.1 Traditional Planning Approaches Fail to Deal with the VUCA World

As we have seen in the preceding sections, traditional planning approaches and systems, either ERP- or APS-based, have two main limitations: first, their enormous dependency on forecasts and high forecast accuracy; and second, the ineffective use of inventory and capacity buffers when dealing with variability in supply chain planning. These planning system deficiencies result in three major root causes of poor supply chain performance:

- The planning loop trap
- The bullwhip effect
- One-sided variability management

These three key issues of traditional planning are the main reasons for insufficient capabilities for managing supply chains in the VUCA world. A substantial number of cost and service issues in supply chains can be traced back to them.

1.3.1.1 The Planning Loop Trap: The Spiral to Inefficiency

Traditional planning concepts pivot around the precondition of planning and producing in accordance with long term but detailed product forecasts, whose expected quality is generally far beyond the required accuracy. As a consequence, companies following these approaches quickly

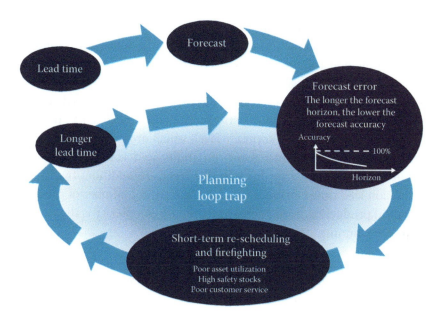

FIGURE 1.10
Many companies are caught in the planning loop trap.

become trapped in the planning loop (see Figure 1.10). This phenomenon describes the vicious relationship between long lead times and the resulting need for far-reaching forecasting horizons, directly causing decreasing forecasting accuracy as planners must look far into the future and expand planning horizons.

Especially in process industries, companies often face long lead times in manufacturing, and therefore have to plan and schedule for make-to-stock (MTS) production. The long lead times, in turn, require that product forecasts guide planning. Naturally, as lead times increase, the accuracy of the product forecasts deteriorates. With more forecasting errors, the need for safety stock and excess capacity increases along the entire supply chain, and inventories swell.

Forecasting errors also result in more rescheduling of orders. However, such short-term changes are little more than firefighting to adjust production schedules to actual demand, also implying that more rush orders need to be squeezed in. Overall, this reduces throughput on all assets due to sub-optimal manufacturing sequences and increased changeover times, thus resulting in reduced asset utilization. In other words, decreasing forecast accuracy ultimately increases production lead times due to reduced asset utilization and inefficient manufacturing sequences. In return, such extended production lead times require even longer forecasting horizons, further decreasing forecasting accuracy. And now, planners are trapped in the planning loop and the same cycle recurs. Lower forecasting accuracy further increases product lead times and the planning loop trap expands further—driving up costs, increasing delays, and creating inefficiencies.

This spiral into decreasing performance and inefficiency depends strongly on the forecasting accuracy level, which drops as planning horizons are extended. The increasing variability and volatility of today's VUCA world only amplify this negative trend. In addition, the overall production lead time is increasing with the growing number of products on typical production lines. Increasing product complexity is thus aggravating the planning loop trap.

1.3.1.2 The Bullwhip Effect: Amplifying Variability along the Value Chain

Your company and its SCM staff might recognize this situation. Despite very stable customer demand for one of your company's "A" products, your manufacturing department may be telling you that production for the very same products constantly swings, seemingly more and more widely. For sure, your company is not alone. The phenomenon behind this situation is called the bullwhip effect and it jeopardizes performance along many supply chains.

The bullwhip effect describes the amplification of demand variability along the supply chain. Even if the end-customer demand is quite stable, upstream demand often fluctuates widely (see Figure 1.11). It is very important to recognize that the bullwhip effect causes substantial supply chain costs. Greater variability means higher inventories, lower utilization, and less efficient manufacturing, all of which increase working capital and costs of goods sold.

However, the bullwhip effect is not inevitable. In many supply chains, much of the variability is not given, but self-made. Anyway, do not blame your company's SCM organization. Today most companies have only limited transparency along their global supply chains, which prevents supply chain planners from effective planning. In addition, their planning activities rely on mostly unreliable demand forecasts, which make effective planning virtually impossible.

Owing to such limited global transparency, every plant or distribution center finds itself optimizing its operations in isolation. Every supply point

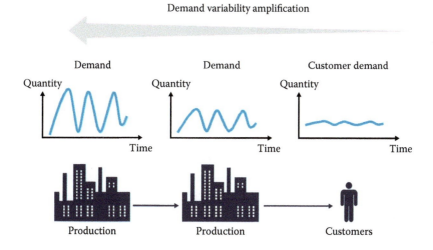

FIGURE 1.11
The bullwhip effect describes the phenomenon of demand variability amplification.

along the value chain is batching demand input to a locally cost-optimized lot size. However, with isolated lot sizing at every value chain step, variability is amplified systematically upstream along the supply chain.

No one should be surprised that planners often overreact to demand signals given the poor accuracy of the forecasts they receive every day. It seems only logical that they inflate replenishment and production quantities in order to be on the safe side. However, this missing trust in demand forecasts, albeit often well justified, creates considerable nervousness and costly variability along a supply chain. If supply chain planning is not capable of handling this variability, but instead amplifies it internally due to ineffective decision making, substantial issues with service and total costs arise.

1.3.1.3 One-Sided Variability Management: Increasing Inventories and Supply Chain Nervousness

Traditional supply chain planning is based primarily on the logic of the MRP run. In this planning approach, estimated demand and planned production are balanced in light of current inventory levels. Based on these levels, an inventory projection is calculated. As soon as minimum inventory levels—the safety stocks—are reached, the MRP logic will lead to the planning of further production runs to stay above the minimum levels and create stock to meet anticipated further demand. The safety stock levels are not touched in this logic.

Because they omit consideration of safety stocks in tactical supply planning, MRP-based planning runs fail to include safety stocks when calculating available stock for all inventory plans. In contrast, the systems trigger replenishment orders immediately once the safety stock levels are reached or even *expected* to be reached. Such a response to expected demand is in this situation especially dangerous. Although we all know that forecasts are typically highly unreliable, even long-term forecasts—those with the greatest inaccuracy—are used by MRP systems to plan orders and avoid touching the safety stocks. As a consequence, even a single peak in demand forecasts will cause a substantial amount of replanning and rescheduling activities.

As shown in the middle part of Figure 1.12, with one-sided variability management, capacity utilization directly follows demand. Inventories, in contrast, are not used actively in planning as a means of dampening variability. Instead, all variability must be buffered by asset capacity, causing much of the nervousness observed in today's supply chains. As a direct consequence of not touching safety stocks, all the market variability and forecast uncertainty are passed one-to-one to production planning and manufacturing capacities through planned stock replenishment signals. Consequently, the capacity side has to manage all the variability—one-sided only—through perpetual replanning and frequent rescheduling activities.

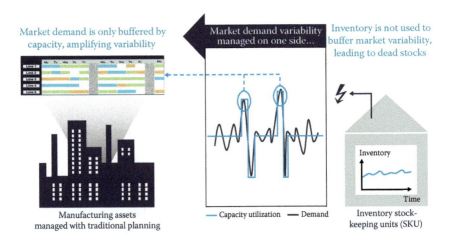

FIGURE 1.12
One-sided variability management leads to ineffective use of assets and inventories.

Even worse, this also implies that planning runs are creating stock replenishment orders with the same one-to-one quantity oscillation as seen in the forecasted (market) demand with all the forecast inaccuracy included. In tactical planning, demand peaks are thus not absorbed through inventory due to the fixed-maintained inventory targets, nor are they dampened through the active use of safety stocks—although dampening variability is the intended purpose of safety stock. Instead, dead stocks accumulate and weaken the working capital position of the whole company.

When demand begins drifting in the VUCA world, it is really a case of two worlds colliding as the traditional MRP planning mode has not foreseen a mechanism for smoothing the increased demand volatility. Similar to the bullwhip effect—and aggravated by it—demand variability is propagated one-to-one from one production stage to the other. Failing to execute variability control by actively using inventory prevents companies from dampening variability and forces supply chains to manage all the variability on the capacity side.

At the operational level, this requires constant production planning adjustments and creates an awkward situation for manufacturing units and raw material suppliers. Perpetual replanning and rescheduling cause excessive effort on the part of planners, but typically result in unsatisfactory results due to the inherent design flaws in traditional planning modes. Although this challenge is well known to supply chain practitioners, it is hard to change this practice when using traditional planning systems and the accompanying supply chain planning processes that were established in the past.

1.3.2 Common Lean Approaches Are Insufficient for Global Supply Chain Synchronization

To overcome the shortcomings of traditional supply chain planning, companies seek alternative planning approaches. In recent years, lean concepts such as value stream mapping or kanban have become popular across a range of industries as potential alternatives to traditional planning and coordination in operations. Originating in the automotive industry, lean concepts have inspired considerable effort to adapt them to the specific requirements of process industries. At the shop floor level, lean approaches have surely helped to improve flow and establish more demand-centric operations.

When extending the lean approach along the entire supply chain, however, many companies find that lean practices tend to over-simplify global supply chain planning. Because lean concepts do not involve integration with corporate planning systems or planning across multiple plants and assets, these concepts appear insufficient for improving flow at the supply chain level and for supporting global supply chain synchronization.

1.3.2.1 The Promise of Lean Principles in Supply Chains

Becoming lean is widely regarded as a promising approach to SCM that makes it possible to achieve a steady flow of goods, driven only by what customers really want—a true pull-driven supply chain. Using lean principles, the automotive and discrete manufacturing industries have found major success in establishing stable material and information flows. Less surprisingly, many companies in process industries today are also on a quest for improved flow. Now more than ever, industry leaders are convinced that a decisive competitive advantage can be achieved for supply chains with a high degree of flow through all stages and to customers. Moreover, observations of process industries confirm this belief: supply chains with stable flows indeed exhibit better service levels, higher asset utilization, and lower inventories.

But would traditional planning practitioners deny the benefits of flow? As we hear all the time, certainly not! Every planner knows that material and products that flow reliably are the easiest to plan for and manage. As lean advocates, all users of traditional planning concepts agree that reliable flow indeed improves supply chain performance and simplifies daily planning tasks. However, are traditional planning systems such as MRP capable of establishing demand-driven flow along the entire supply chain? When considering all the planning issues supply chain managers face today, it must be admitted that they are not. Strong dependency on (accurate) forecasts prevents ERP and APS systems from pulling products based on customer demand and establishing stable and reliable material flows.

1.3.2.2 Cyclic Scheduling: Lean Scheduling for Process
Manufacturers

Many companies have benefited from applying lean manufacturing, enjoying lower inventories, reduced variability, and high acceptance on the part of the people who actually work with lean principles and tools. However, practitioners in process industries also face significant limitations when attempting to implement lean tools such as kanban or heijunka beyond the shop-floor level. Traditional lean principles and tools are deeply rooted in the automotive industry, which does not operate under the manufacturing constraints that are present in process industries. Especially when it comes to planning and scheduling, lean concepts thus often struggle in these industries.

In process industries, the manufacturing restrictions are much more manifold and complex, demanding intensive supply chain and operations planning activities. Planning must address a variety of industry-specific supply and manufacturing restrictions: long supply lead times, variable product yields, and sequence-dependent changeover times, to mention just a few. Minimum campaign sizes need to be taken into account due to technical or economical restrictions. Furthermore, production sequences determine the efficiency of changeover operations and the amount of waste in production. Thus, an optimal sequence for production—for example, from bright-to-dark colors, or low-to-high additive concentrations—must be defined and maintained.

In recent years, the product wheel has evolved as a lean approach for achieving leveled production, smooth production flows, and reduced cycle times in light of relevant manufacturing restrictions and challenges inherent to process industries. To maximize equipment utilization and labor productivity—two major competitive advantages in process industries—product wheels aim to accomplish the dual objectives of production leveling and multi-product scheduling to an optimum practical extent. They combine these two objectives with a third advantage: the optimization of multi-product schedules. Changeover time, costs, and difficulty depend on the sequence in which products are manufactured; the optimization of the production sequence is thus a critical feature for process industries.

As shown in Figure 1.13, a product wheel is a visual metaphor for a structured, regularly repeating sequence of all products to be made on a

FIGURE 1.13
The idea behind product wheels.

certain asset or piece of equipment; it is designed to manufacture products in a cyclic sequence in short cycle times without sacrificing customer service. In this way, product wheels constitute an effective way to bring lean planning and scheduling principles into process industries. They combine the simplicity and flow thinking of the lean approach with the requirements that are characteristic of process industries.

The length of every spoke of a product wheel represents the production quantity of a certain product, so the spokes need to be arranged according to the optimal production sequence. The length of each spoke—the production quantity within one cycle—is based on the demand takt rate, that is, demand averaged over some period. When using product wheels, production will be triggered only if it is a product's turn, ensuring stable manufacturing flow and minimizing (work-in-process) inventories. In this way, product wheels are an important measure for establishing pull principles in the highly constrained process industry environment. Product wheels can be employed in both make-to-order (MTO) and MTS environments. In fact, MTO and MTS products can be scheduled on the same wheel.

Still, can we resolve all planning challenges in the VUCA world by adopting product wheels in process industries? The first implementations of product wheels were expected to produce every product in every cycle of the wheel. As industry experts such as Peter L. King recognize, this limits their application to stable product portfolios on locally planned assets. Although some extensions of product wheels have been discussed in the past, all standard product wheels presented so far struggle to manage short-term volatility and fail to ensure global synchronization along supply chains. This is the reason why we introduce Rhythm Wheels in this book as approach to overcome such limitations of standard product wheels.

1.3.2.3 Limitations of the "Product Wheel" Approach for Managing End-to-End Supply Chains

When it comes to supply chain planning and execution, lean advocates are often accused of being antitechnology. Of course, it is reasonable to avoid using IT in applications in which it is wasteful, confusing, or not reflective of reality. Unfortunately, this has been true for quite some time with regard to traditional planning approaches such as MRP or APS systems. Consequently, most product wheels that have been introduced into industry practice are locally managed and are mostly applied manually or with simple Excel-based solutions outside existing corporate information systems. Lacking support by corporate planning systems combined with a lack of tailored business concepts has, however, had a very undesirable consequence: today, the application of standard product wheels is limited to rather stable and fast-moving product portfolios. They cannot be used for slow-moving or highly volatile products.

However, as value chain networks in process industries encompass several manufacturing steps that are distributed across plants around the globe, end-to-end supply chain synchronization is crucial to ensuring efficiency. When attempting to apply the standard product wheel to separate but interlinked plants at the same time, three major shortcomings have been identified by many practitioners:

- The rigid design of standard product wheels limits their application to fast-moving and stable product portfolios. However, when operating in a VUCA world with an increasing share of volatile products, this constitutes a major limitation. As standard wheels do not incorporate concepts for mitigating short-term volatility or adjusting to changing conditions, they quickly lose effectiveness when exposed to highly dynamic business changes.
- If product wheels are locally designed and maintained without integration with corporate planning systems, the designs and techniques typically in use tend to differ significantly across sites or even across manufacturing lines of the very same plant. Lacking well-governed standards runs the risk that product wheels may be locally optimized, but leads to sub-optimal solutions for the entire supply chain. Without proper global synchronization, companies face the risk of increasing cycle times, tying up more working capital, and prolonging customer response times.
- The proliferation and sustainability of product wheel implementations have been negatively impacted by the lack of appropriate planning and execution technology. Site-specific product wheels do not foresee real-time sharing of detailed information as an ERP or APS system would do. However, low demand visibility and planned production and inventory levels along the supply chain reduce the flexibility needed to react to changes and make global synchronization virtually impossible.

Consequently, well-intentioned lean initiatives based on product wheels—the most suitable approach for lean planning and scheduling introduced to process industries so far—very often lack process standardization and global visibility in corporate planning systems, hindering global synchronization along supply chains.

1.3.3 How to Back Out of the Dead End of Today's Planning

The lessons learned from observing today's approach to supply chain planning in process industries are clear: both traditional planning and common lean approaches, even when adapted to the requirements of process industry manufacturing, fail to a considerable degree when addressing increasing VUCA challenges. Thus, we believe that it is time to back out of the dead end in global supply chain planning—or avoid it altogether—by taking the best of traditional and lean planning and tailor it to the new realities in process industries. This is what we call LEAN Supply Chain Management.

1.3.3.1 Both Common Lean Approaches and Traditional Planning Fail

Supply chain planning has to find answers to the growing VUCA challenges in the business world. Without the capability of managing those factors, costs and inventories will surely continue to rise and service issues as well as shortages will occur even more frequently. As we have seen, when it comes to planning and execution, both common lean approaches and traditional planning suffer from serious limitations when it comes to solving the VUCA challenges.

Traditional planning is subject to several vulnerabilities when addressing uncertainty and volatility; in particular, three root causes of weak performance can be identified that are mutually amplifying:

- The planning loop trap increases forecast errors.
- The bullwhip effect increases volatility and variability due to over-reaction to forecast errors and too little consideration of actual costumer demand.
- The traditional planning approach in ERP and APS systems transfers all the volatility—with no smoothing—to production planning.

On the other hand, even product wheels, the most sophisticated lean concept for process industries, fail, as do other conventional lean planning approaches, to manage complexity:

- Product wheels, albeit tailored to process industries, struggle to manage high-mix and highly volatile product portfolios and to cope with business dynamics.
- The lack of system support and standardization prevents lean planning from providing visibility beyond local manufacturing sites into global SCM.
- Common lean concepts fail to achieve global supply chain synchronization.

Insofar as both traditional and conventional lean concepts struggle in the VUCA world, the key question for SCM today is how to go forward. The verdict is clear: neither traditional planning nor conventional lean concepts alone can solve the current challenges. Traditional planning suffers from its over-complexity and insufficient capability for managing volatility and variability, while over-simplified lean approaches help to solve some local issues but fail to coordinate and synchronize global value chains.

1.3.3.2 Resolving the Traditional Planning versus Lean Conflict

In the face of all the above-mentioned planning issues in the VUCA world, many companies are experiencing intense deliberations while trying to determine the right way forward. The typical conflict between traditional versus lean planning goes something like this: many lean initiatives attempt to abandon traditional supply chain planning, which uses ERP or APS systems. Why? In the eyes of lean advocates, they are inappropriate, transaction-intensive, and nonvalue-added when compared with what lean planning and execution tries to accomplish. In the eyes of the most lean facilitators and advocates, ERP and APS are just overcomplicated and wasteful dinosaurs that prevent companies from establishing demand-driven supply chains.

This, however, often causes tremendous friction between supply chain planners and the lean advocates who are pushing to eliminate formal planning systems. Supply chain planners see every day that they need to be able to see the total demand picture and develop global corporate capabilities for inventory and capacity planning, since otherwise blind spots will exist in the planning process, resulting in shortages, firefighting, and even excessive inventory positions. Although many planners favor lean principles, they regard the simple pull approach of lean planning as a gross oversimplification of the complex planning and scheduling scenarios that are the norm in today's VUCA environment.

To achieve a true step change in supply chain performance and bring greater agility and resilience into SCM, we believe that it is counterproductive to choose between lean or traditional planning as though they are mutually exclusive. Both approaches are burdened by serious limitations when dealing with the VUCA world, so we should think more about taking the best of both worlds.

1.3.3.3 Prepare for the VUCA World by Opting for LEAN SCM

To back out of, or avoid altogether, the dead end of today's supply chain planning practices and regain lost performance by establishing flow and demand-driven supply chains to withstand the turbulence of the VUCA world, we have invented LEAN Supply Chain Management—LEAN SCM—a planning paradigm that brings formerly disconnected planning approaches together and tailors them to the specific requirements of process industries.

As highlighted in Figure 1.14, LEAN SCM is, first and foremost, designed to eliminate the root causes of poor supply chain performance that result from today's VUCA challenges. In particular, it eliminates the strong dependency on accurate forecasts that is the Achilles heel of any

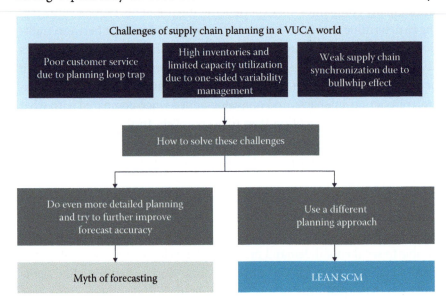

FIGURE 1.14
LEAN SCM provides a new planning paradigm.

MRP-based planning approach. Through LEAN SCM, your company can also rid itself of the antiquated and complex rules governing demand and supply order generation in traditional supply chain planning, rules that lead to unrealistic schedules and constitute a huge problem for flow. Also, LEAN SCM avoids reliance on concepts with little or no connectivity at the plant, enterprise, or supply chain level of conventional lean planning, which is a reason for poor global supply chain synchronization. We thus introduce an innovative Rhythm Wheel approach with formalized variability control mechanisms that extend standard product wheels to high-mix and volatile product portfolios and enable the synchronization of Rhythm Wheels along the global value chain.

In the face of the VUCA challenges, there are now two ways to conduct SCM: first, SCM can accept the drawbacks of the traditional planning and conventional lean practices as they exist today in organization, processes, and planning systems. But all too often this results in high inventories, poor asset utilization, and service issues—all of them jeopardizing top-line margins. Or, second, SCM can challenge and fundamentally change how your company approaches supply chain planning and global coordination. From our point of view and in the context of this book, we show you how to take the second way by going for LEAN SCM.

CHAPTER SUMMARY

Virtually all industry leaders in process industries emphasize that mastering the VUCA world and its impacts is the top SCM priority. Strong pressure on inventories, service, and top-line margins urgently demand a rapid but sustainable solution.

Especially the management of volatility and variability is a key challenge for supply chain planning, the backbone of any supply chain. Traditional planning systems such as MRP and APS struggle in this effort due to their strong dependency on unrealistically high levels of forecast accuracy, trapping them in the planning loop and subjecting them to the infamous bullwhip effect. Furthermore, these traditional planning approaches lack efficiency due to their one-sided approach to variability management, which causes ineffective use of both capacity and inventory.

However, conventional lean approaches, such as kanban and heijunka, often considered as viable alternatives to traditional planning, are also unable to overcome the challenges of the VUCA world. These approaches provide some advantages when it comes to variability management, but they fail to establish end-to-end synchronization along supply chains—a prerequisite for successfully managing the global value chain networking in process industries.

To help your company back out of or avoid a dead end in supply chain planning, we propose in this book a new way: LEAN SCM. We combine elements of both lean and traditional SCM to develop a holistic concept that adapts planning to the VUCA world while meeting all specific requirements of process industries. Along these lines, we guide you through a paradigm change that will enable your company to achieve a step change in supply chain performance.

2

Guiding Principles of LEAN SCM Planning: Facing VUCA Challenges

Before we lay out the fundamentals and building blocks of the LEAN SCM Planning framework later in the book, we must consolidate major planning requirements we have identified. In this chapter, we formulate central *principles to adopt* in the design of the new supply chain planning paradigm. They will guide us in framework development and provide a first impression of the conceptual basis on which we build our system. Those principles provide first thoughts leading to how we recommend facing VUCA challenges in tactical and operational supply chain planning. We consolidate these principles and group them into three areas, offering an opportunity for some early reflection on several critical questions:

- LEAN demand—How to cope with rising demand variability and uncertainty in planning?
- LEAN supply—How to establish firmer control of supply volatility and supply reliability?
- LEAN synchronization—How to master complexity and reduce ambiguity?

We begin with the demand view, because it triggers all subsequent planning work.

2.1 LEAN DEMAND: HOW TO COPE WITH RISING DEMAND VARIABILITY

Predicting the future has always been a challenge. Reflecting our own experience and looking back on current demand management practice in companies we have worked with, we have to admit that most of these companies seek perpetually to predict future demand but continually fail to do so with sufficient accuracy. Nevertheless, all planning activities today start with sales forecast input—predictions of future demand. However, in today's VUCA world forecasting accuracy is more difficult to achieve than ever. The search for new ways to approach planning and forecasting is therefore on many supply chain planners' agendas. So how can companies cope with this forecasting dilemma? We suggest the following three principles to guide SCM into the future:

1. Accept uncertainty and eliminate the need for certainty—put the forecast accuracy myth aside.
2. Obtain better aggregated demand views—be better prepared for consumption-driven supply.
3. Stop using forecasts to trigger manufacturing—it is better to respond to real consumption based on pull.

These principles would, of course, have far-reaching consequences for any traditional or conventional supply chain organization; we now explain them shortly so that your company can put them into practice.

2.1.1 Accept Uncertainty and Eliminate the Need for Certainty in Execution

In process industries, lead times of up to one year are not unusual. But can a sales organization accurately forecast all their products up to 12 months ahead at the detailed SKU level? As experience shows, they cannot. Most sales forecasts turn out to be completely wrong and the question is: Should we really go ahead and manufacture products, building up inventory according to those forecasts?

Forecasting accuracy is closely connected to the sales volumes and variability associated with each product. As a rule of thumb, the higher the sales quantity of a product the lower is its variability, and thus the lower the forecasting error (see Figure 2.1). For high-volume products with stable demand, most forecasts fall therefore into an acceptable range of accuracy, something planners can live with. For a large number of products in the portfolio, however, demand variability makes accurate forecasting impossible: the lower the volume, the poorer the forecasting accuracy.

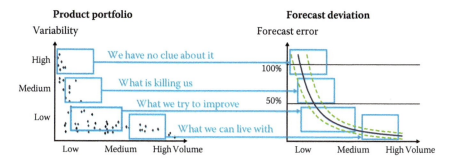

FIGURE 2.1

Forecasting accuracy increases with product volume and decreases with demand variability.

To reduce the increasingly negative effects of poor SKU-level forecasting, companies still desperately try to improve their SKU forecasting accuracy. But let us face facts: all those accuracy improvement initiatives at the detailed SKU level have failed to significantly change results in the past and in the face of even greater VUCA challenges they are even less likely to succeed.

So, what can companies do to solve this obvious dilemma in SCM? Instead of continuing to try to improve SKU-level forecast accuracy, companies should take a different way: they should accept uncertainty and eliminate the need for exact and detailed SKU-level forecasts.

This sounds easy, perhaps, but it is likely to cause conflicts given how planning processes and systems work today. In MRP, for instance, today's systems require detailed SKU-level input to make net demand calculations, even for long-term stock replenishment plans. Consequently, the resulting replenishment signals include all the variability and are distorted by forecast errors—but are still used to trigger production.

We hope to change this practice in order to ensure that no forecast errors are included in production orders for MTS products your company manufactures. By applying this principle, your company can dispense with the need for detailed but typically inaccurate SKU-level forecasts and significantly reduce planning complexity.

2.1.2 A View of Aggregated Demand: Be Prepared for Consumption-Driven Supply

If it is no longer necessary to use detailed SKU-level forecasting in operations, it is natural to ask: what then is the future role of forecasting in

SCM? To master the VUCA world, supply chains have to rely more heavily on pull strategies in order to realize consumption-driven supply management. To do so effectively, there is still a need for improved aggregated forecasts to enable the right configuration and parameterization along the supply chain.

As many supply chain practitioners have experienced, it is usually possible to accurately estimate total demand three months in advance, while it is virtually impossible to forecast precisely on which day or even during which week a particular customer order will arrive. Aggregate forecasts have the advantage that their accuracy is higher due to statistical pooling effects. Therefore, we propose using aggregate forecasts for parameter setting: the higher the level of aggregation, the better the forecast accuracy. A monthly forecast, for example, is more accurate than a daily forecast (see Figure 2.2). An aggregated forecast can effectively be used to configure supply chain parameters and prepare the supply chain for later fulfillment of final customer orders.

Acknowledging that forecasts at the detailed SKU level have not been accurate in the past and are likely to be even less accurate in the VUCA world suggests the advantage of replacing or complementing them with aggregated forecasts. For instance, it is possible to improve the supply chain configuration considerably by designing it based on monthly forecast information instead of using forecasted daily SKU demand, which is not reliable. As shown in Figure 2.2, using monthly instead of daily or even weekly forecasting can improve planning accuracy significantly. To complement these efforts, forecasts at the product family level or based on postponed value chain steps can also be used.

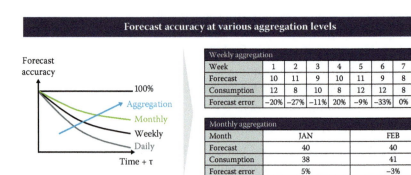

FIGURE 2.2
How forecast accuracy increases with time-wise aggregation levels.

Overall, aggregated forecasts are more stable and accurate. Therefore, aggregated forecasts provide a better basis for fundamental supply chain decisions to improve the supply chain configuration like stock allocation, asset investment, and sourcing mode definition. Although such configuration decisions are likely to be better when based on aggregated forecasts, supply chain parameterization for inventories must be designed on a SKU-by-SKU basis, at least for target setting and replenishment calculations; parameterization for production-leveled schedules using the Rhythm Wheel design can still be achieved on a product family basis. Therefore, product family forecasts should be disaggregated to the SKU level using proportional factors, but such disaggregates still achieve higher accuracy and contain less variability than, for example, any of the individually planned SKU-based forecasts that project 24 months into the future.

With higher forecasting accuracy, inventory target levels at the tactical level can be adjusted regularly up or down in accordance with the forecasts, so that the range of demand variability that must be covered by safety stocks will be narrower by far and the overall cycle stock can be optimized.

2.1.3 Stop Using Forecasts to Trigger Manufacturing: Respond to Real Consumption

To solve problems caused by the increasing demand uncertainty and variability that make SKU-level forecasts very inaccurate, with the supply chain configuration based on forecasts aggregated at some level, we recommend applying real demand-driven, pull-replenishment principles to move beyond the "forecast accuracy myth" in customer order fulfillment and operations. In pull mode, replenishment is triggered by real customer demand, not by forecasts. In other words, production processes are initiated in response to actual customer orders, not in anticipation of those orders. Pull thereby directly reacts to the voice of the customer.

With consumption pull concepts, aggregated forecasts are used only to set inventory targets and inventory replenishment levels as part of supply chain planning parameterization. Planning runs using those parameters later create pull replenishment signals and trigger actual production by comparing current inventory levels with those targets. Thus, although the inventory replenishment target levels are based on and adjusted to forecasts across the planning horizon, on a daily basis final production

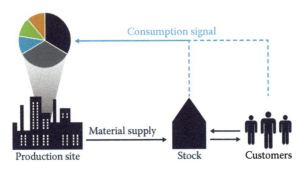

FIGURE 2.3
Real consumption should trigger/pull production.

will always be based on real consumption, current inventory status, and a "production-leveled" schedule (see Figure 2.3).

It would perhaps be natural to object here that using any forecasts in planning seems more like applying a push strategy than the postulated pull strategy. As a response to this obvious first reaction, we observe that even Toyota, widely known as a major promoter of lean and pull principles, expends considerable effort on forecasting. But Toyota does not use forecast information to trigger orders. Instead, they use them for configuring the supply chain, for example, for determining how many kanban cards are needed for each product or part type at various inventory locations. Reflecting this and based on our own experience in process industries, we follow the pull principle; while supply chain configuration with inventory planning and target setting should be based on best-available aggregated forecasts, actual replenishment and production should always be based on current inventory status and real consumption needs.

In this vein, it is possible to avoid constant operations re-scheduling due to variability and uncertainty in forecasts and calm supply chain nervousness. We thus recommend implementing pull consumption whenever possible.

2.1.3.1 Summary of LEAN Demand Principles

The "planning loop trap" in traditional, forecast-oriented supply chain planning is a major challenge for every company in process industries, resulting in high inventory levels, widely fluctuating capacity utilization, and, in spite of it all, unsatisfactory customer service. Nevertheless, undertaking additional forecast improvement initiatives or blaming the lack of "market intelligence" on sales does not solve the root cause of the problem.

The simple truth is that demand for low- and medium-volume products will always remain hard to predict and thus be linked with low forecast accuracy. To move beyond the "forecasting accuracy myth," companies must change how they create the relevant outlook on future customer demand in SCM. On the one hand, they should use aggregated forecasts with higher accuracy (through statistical pooling effects) for the tactical configuration of right-sized buffers in inventory and capacity in order to prepare for the ideally leveled product flow along supply chains. On the other, they should respond only to real consumption-triggered pull signals when fulfilling customer orders in operations.

2.2 LEAN SUPPLY: HOW TO GET A GRIP ON SUPPLY UNCERTAINTY AND RELIABILITY

As highlighted above, today every supply chain faces increasing variability and uncertainty. As a consequence, companies have to plan much more thoroughly for these factors on the demand side. Furthermore, they need to create and utilize appropriate buffers in inventory and capacity on the supply side. Yet most companies lack the required processes and systems for developing adequate safety stocks and an effective capacity buffer strategy. Nevertheless, companies can either choose for themselves how to buffer against variability or this will be "chosen" for them; when it is "chosen," it will show up in the form of lost sales, late shipment penalties, and a more chaotic supply chain. So, we recommend being better prepared for increasing variability by applying LEAN Supply principles:

1. Manage demand spikes with planned and right-sized safety stock buffers.
2. Level production plans to create flow and stabilize utilization.
3. Use cyclic production patterns to achieve a common takt and regularity in operations.

2.2.1 Manage Demand Spikes with Planned and Right-Sized Safety Stock Buffers

In process industries, forecast variability and order volatility enter supply chains typically through MTS products due to long supply lead times.

Such MTS products are usually assigned fixed inventory target levels and associated significant safety stocks as a buffer against uncertainty.

Unfortunately, however, safety stocks are used only for short-term volatility impacts in fulfillment and manufacturing and for reactions after a short-term demand peak hits the supply chain in an unexpected way that creates variation between forecast and real consumption, causing inventory to drop below a pre-determined (safety stock) level. In this case, immediately and often hectically, a new replenishment order is triggered and flagged for expediting before being passed along to production (see the left side of Figure 2.4).

We are however convinced that safety stocks should be put into place also to prevent the upstream propagation of demand spikes and further amplification of variability. Therefore, within the tactical planning horizon, new replenishment orders should not be triggered immediately when inventory falls below the safety stock level. Safety stocks have been designed and configured for those variability peaks. In this sense, inventory buffers should be used to dampen demand variability and reduce the infamous bullwhip effect upstream along the supply chain.

For effective supply chain planning, we recommend changing the current practice of buffering variability in two ways. First, safety stocks should be used actively as a buffer against demand spikes (also for forecasted demand); in this way market variability can be dampened to some extent. Second, inventory replenishment levels and the included safety stock buffers should be adapted dynamically to actual demand.

This prevents demand variability from entering the supply chain by buffering it in inventory that has been designed for variability management (see the right side of Figure 2.4). In this sense, safety stocks are actively used to cover demand spikes. When configured correctly and dynamically

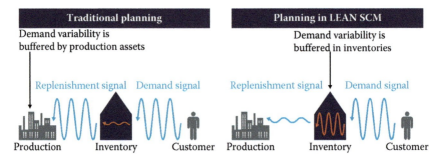

FIGURE 2.4

In LEAN SCM, demand variability is buffered in inventories and not passed on into the supply chain.

adjusted, safety stocks become a very powerful variability control mechanism indeed, and act as a first firewall against variability within SCM, also offering a means of avoiding that demand variability is propagated one-to-one or even amplified upstream. This built-in variability control actively counteracts the bullwhip effect and enables smoother capacity utilization.

2.2.2 Level Production Plans to Create Flow and Stabilize Utilization

If manufacturing commits to production campaigns that consist of MTS products, the supply chain obviously cannot be tied to real customer demand, which makes it difficult to meet foreseen or unexpected demand changes that occur frequently in the VUCA world. Therefore, most companies invested substantially in ERP or APS technology to cope more adequately with customer demand variability and volatility in their manufacturing schedules. Unfortunately, however, an increasing number of "rush orders" and replenishment orders caused by demand changes have to be scheduled—and even more often re-scheduled—in order to meet real or assumed demand (forecasts). This might suggest that manufacturing plans and schedules are perfectly tied to customer demand but in fact we see in most cases higher changeover costs and lower capacity utilization (see Figure 2.5). Furthermore, when this happens, product flow is thrown out of balance and overall throughput decreases.

To be sure, companies in process industries should, and typically do, aim for balanced flow and smooth utilization with regard to their capital-intensive assets. Therefore, production sequences and quantities should be leveled over time to improve flow and achieve more stable capacity utilization. Figure 2.5 illustrates how production quantities will be leveled by

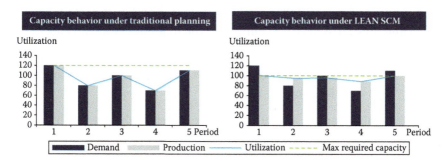

FIGURE 2.5
LEAN SCM enables stable capacity utilization and reduced excess capacity buffers.

FIGURE 2.6
Leveling of production quantities reverses the bullwhip effect.

LEAN SCM. In contrast to traditional planning, production-leveled schedules avoid overstressing of production capacity by adapting to demand peaks. Instead, demand peaks are now buffered actively in safety stocks and replenished with a time delay. As a direct consequence, capacity utilization becomes smoother and maximum capacity is lower compared with traditional capacity planning. Finally, leveling production reduces the need for capacity slacks, which means lower overall capital expenditures.

Moreover, by following LEAN Supply principles, upstream production stages can benefit from leveled production quantities as well. If production at any supply chain stage is leveled, the required input materials are leveled as well. This creates smooth replenishment signals from downstream to upstream stages, which makes it possible to reduce inventory levels. This effect can be regarded as the reverse of the bullwhip effect (see Figure 2.6).

A major objective of LEAN SCM Planning is providing formalized variability control mechanisms for inventories as well as manufacturing capacities. Therefore, we recommend building on the product wheel concept but evolving it to provide greater flexibility in high product-mix environments that face high demand variability and order volatility. In addition, a formalized variability control mechanism is required for production-leveled schedules to provide SCM with the key to controlling variability propagation upstream along the supply chain. In this sense supply chain planning can funnel and dampen the variability at each Rhythm Wheel-managed asset and, in sum, reduce variability and the infamous bullwhip effect.

2.2.3 Use Cyclic Production Patterns to Achieve a Common Takt and Regularity

Today, manufacturing schedules react very nervously due to the one-sided variability management approach of traditional planning. This creates significant variations in production rates and fluctuations in product flows, both ultimately leading to waste because manufacturing lines, equipment, workers, and required inventory must always be prepared for

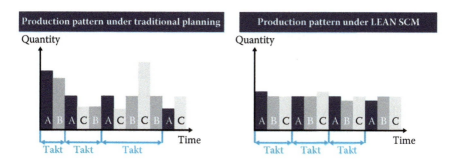

FIGURE 2.7

Cyclic production patterns lead to constant takt.

peak production. Traditional planning has responded to this by reserving additional capacity buffers on production lines. But even with these additional buffers, firefighting is still a daily routine that leads to burdensome planning efforts. Erratic production schedules and unfavorable order sequences due to the high variability of order quantities and completion dates remain common phenomena (see the left side of Figure 2.7).

Therefore, LEAN SCM focuses on implementing cyclic production patterns to achieve more favorable production sequences with regard to changeover costs and planning effort. Cyclic production patterns provide greater transparency in manufacturing as they follow a regular takt and mostly repeat the same production sequence over and over.

Cyclic schedules also contribute to production leveling and reduced planning complexity. The entire planning concept and its outcomes are easy to understand and thus are readily accepted by planners at the shop floor level. Repetitive patterns also facilitate organizational learning as shop floor employees are more certain about which products are to be manufactured next and thus can better master the production process, leading to economies of repetition. In light of better overall supply chain synchronization, upstream supply chain stages can better oversee the needs of their customers and what needs to be supplied to downstream stages. In this sense, cyclic scheduling enables a more effective coordination of activities between production stages to be achieved.

2.2.3.1 Summary of LEAN Supply Principles

Without doubt, the goal of achieving a steady and constant material flow on the supply side is of utmost importance. With the LEAN Supply principles we have introduced, your company can meet this objective while also buffering variability in inventory to a greater extent, not purely and one-sidedly on production assets. Therefore, dynamically adjusted safety stocks and inventory replenishment targets should become a variability control mechanism and act as the first firewall against variability. In this way, demand peaks can be kept under control and the bullwhip effect is more effectively prevented.

In addition, manufacturing quantities should be actively leveled with cyclic patterns to create robust production schedules that require lower capacity buffers. This leads to lower capital expenditures, which is a crucial competitive factor in capital-intensive process industries.

Moreover, cyclic production patterns introduce local takt in manufacturing. The transparency achieved in the production sequence and more predictable takt provide a new level of coordination between production stages. For your company's employees, cyclic patterns provide simple and comprehensible rules for managing production sequences and quantities, enabling rapid improvements in the production process.

2.3 LEAN SYNCHRONIZATION: HOW TO MASTER COMPLEXITY AND AMBIGUITY

In a LEAN end-to-end supply chain, demand and supply are synchronized. This ensures that the right product goes to the right place, in the right quantity, and at the right time. Synchronization also implies coordination across supply chain stages. By contrast, with the traditional planning approach, high inventory levels and widely fluctuating capacity requirements often occur due to inadequate coordination between production stages. By creating transparency of demand information across all supply chain stages and furthermore synchronizing production stages, a company can reduce inventories significantly while simultaneously balancing capacity utilization.

The following principles will enable a LEAN synchronized supply chain:

1. Separate planning activities and slice complexity for global synchronization.
2. Use "parameter-driven" end-to-end supply chain planning.
3. Establish visibility and a collaborative environment for end-to-end synchronization.

2.3.1 Separate Planning to Slice Complexity for End-to-End Synchronization

Traditional ERP- and APS-based planning concepts have blurred the line between planning parameterization—the way we configure all activities that need to be conducted during a planning run—and planning execution—the way we create replenishment plans or daily production orders. Today's planning practice, with no separation between planning parameterization and execution, results in perpetual planning, replanning, and order chasing as systems "shuffle the whole order pack" each time they run, requiring constant attention from all planners. Technically speaking, traditional planning slices corporate supply chain planning activities vertically into location-specific MRP runs to cope with the complexity involved, consequently increasing the need for synchronization across site locations.

Even within the local MRP, planners are barely able to distinguish forecasted demand from real consumption during a multi-level planning run. Such planning practices constantly push manufacturing to produce orders that are based on volatile and inaccurate forecast data; in most cases, they are "planning operations" to either over- or under-supply the supply chain.

In supply chain organizations that face ever-increasing network complexity and demand variability, it is vital to separate activities associated with the planning configuration from the daily generation of replenishment and manufacturing orders. As shown in Figure 2.8, end-to-end supply chain planning is sliced into a global tactical planning configuration and local operational planning that triggers execution. At the tactical planning level, all production and replenishment parameters such as inventory and replenishment levels or cycle times are configured as

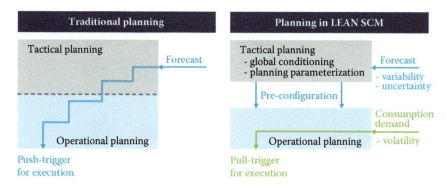

FIGURE 2.8
Tactical and operational planning are separated.

replenishment and manufacturing orders are released on the operational level during order execution.

The tactical configuration and its renewal should cover all planning activities designed to support the global S&OP process in order to build, synchronize, and agree on common supply chain plans. Best practice would be to align those plans with the frequency of the S&OP (in practice often monthly). Once the tactical planning cycle is complete, the time in-between will be spent in operational planning: raising orders against the tactical configuration in line with real customer demand pull.

2.3.2 "Parameter-Driven" End-to-End Supply Chain Planning

Configuration or parameterization in tactical supply chain planning is about building and agreeing on network-wide capacity plans and corporate inventory plans. In traditional supply chain planning, this is done sequentially and typically distorted by forecast variability. A forecast is used to determine the plans for the finished good stage; the dependent demand resulting from the underlying production plan is propagated upstream to the next supply chain stage. This stage again determines its own production plan and propagates the dependent demand upstream (see the left side of Figure 2.9). The drawback of this planning approach is the enormous planning effort and heightened planning nervousness, since all capacity and inventory plans need to be adjusted whenever a single production plan changes.

Consequently, most companies see the adoption of global parameterization and synchronization processes—undertaken to define the right level of inventory and capacity along the supply chain with the right buffer

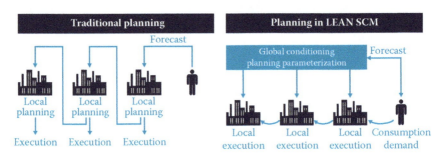

FIGURE 2.9
Global pre-configuration of supply chain parameters reduces planning complexity and planning nervousness.

against variability and uncertainty—as itself a critical business process. If SCM does not establish those processes that enable it to define a robust parameter and buffer strategy and some measure against which to pre-configure supply to address variability, the entire business might struggle.

In global networks, buffer and safety stock parameter setting plays a critical role in helping to protect the supply chain from variability and disruptions. Today, however, this is not an isolated task carried out at each plant or warehouse. Instead, a formalized global planning approach to defining buffers in inventory is required to ensure that the right balance between the benefits and costs is targeted. To achieve this, supply chain planners must determine both the location and size of inventory buffers at the same time, which requires the use of multi-echelon inventory optimization approaches. To make these decisions as robust as possible in the face of intensifying business dynamics, companies need to establish processes and rules for dynamic inventory target setting.

Tactical parameterization also requires focusing on capacity models that result from cyclic production patterns based on flexible Rhythm Wheels. The capacity models in place should be aligned with the established inventory rules and should also contribute to mitigating demand variability. Therefore, *variability control tactics* should be formalized at each Rhythm Wheel-managed asset.

Global parameterization uses forecasts for determining just a few global control parameters (e.g., target inventory levels, Rhythm Wheel cycle times), which constitute the framework for operational planning and execution (see the right side of Figure 2.9). Each supply chain stage simply follows these globally pre-configured parameter settings. Since only a few parameters need to be determined, planning effort can be significantly reduced. Furthermore, separated tactical pre-conditioning enables efficient planning execution, since parameter pre-configuration can be held constant for a certain period.

2.3.3 Establish Visibility and a Collaborative Environment for Synchronization

To optimize inventory levels and capacity utilization globally, each supply chain step should be synchronized with the upstream and downstream stages—and with customer demand. In the simplified and parameter-driven LEAN SCM Planning framework, synchronization of operations is achieved by global synchronization of the supply chain parameters. The mechanisms and rules for variability control of inventories and capacities introduced above are the central keys for supply chain synchronization in today's highly volatile and complex supply chains.

To achieve optimal flow and throughput along the supply chain, companies need to establish a global takt, which involves orchestrating and guiding all tactical parameterization and renewal processes. Aligning all manufacturing steps to a global takt ensures that all stages always produce the right quantities as multiple Rhythm Wheels mesh with each other like gears at the right time to serve global customer demand (see Figure 2.10).

However, functional silos in an organization might limit opportunities for systematically reducing complexity and prevent optimization of the end-to-end supply chain. Often, managers tend to focus on their own functions or departments. Without a cross-functional, end-to-end perspective along the supply chain, the required visibility and trust within the organization are lacking, keeping it stuck in a pattern of local silo decisions (Figure 2.11).

A global organization must transcend local interests to achieve global coordination and maximize performance. This requires breaking through functional silos and departmental walls in the minds of employees. However, employees are usually incentivized to aim at potentially conflicting (local) targets. To get people aligned and committed to common corporate objectives, it is essential to establish an end-to-end mindset and effective cross-organizational collaboration. However, collaboration starts with the fundamentals known to successful supply chain professionals: transparency, timeliness, and discipline.

To overcome collaboration barriers, people need to understand their individual impact along the entire value chain and the role of their own

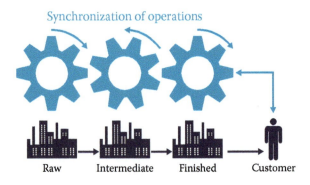

Synchronization of operations

Raw — Intermediate — Finished — Customer

FIGURE 2.10
Operations are synchronized by the synchronization of supply chain parameters.

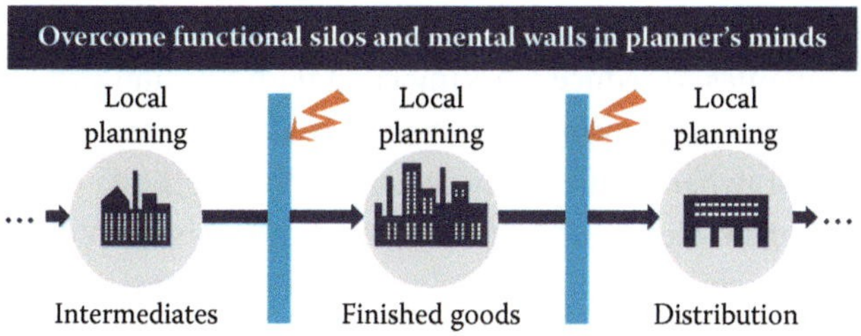

FIGURE 2.11
Create visibility and trust and tear down collaboration barriers.

work within the overall network. This allows them to evaluate the impact of their own local decisions on the rest of the supply chain.

However, to reach such important corporate goals, cross-functional empowerment is crucial to SCM; it ensures the foundation for successfully introducing a collaborative mode within an organization. Such empowerment enables global coordination of targets and incentives to guide the organization in the right and synchronized direction. This allows all stages of the supply chain to share the benefits of end-to-end collaboration activities.

2.3.3.1 Summary of LEAN Synchronization Principles

LEAN SCM calls for synchronizing demand and supply such that high customer service levels can be achieved at the lowest possible cost. To coordinate information and material flows efficiently along a supply chain, an organization needs to establish end-to-end collaboration. Only in this way can it achieve a true global optimum for the entire supply chain. By working with a parameter-driven framework, the planning effort is significantly reduced. Execution has only to adhere to the pre-configured supply chain. By synchronizing parameters globally, supply chain operations can be synchronized to customer demand, enabling stable material flows and ensuring that the right quantity is produced at the right time.

CHAPTER SUMMARY

Process industries face growing complexity and variability in the VUCA world. Traditional planning methods typically result in high inventory levels and the need for excess capacity since they cannot manage these challenges efficiently. At the same time, competition and capital markets pressure companies to reduce working capital and capital expenditures.

Companies can now choose between two alternatives: either learn to live with the drawbacks of traditional planning methods or change the fundamental principles on which SCM operates. This chapter introduced LEAN SCM principles for achieving an end-to-end synchronized supply chain. By following LEAN demand, LEAN supply, and LEAN synchronization principles, variability and complexity can be managed efficiently. The benefits on the demand side are high customer service levels and short lead times. On the supply side, LEAN SCM brings stable capacity utilization and improved asset performance. From an end-to-end perspective, total lead time is reduced and inventory levels are minimized. To explore how thought-leader companies of the process industries put these principles into practice, we recommend reading Chapter 12. There you will find the industry cases of your peers who already made big steps toward LEAN SCM.

3

Fundamentals of LEAN SCM Planning: A Paradigm Shift in Planning

Many companies struggle to adapt supply chain planning to the new business realities of the VUCA world. While the realities have changed, however, supply chain planning still all too often follows the principles and recipes of the past, when stability and predictability were still predominant characteristics in process industries.

Here, we describe a paradigm shift in supply chain planning, offering a new paradigm that allows companies in process industries to adapt their supply chain planning and design to the realities of the VUCA world more effectively. The essence of the new planning paradigm introduced in this book changes how your company should address the root causes of poor customer service and unsatisfactory results in SCM. With LEAN SCM, your company will address variability and uncertainty directly and actively manage these factors, while considering the unique manufacturing constraints on supply chains in process industries. In this chapter, we outline the fundamentals of the LEAN SCM Planning framework, answering the following questions:

- What is the best supply chain paradigm to follow in the VUCA world?
- What are the core building blocks of the new LEAN SCM Planning approach?
- How does LEAN SCM provide a basis for achieving a step change in performance and results?

3.1 WHAT IS THE MOST SUITABLE SUPPLY CHAIN PLANNING APPROACH TO FOLLOW?

A frequent topic of discussion is whether a lean, agile, or resilient supply chain strategy best fits a particular company's business model. *Lean* strategies are designed to reduce costs and not to "waste" valuable resources that are necessary to meet customer demand at any stage along the supply chain. On the other hand, *agile* supply chain strategies enable companies to react to and adapt rapidly and effectively to variations in customer demand. By following the *resilient* strategy, a supply chain becomes more robust to major disruptions in supply that might result from natural disasters, geopolitical developments, or other unexpected events (see Figure 3.1).

Many publications focus on the supply chain trend of the time, typically mirroring general economic trends. It seems that during economic downturns a greater number of organizations favor *lean* manufacturing and supply chain principles, focusing their activities on inventory reduction and optimization of base costs. In contrast, during economic growth phases, many companies favor *agile* and *reliable* supply chain principles, tolerating higher inventory levels to avoid missing any new market opportunity. Before we draw any conclusions about which is the right paradigm to follow, we briefly characterize all three through the lens of supply chain planning.

3.1.1 The Lean Supply Chain Is More about Waste Elimination and Cost Efficiency

The roots—and broad acceptance—of *lean management* principles go back decades. Despite the huge number of *lean principles* and supporting

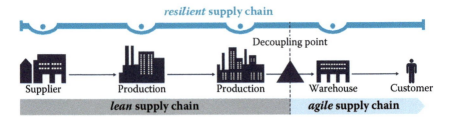

FIGURE 3.1
The traditional view of the lean, agile, and resilient supply chain paradigms.

lean tools, we will keep the story short and refer the reader to one of the numberless publications in the field of lean manufacturing or lean management. Furthermore, most companies already have substantial experience in this area through lean manufacturing initiatives, which aim at increasing value added by eliminating waste and improving operational excellence at the shop-floor level. Beyond the shop floor, lean principles have already been extended in many companies through a variety of lean supply chain improvement programs. Starting typically with a value stream mapping initiative, waste and excess resources (measured in costs, time, and inventory) are eliminated first. In the second step, the major intention is to create *leveled flow* across manufacturing sites and along the entire value chain, to smooth resource utilization, increase throughput, and improve cost efficiency. For a *lean* and *leveled flow* along the supply chain, it seems obvious, a matter of common sense, to move products downstream in the fastest possible time with a minimum of waste, and to perform these activities without interruption when a customer requests them.

However, most practitioners in process industries consider the *lean* and *leveled flow* paradigm to be the most suitable for the supply chain steps upstream of the decoupling point. At the decoupling point, companies have typically held their strategic stocks, as demand at this stage is typically smoother and higher product volumes flow through a manageable number of value streams. In contrast, *leveled flow* principles are rarely applied downstream of the decoupling point in process industries, which can be explained by the typically divergent supply chain structures and high-mix product "fan-outs."

3.1.2 The Agile Supply Chain Is More about Responsiveness and Customer Service

Many companies have developed capabilities for responding more rapidly to changes in customer needs in terms of volume and variety. The main characteristics of such *agility* capabilities are considered to be flexibility, responsiveness, and speed in mobilization of global resources. These qualities allow companies to keep up with changes in technologies, supply markets, and customer expectations. A key enabler of an agile supply chain is "strategic flexibility" to enable prompt adaptation of network structures and relationships between all network entities, including suppliers and contract manufacturers. On the tactical planning level, vital *agility* enablers include supply chain visibility and integrated processes. Responsive tactical management decisions, such as shifting production volumes from one plant to another or to a contract manufacturer, are possible only when SCM can access information in real time. Furthermore, agile supply chains are characterized through their market sensitivity and are closely connected to end-user trends and demand signals. An agile supply chain is therefore the key enabler for companies to exploit opportunities in volatile and dynamic marketplaces.

The *agile* paradigm is widely accepted, but the common understanding of SCM practitioners in process industries is that it should ideally be applied downstream from the decoupling point, where product specifications vary more widely and demand variability regarding a particular product variant is high. This attitude can be explained to some extent by the experience these practitioners have with planning processes and systems of the past. Because frequent planning and rescheduling of upstream operations, which are highly dependent on product sequences, utilization, and cost-sensitive assets, cause disturbance they would like to avoid.

3.1.3 The Resilient Supply Chain Is More about Risk-Avoidance and Robustness

Today's global marketplaces are subject to increasing levels of turbulence and risk exposure, often rooted in unpredictable supply chain events, for example, natural disasters, geo-political developments, and economic or legal developments. But even as companies have labored mightily to squeeze costs and reduce variability buffers (inventory and lead times) in the past, their supply chains have become more vulnerable to such events. When major disruptions occur in such "buffer-stripped" networks, supply chain activities tend to be interrupted and require a long time to recover. Therefore, more and more companies have recently been adopting countermeasures such as systematic supply chain risk management and promoting best practices to increase robustness and resilience.

During the financial crisis of 2008–2009, many companies experienced bankruptcies among their major suppliers and contract manufacturers. When a company follows a single sourcing concept, such an extraordinary event can lead to substantial issues for customers and thus for the entire supply chain. To master such a situation, the prompt *strategic* reconfiguration of a company's own network structures is essential to achieve supply chain resilience.

On the *tactical* level, the key imperative for developing resilient capabilities is global visibility. Take for example the eruption of the "E15" volcano in Iceland, an unforeseen natural event that disrupted air travel in Europe for 6 days in 2010, or Hurricane Sandy of 2012, whose 110 mph winds caused $74 billion in damage in a swath that it cut from the Caribbean to Cape Cod. Both events paralyzed air, rail, and road networks. In such cases, the complexity involved in revising integrated supply chain, production, and distribution schedules cannot be overstated. Yet creating and evaluating feasible recovery scenarios are impossible without accurate and timely data. In SCM, several urgent questions arise: how can product supply be rerouted from air traffic to available trucks? How can product confirmations be reallocated to serve markets short on stock first? Many companies in these situations could not apply appropriate countermeasures to manage the disruptions adequately due to the lack of end-to-end supply chain visibility.

Finally, on the *operational* level, short-term resilience is more about organizational behavior, and enables a company to stand firm while conducting everyday business. In this case, stakeholders in the supply chain and operations units need to be aligned with the organizational attitude that determines how effectively they react to such disruptions in supply. They need to be able to adequately manage and buffer delays in production and transports. For example, when touching safety stock levels, employees in operations tend to overreact and usually create more turbulence at the top—even though safety stocks have been calculated for exactly such disruptions.

The *resilient* supply chain paradigm focuses on activities that are involved in meeting these challenges. Resiliency determines how well the supply chain responds to disturbances and how quickly it returns to the original state of business activity. Your company's potential to exhibit resilience can be a competitive advantage if its supply chain can respond more quickly to disruptions than its competitors.

3.1.4 Trade-Offs among the Common Paradigms in Supply Chain Management

As demand variability and volatility have been increasing in recent years, many companies have been discussing whether to shift their SCM strategies toward agility and reliability, with the implication that any progress made toward lean and leveled flow no longer seems impressive. However, based on our experience with leading companies in process industries, it should not be a question of choosing agility OR reliability, but rather a question of how to achieve agility AND reliability in accordance with market needs. The question then is where the optimal balance between these paradigms lies.

In other words, how can your company's SCM organization ensure that it can maintain the necessary agility in moving goods and services through its extended supply networks, not only to satisfy ever-changing customer demand or manage threats posed by disrupting events, but also to effectively manage the supply chain in an era of lean practices carried out to achieve maximum efficiency?

Obviously there are trade-offs among the broader concepts involved in SCM (see Figure 3.2); the effects of which strongly depend on the business environment. So, first, it seems important to increase a company's agility to ensure that it can operate effectively in a more dynamic environment. In general, the benefits of agility increase substantially in the face of variability, but this strategy is typically accompanied by an increasing need for working capital and often more costly operational processes.

On the other hand, it is more imperative than ever for companies to accomplish better results while consuming fewer resources. Thus, many companies aim for the benefits of lean principles in their supply chains. As a consequence, the entire supply network requires leveled product flow and synchronization of activities to ensure minimal buffering in inventory and capacity. Finally, any supply chain, whether lean or agile, must

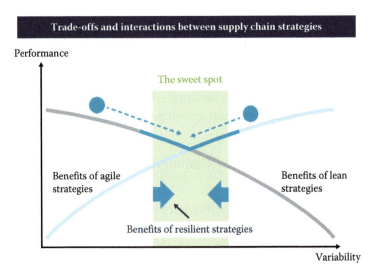

FIGURE 3.2
How to find and stay in the sweet spot of supply chain performance.

have the capabilities it needs to manage risk and ensure resilience following disruptions. As all strategies offer appealing benefits, a balanced supply chain strategy is required to manage VUCA challenges successfully. But the key question is then: how can your company find the "sweet spot" that balances the different strategic directions?

Here it is important first to recognize that this requires a strategic direction that combines both leanness and agility. As outlined above, a company must "plan for variability" across the entire planning horizon and prepare its operations to "respond to volatility" as the plans are executed. Ideally, an SCM organization is able to plan for variability in a lean way, but when it comes to executing its plans it must also be able to respond to volatility with great agility. A strategy that combines these two elements effectively will certainly find the sweet spot for supply chain performance.

To remain in the sweet spot, however, requires resilience capabilities to bring supply chains back into a desired level when out of balance. Once SCM has set a planned direction (plans), ideally around the "sweet spot," velocity is required to get there quickly. Velocity is defined as speed to rebalance with the direction set in supply chain plans and therefore a major resilience capability.

To be sure, most supply chain professionals would agree with us to this point in the discussion. However, most would now wonder how they can find this sweet spot of targeted supply chain performance and, even more importantly, how can they remain in this area of heightened efficiency when in the face of increasing demand variability one of these unpredictable events affects the company.

So how can your company keep going in the "planned" supply chain direction while adjusting to the perpetually changing supply chain reality in execution? Obviously, this means setting the correctly balanced direction for supply chain performance. But there is also a need to structure supply chains and planning capabilities to incorporate formalized principles and processes that support continuously renewed and resilient plans. Responding with the necessary speed that the VUCA world demands from companies today creates the needed "velocity to resilience."

3.1.5 How LEAN SCM Combines and Builds upon a New Planning Paradigm

To help your company find the sweet spot of supply chain performance, we have developed the *LEAN SCM Planning framework* for process industries that actively addresses the challenges of the VUCA world. Therefore, we hark back to the LEAN SCM Planning principles and a combination of the conceptual elements of the *agile* and *lean* supply chain paradigms highlighted above. One of the central elements of this new planning framework is taking an integrated approach to "closing the loop" between variability management and volatility management. This means that SCM planning processes will be integrated through continuously ongoing renewal activities (on the strategic, tactical, and operational levels) with effective response capabilities in execution (see Figure 3.3).

We are convinced that efforts made to build "agile response capabilities" in supply chains—especially in the capital-intensive process industries—should not be limited only to the downstream or customer-facing value chain steps. In fact, preparing response capabilities is the first step toward creating "strategic flexibility" in supply chains. The results of a flexible and prompt reconfiguration of supply chain networks and relationships (see overview in Chapter 4) are essential to adapting quickly to new market developments or serious supply disruptions. Additionally, we have incorporated automated segmentation and related classification activities into a *regular strategic capability review* for "supply mode selection" in production and replenishment to better keep pace with the new market dynamics (see Chapter 5). Finally, we have embedded operational agility into the framework using the best way to "respond to volatility" by driving supply chains based on real customer demand. Propagation of the "customer voice" is enabled through *demand-driven* and consumption-based pull replenishment signals (see Chapter 7)—from the first customer-facing value chain step upstream to the last upstream stage. However, to operate a supply chain in this way, the SCM organization needs to prepare and plan for right-sizing and parameterization of buffering capabilities in inventory and capacity (see Chapter 6).

In the face of the long product lead times that are typical in process industries, companies must create tactical plans for 12 months to as many as 36 months ahead to be able to synchronize every last stakeholder upstream along the supply chain. To do so effectively, the best possible demand forecasts of customer expectations must be generated and used for supply chain parameterization. There are, however, two major differences regarding the use of forecasting information in LEAN SCM Planning compared with traditional planning modes. First,

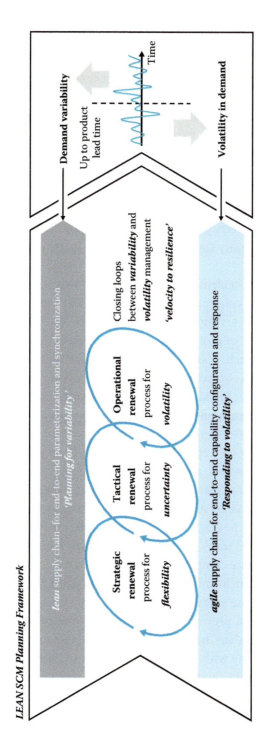

FIGURE 3.3

LEAN SCM Planning framework for process industries facing the VUCA world.

forecast information should never be translated and released directly into operational manufacturing orders, because there is too much uncertainty around those forecast data. In contrast, real consumption data that becomes available later in the process should be used to trigger manufacturing orders. Therefore, product-level forecasts are no longer needed for tactical planning, as always been assumed in SCM, because we do not need them for operational order creation. The second difference is consequently the replacement of product-level forecasts at the tactical level by more aggregated forecasts. The basis for such aggregation is for instance time or product hierarchies. Overall, this is an efficient way to reduce demand variability along supply chains as much as possible. Simultaneously, this also helps to significantly reduce the workload and complexity involved in the monthly demand forecasting process.

The *lean* supply chain elements we introduce in this book will provide your company with a conceptual framework for clear (and, as far as possible, stable) direction-setting across the entire end-to-end supply chain organization and help to avoid ambiguity among roles in the new LEAN SCM Planning approach. Starting with the best available data for future demand expectations, we offer, on the tactical level, *integrated supply chain parameterization* based on only two manageable parameters, thus providing simplicity and reducing planning complexity. These parameters are product- and asset-specific cycle times combined with right-sized inventory targets. The conceptual foundation for cycle time determination is the Rhythm Wheel approach; while inventory target setting is based on multi-echelon planning approaches that consider variability along the entire end-to-end supply chain (see Chapter 6). All those supply chain parameters will be calculated to provide the best possible *synchronization* to the "global takt" of each value chain to ensure overall cost-efficiency and corporate competitiveness. However, a supply chain that is parameterized and synchronized in this way is in effect a snapshot of the "*lean* ideal state." Since every "tactical plan" needs continuous renewal, we provide formalized processes for adaptation to anomalous or unexpected situations (see also Chapter 6). In this sense, our approach not only provides your company with a tactical "plan for variability," but also the acceptance of uncertainty and the preparation for changes in the business environment.

The concept of LEAN SCM Planning therefore features continuous control loops. In such closed-control loops, there is an objective function (the tactical plan) that aims for the best possible end-to-end leveled flow, along with continuous sensing to recognize deviations from the objective function (monitoring and feedback) with formalized renewal processes on different time horizons and decision levels (adaptation):

1. The *strategic renewal process* (described in Chapter 5) reviews decisions involving the redesign of the supply chain network and supply chain mode configuration, accounting for flexibility in capabilities.

2. The *tactical renewal process* (described in Chapter 6) reviews and regularly updates the planning parameters, accounting for uncertainty in plans. It is best embedded in an operational, monthly S&OP cycle. Here, the supply chain is also prepared for exceptional demand and supply events such as building up stock for marketing campaigns or plant shutdowns.

3. The *operational renewal process* (described in Chapter 7) involves short-term parameter "factoring" to adapt preconfigured operations to short-term changes in consumption demand or supply deviations, enabling volatility management in execution.

Renewal processes in LEAN SCM should be built up in a formalized way to enable rapid adaptation to changing conditions or disruptive events at any supply chain planning level. By adopting such processes, your company will be able to achieve greater velocity in SCM. Furthermore, the associated *resilience* capabilities will increase SCM capabilities at the strategic, tactical, and operational levels.

In the LEAN SCM Planning framework, demand variability is effectively managed through robust planning processes for parameterization and synchronization, and demand volatility is effectively managed through robust response capabilities that are configured and prepared in supply chain execution. The implications of this new SCM paradigm are far reaching. The most fundamental implication is that tactical planning and operational execution are set up in a completely new way. However, by separating these functions, LEAN SCM actually synchronizes them much more efficiently. In this vein, LEAN SCM takes a different course from supply chain planning systems currently in place in order to cope with the new realities of the VUCA world. In the end, this will allow your company to make more out of its supply chain and thus lead to new competitive advantages (see Section 3.3 for benefits of LEAN SCM).

3.2 THE BUILDING BLOCKS FOR LEAN SCM PLANNING: CONCEPTS AND HIGHLIGHTS

Given the ongoing evolution of SCM, new ideas and concepts are constantly emerging. In many cases, old concepts are dusted off and reintroduced as something new. Of course, it might be fair to ask yourself if LEAN SCM really does provide something valuable and new for you and your organization. Therefore, we now highlight the five core building blocks of the *new* LEAN SCM Planning concept we introduce in this book. In some cases, you might recognize familiar elements, but you will also see that each building block goes far beyond what has been used before for planning in process industries.

The paradigm shift entailed by the LEAN SCM Planning concept will result in meaningful changes at various areas and levels of supply chain planning—on the demand side, the supply side, and in supply chain synchronization as well. In each of these areas, some fundamentally new conceptual elements will be introduced that are significantly different from today's traditional planning processes and systems. These conceptual enhancements for planning in process industries will help your company overcome the obstacles of today's VUCA world.

Before introducing each element in detail in the course of this book, we will first introduce the major highlights to explain the motivation and fundamentals of the supply chain planning architecture we are laying out. In detail, those building blocks provide answers to the following questions:

- How can cyclic planning be applied in high-mix and high volatile manufacturing environments?
- How can dynamic inventory targets and active use of safety stocks in planning reduce variability?
- How can Rhythm Wheel planning and global takt synchronization be aligned in a flexible way?
- How can tactical pre-parameterization and planning execution reduce complexity?
- How can visibility be created with LEAN SCM Planning in existing IT systems for sustainability?

3.2.1 Flexible Rhythm Wheels Enable Cyclic Planning while Responding to Variability

We must admit that the conceptual roots and foundations of Rhythm Wheel-enabled planning are not absolutely new. However, LEAN SCM Planning extends their applicability to high-mix product manufacturing environments that operate under high variability; it further makes it possible to achieve end-to-end synchronization of multiple Rhythm Wheels that operate in interlinked but globally dispersed production sites.

Industry experts such as Ian F. Glenday, Peter L. King, and Raymond C. Floyd have already introduced "cyclic scheduling" and "product wheel" applications in process industries. They were able to connect the general *lean* (manufacturing) concepts and *lean principles*, as well as the underlying elements of simplicity, flow, and pull, with the physical restrictions of production planning and control in process industry manufacturing.

They first started to apply cyclic schedules with the ultimate ideal of producing every product every cycle (EPEC), with the dominant objective of achieving simplicity and stability in manufacturing. But they also recognized that an EPEC schedule is not feasible in all manufacturing environments and across the entire product range running on a specific production line. Later they developed methods for differentiating, for example, fast-mover and slow-mover portfolios in product wheels. Although they showed how to identify products that are suitable for specific EPEC schedules, product wheels would still be propagated for a more stable environment, typically with fixed product wheel cycle times. Nevertheless, product wheels within such a stable cyclic scheduling environment have already been applied at various chemical manufacturing sites of industry giants such as DuPont and Dow Chemical. At these facilities, the companies apply a simple product wheel design process, based on practical methods for finding the optimum manufacturing sequence, minimizing changeover costs, and freeing up useful capacity.

However, in discussions with various companies from process industries, we identified three unanswered questions that were highly relevant for most of the companies:

- How can product wheels be applied in a high-mix production environment with a high mix of variability patterns in demand, as often seen with pharmaceutical manufacturers, in specialty chemicals, or food production?

- How can multi-step supply chains that use product wheels at more than one site be synchronized, and how can that synchronization be adapted in a flexible way to cope with variability and unexpected events?
- How can global visibility of supply and capacity be achieved for end-to-end capability synchronization and how can product wheels be integrated into corporate supply chain planning?

In recent years, we have consolidated practical product wheel experiences across process industries—using our own approach and those of various industry experts. To achieve the vision of a transformation of the product wheel concept into a high-product-mix and high-demand-volatility manufacturing environment, we refined the existing industry approaches to product wheels along several dimensions. These efforts allowed us to introduce the "Breathing" and "High-Mix" Rhythm Wheels concepts (see Section 6.1 and Figure 3.4).

The most important conceptual advancement is our approach regarding management of cycle time and cycle time boundaries (upper CT+ and lower CT–) for Rhythm Wheel-based planning. The concept of cycle time boundaries and a formalized process for calculating and using them now provides the required manufacturing flexibility to manage increasing demand volatility. And this can be done without jeopardizing the benefits of cyclic planning that makes the use of product wheels so appealing.

The effect of "breathing" in Rhythm Wheels within cycle time boundaries can be explained and visualized as a ballooning tube, providing production output in the product sequence as designed, but with more or less throughput allowed within the parameterized volatility boundaries. This increased flexibility on a cyclic-takted production line should be

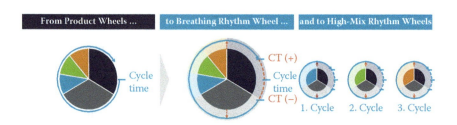

FIGURE 3.4

Breathing and High-Mix Rhythm Wheels enable cyclic planning while responding to variability.

monitored and managed carefully (see Section 7.3). Permanent measuring and visualizing of cycle time developments over time-phased Rhythm Wheel cycles are central and necessary for monitoring the pace or takt of a production asset. It is similar to monitoring the heartbeat with a cardiograph; we must constantly monitor the drumbeat in manufacturing to understand and act in light of implications along the supply chain.

In case tactical planning variability or any unexpected demand peaks occur above the set cycle time boundaries—above the calculated and therefore allowed volatility boundaries in manufacturing—then LEAN SCM makes it possible to mitigate such volatilities with designed inventory buffers. We explain this in greater detail in the following section, which covers dynamic inventory target setting and the effective use of safety stocks in planning. We should emphasize here that the concept of cycle time boundaries for supply chain synchronization in cyclic planning provides for the very first time a formalized key to smoothing variability and limiting volatility propagation upstream along the supply chain.

Even as the Rhythm Wheel approach is designed to be more flexible and to "breathe" within the cycle time spans, at the same time (replenishment) demand propagation upstream is restricted within the defined upper and lower cycle time boundaries. This achieves significantly reduced variability upstream along the Supply Chain. Effectively, demand variability is cut off, above and below the agreed cycle time boundaries as defined in the Rhythm Wheel design. Therefore, at every Rhythm Wheel-managed stage there is "funneling" and "dampening" of variability in a controlled way up to the next sourcing site (see Figure 3.5). Instead of incurring the infamous bullwhip effect through upstream propagation and internal amplification of internal demand, it is possible now to reduce variability propagation actively and in a highly controlled way. Without doubt, this new and more flexible planning approach will allow for a step change in supply chain performance as it enables much more efficient use of the working capital tied up in safety stock and variability buffers.

One of the central challenges for cyclic planning is finding the best cycle time for each manufacturing line. Companies that have already applied product wheels will know how complex and cumbersome this is since typically only self-made Excel files are available for decision support. And this becomes even more difficult when it comes to high-mix manufacturing lines with several low-volume products needing a much lower manufacturing frequency than those defined by the product wheel cycle time. For these cases, we have developed a High-Mix Rhythm Wheel concept with

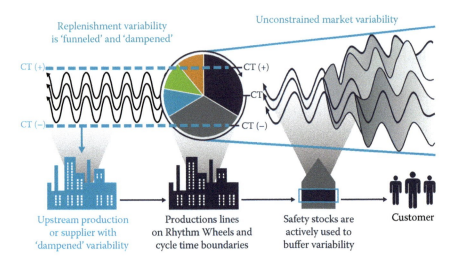

FIGURE 3.5
Upstream variability is actively funneled and dampened.

dynamic calculation and balancing of product manufacturing frequencies over the planning horizon (see Section 6.1). This falls very much in line with the principles of "campaign planning" in process industries, but is now aligned with the concept of cyclic planning with Rhythm Wheels.

When necessary, Rhythm Wheel parameters should be adapted through the tactical parameter renewal process (described in Section 6.4), which is typically integrated into the monthly S&OP process. Short-term adjustments are made within the short-term operational renewal process, which we call parameter factoring (see Section 7.2).

Traditional product wheels are typically locally managed manually or with simple Excel-based solutions outside of the existing corporate information systems; their planning results are then transferred manually into local ERP systems for execution. As a consequence, product wheel principles and associated techniques tend to significantly differ across sites or even manufacturing lines. Therefore, such well-intentioned initiatives very often ignore process standardizations and lose global visibility in corporate planning systems, hindering global synchronization of supply chains.

To avoid such issues in Rhythm Wheel-empowered planning, we developed the LEAN Rhythm Wheel Planning Suite inside the corporate IT systems platform for supply chain planning (see Chapter 10.3). The LEAN Planning Suite includes the "Rhythm Wheel Designer," "Rhythm Wheel Scheduling," and the "Rhythm Wheel Monitor" components that make it possible finally to achieve the required global visibility for Rhythm Wheel-cycled product supply and takted capacity utilization for synchronization of supply chains.

3.2.2 Dynamic Safety Buffers in Planning for Two-Sided Variability Management

A major improvement in LEAN SCM Planning is the use of safety buffers and, specifically, safety stocks that are used to hedge against variability. In most companies, this alone could involve a paradigm change in supply chain planning practice, because in traditional supply chain and tactical production planning processes, variability is managed and buffered one-sidedly, through capacity. With LEAN SCM, we want to change this and improve the ability to "plan for variability."

So how does this work today? In supply chain planning practice, inventory target levels are carefully calculated and recalculated. In particular, these calculations include various safety stock elements that account for customer demand fluctuations, forecast inaccuracy, and lead time variations.

The calculated safety stocks are later used only for unexpected supply and fulfillment situations, responding to short-term volatility (see Figure 3.5). So far this seems quite logical and should not be a cause for complaint, and definitely not a basis for the sort of "panic behavior" or "nervous reactions" that arise in SCM organizations when safety stock must actually be consumed in the face of a potential shortage. Essentially, safety stocks are designed for such situations and should not give the planner a bad conscience, as it often does. However, although this is still not the supply chain "paradigm change" we have been describing, it provides a good indication of how planners typically react in tactical planning as well.

In contrast, safety stocks are never used in the tactical planning process despite the fact that they have been planned for variability. The previously calculated safety stocks that are built up to meet inventory replenishment targets in the calculation are never touched in those tactical planning runs, despite the fact that they have been designed for, and are intended to be used for, specific demand peaks. As a consequence, resulting from this practice, very often a large amount of dead stocks occur in overall inventory, and this share will further increase when the variability increases (see Figure 3.6).

But even more painful is the fact that planning runs are creating stock replenishment orders with the same one-to-one quantity oscillation as in

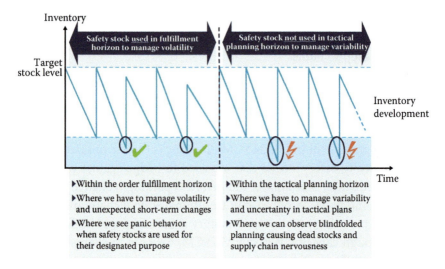

FIGURE 3.6
Safety stocks in the fulfillment and the tactical planning time horizon.

the forecasted (market) demand signals with all their forecasting inaccuracy. On the tactical planning horizon, demand oscillations are therefore not absorbed through inventory due to the fix-maintained inventory targets, nor are they dampened by the safety stocks included in the tactical plans. This means in consequence that all the market variability and forecast uncertainty are passed along via the planned stock replenishment signals, one-to-one into production planning and manufacturing capacities. Consequently, the capacity side has to manage all the variability one-sidedly through permanent replanning and rescheduling activities.

Such planning behavior typically results in lower capacity utilization, since production and asset planners have to violate the best possible production sequence to maintain sufficient capacity buffers to manage increasing demand variability. But with ever-increasing market dynamics on one hand and the need to accomplish better results with fewer resources in less time on the other, the traditional planning approach must be seen as a conceptual dead-end for today's variability management. This therefore calls for a paradigm change in this supply chain practice.

The new LEAN SCM Planning concept masters variability with a two-sided approach (see Figure 3.7). Demand variability in supply chain planning is managed on both sides, on manufacturing capacities as well as on the SKU. In particular, safety stock elements in all SKU inventories are now actively used in tactical planning runs, as they have been designed to

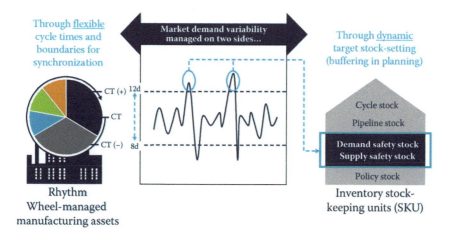

FIGURE 3.7
Market demand variability is managed on two sides.

do, to level the replenishment trigger signals and keep market noise out of manufacturing, at least to some extent. To make this happen, we have developed and designed an additional renewal process as a structured and disciplined approach for the dynamic adaptation of target inventory replenishment levels (IRLs).

This now allows SCM to keep some parts of demand variability—peaks in the forecasted demand signals—out of manufacturing plans. In this way, planners can actively use the safety stock in tactical "planning for variability" as designed. This will smooth capacity utilization and planners will not need to resolve so many unnecessary changes in production plans and schedules.

3.2.3 Cycle Times and Inventory Targets Aligned to Global Takt for Synchronization

We have now introduced several concepts and tools that can be used to "plan for variability on the supply side" and create ideally leveled plans for manufacturing capacities. We achieve this with cycle times and boundary setting within the flexible Rhythm Wheel design. Consequently, we "plan for variability on the demand side" through dynamic inventory sizing and right-sized target setting to enable real consumption-based replenishment down the line in operations and fulfillment (see Section 6.2). Overall, we now have developed the concepts and instruments your company needs to mitigate market variability in planning. This is achieved through ongoing adaptation of cycle time boundaries and subsequent use of the calculated safety stocks in our tactical plans.

We have emphasized the importance of end-to-end supply chain synchronization in planning. Therefore, all LEAN SCM Planning concepts are tailored to achieve true end-to-end supply chain synchronization. But how can your company achieve multi-echelon synchronization of global supply networks at the tactical level? This question is especially important in process industries due to typically large differences in lead times, starting generally with long chemical conversion processes and moving downstream to the shorter physical formulation and packaging processes.

To see a better illustration of this issue, it is useful to view the various Rhythm Wheel designs as gear wheels of unequal sizes, representing unequal cycle times, which must nevertheless mesh with each other (see Figure 3.8). To create end-to-end leveled flow, it is necessary to formalize a "global takt" for synchronization and to balance the cycle time differences of the interlinked product flows with the right-sized inventory levels from step to step (see Section 6.3).

In a stable supply chain environment, this could be enough for the purpose of synchronization, but not in cases with high demand volatility and high product mixes in the manufacturing portfolio. To see this, recall the above-mentioned picture of ballooning tubes to visualize the "breathing" Rhythm Wheel. Now imagine having many of those "ballooning tubes in a tightly interlinked pipeline." Imagining this might bring to mind a

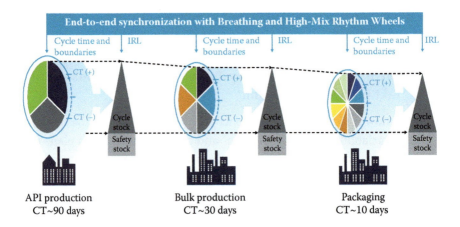

FIGURE 3.8
Unequally sized Rhythm Wheels to be geared with the global takt and dynamic target inventory replenishment level settings.

fragile network. But when they are accurately leveled, Rhythm Wheel-balanced supply chains provide a significant "calming effect" representing a step change in performance and financial benefits.

We have defined two core parameters that are required for synchronization: cycle times and cycle time boundaries for every Rhythm Wheel, and a dynamic inventory target-setting approach for the multi-echelon supply chain. In line with these parameters, we have designed continuous planning control loops with formalized renewal processes on tactical (see Section 6.4) and operational time horizons (see short-term parameter factoring in Section 7.2). These techniques provide structured realignment and decision support for adapting synchronization parameters when production falls out of takt or an imbalance occurs. When these renewal processes are embedded in operating models, the needed capabilities for agile SCM are developed. By adopting such renewal processes, your company will be able to carry out SCM with more direction and therefore more velocity. As a consequence, the supply chain can react rapidly when something falls out of sync through a disruption, providing resilience through a global re-takt or reset of the supply chain.

3.2.4 Separation of Tactical Pre-Parameterization and Planning to Reduce Complexity

At this stage, it is important to understand how to manage demand planning, supply planning, and synchronization following the new supply chain concept. The next logical step is the transfer of these planning concepts into process architecture and later into system architecture for implementation. Although this is a logical consequence of the process we have described, it constitutes the largest barrier to a LEAN SCM Planning transformation in most companies.

Today's process architectures in supply chain planning are very much constrained by the underlying ERP and SCM software tools, as discussed above. These tools simply lack the required differentiation between variability and volatility and lack the capacity to manage both efficiently. Under the LEAN SCM concept, a more differentiated view of volatility and variability is taken. Demand variability has been introduced here in line with future demand forecasts outside customer order lead times, and volatility has been introduced in the form of demand deviations within the lead times of the received order against the plan. This distinction has had far-reaching consequences in our conceptual framework. Therefore, the new approach in supply chain planning must involve a two-step planning architecture (see Figure 3.9) that makes a clear distinction between:

1. The *tactical planning process layer* in which companies "plan for variability" and the demand and supply parameter configuration (lean-leveled parameterization of cycle times in Rhythm Wheels and inventory targets reflecting the long lead times and variability).

2. The *operational planning process layer* in which companies "respond to volatility" in more simple planning runs for consumption-based, pull replenishment within the preconfigured parameters and buffers, while strictly keeping forecasted elements out of manufacturing and fulfillment orders.

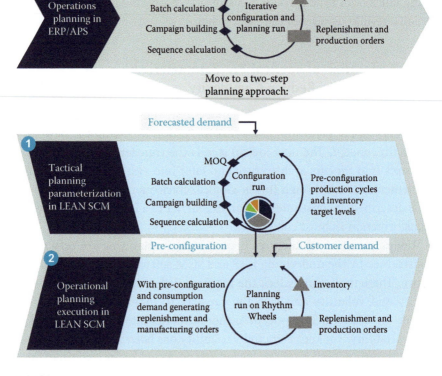

FIGURE 3.9

Two-step separation of tactical configuration and operational planning in LEAN SCM.

Maintaining this distinction is not an easy task; in fact, it is not even possible within the process architectures and planning systems generally in place today. The way in which they manage variability and volatility in the planning processes makes it irrelevant whether they have forecasted future demand variability, volatility associated with incoming customer orders, or just volatility associated with internal stock replenishment orders; those ERP-based or APS-based planning processes manage these factors in the same mixed-up way. In consequence, they give planners no chance to identify the root cause of volatility in his or her typical MTS environment, and that is why they add some "just-in-case stock" to be on the safer side.

As we have outlined in presenting the LEAN SCM Planning framework, long lead times force supply chains to plan for future demand and demand variability. In LEAN SCM Planning, this is done at an aggregated demand level, but in a way that enables planners to prepare the right supply chain plans and parameter configurations for the right level of the needed capacities and inventories later in execution. With leveled and takted plans and right-sized parameter configurations (parameterization at the tactical level), supply chain managers can develop efficient capabilities for agile responses to customer orders and associated order volatility. This makes it unnecessary to guess or forecast product levels for short-term customer orders. Within the operational planning and fulfillment time horizons, it is now possible to work within the buffers and parameters that have been planned on the tactical level. This makes the supply chain well prepared to respond to consumption-driven pull replenishment based on the real customer order.

Regarding eventual implementation of LEAN SCM Planning, we wish to emphasize and summarize the most important aspect of the planning process architecture that is required to make the recommended changes possible. In this respect, it is important to separate the tactical preconfiguration and operational planning layers in your company's existing business process architecture and system platforms. The parameterization or preconfiguration of the supply chain is most efficient when based on aggregated demand forecasts, but actual operational planning is conducted only on actual (and thus reliable) consumption signals. This separation is essential, but it is not possible to transform your company's existing process and system architecture into LEAN SCM Planning without additional IT planning support.

3.2.5 Enabling IT to Create Global Visibility and Staying Power for Sustainability

In SCM, it is not possible to achieve end-to-end visibility of inventories or utilize a company's capabilities for synchronizing capacities and inventory buffers in the global network without an integrated IT planning system. The LEAN SCM Planning framework and process architecture are designed to achieve this visibility and enable end-to-end synchronization and should be seen as the *"conditio sine qua non"* or "must-have" to get LEAN SCM up and running. However, it is important to note here that the designed planning process architecture we have described cannot simply be transferred into existing planning systems.

So, does this mean that your company will need to replace its existing planning system platforms to implement the new LEAN SCM Planning approach? Absolutely not! Obviously, we cannot ask companies to replace their SAP ERP and SCM systems, for example, which are commonly used in process industries. Companies have typically made substantial investments in those systems, and the achieved standardization and consolidation must not be jeopardized.

Consequently, when we first began assisting companies in transforming their planning concepts toward LEAN SCM Planning, we had to do this on the existing system platforms. To allow for an effective transformation to LEAN SCM, we developed additional IT concepts and software add-ons to enhance the existing planning systems for the initial implementation. These enhancements of the existing IT infrastructure address both of the key layers of supply chain planning (see Figure 3.10):

1. **End-to-End (global) LEAN SCM Planning Add-on Suite** for tactical supply chain parameterization, multi-stage cycle time and multi-echelon inventory synchronization with decision support for the global *tactical renewal processes*; the required solutions for implementing new "planning for variability."
2. **Site-Level (Local) LEAN SCM Planning Add-on Suite** for Rhythm Wheel-managed assets, with Rhythm Wheel design and monitoring, Rhythm Wheel planning heuristics with parameter factoring during the *operational renewal processes*, creating resilience against order-injected volatility in manufacturing; the required solutions for creating and implementing an agile "response to volatility."

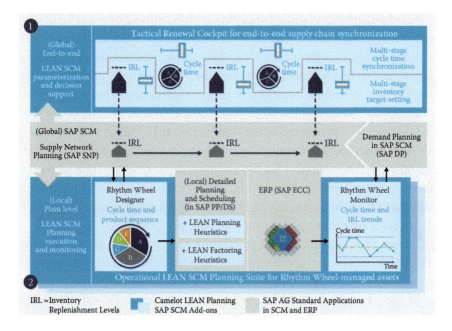

FIGURE 3.10

LEAN SCM Planning embedded in an existing ERP and SCM planning environment with add-ons.

The key challenge regarding end-to-end tactical supply chain planning is to achieve real-time visibility into capacity and inventory. Companies already using integrated global supply network planning (SNP) on top of their ERP systems are well-prepared for the next step: adapting the new LEAN SCM Planning approach to global planning. To provide the right system support, however, we added an IT-enabled planning decision support component for regular tactical supply chain parameter (re-)configurations on top of SNP.

In LEAN SCM Planning, there is also an additional need for IT capabilities to support the evaluation of demand variability. Based on demand variability data, it is necessary to undertake segmentation of the entire product portfolio and classify it according to the most appropriate supply chain channel and supply mode, resulting in a differentiated customer–supply interaction matrix (see Sections 5.1 and 5.2). Once this preparation is in place, the next step is to "plan for variability" and decide on the right buffer size for capacity and inventory according to the identified variability, lead times, and assigned service levels in the individual customer–supply channels.

Therefore, the first planning application built as an add-on was an inventory target-setting tool to enable dynamic IRL calculations. This was further supported by optimization capabilities for multi-echelon inventory target-setting. In addition, the optimal inventory levels depend directly on the Rhythm Wheel-managed cycle times. Furthermore, both cycle time and inventory targets must align with the global takt to achieve end-to-end product flow synchronization (see Section 6.3). We have consolidated all these functionalities under the umbrella of a *tactical renewal cockpit*, providing decision support for the end-to-end parameterization of supply planning processes. Simulation and what–if analysis features have been embedded into the tool to make optimal evaluation of scenarios possible and provide options regarding supply chain performance.

Companywide visibility and standardization remain a barrier for the classical product wheel approach that is applied in many companies; product wheels are managed either manually or based on individually developed local Excel solutions. In contrast to those local and often isolated solutions, we have envisioned leveled flow with multiple Rhythm Wheels that are synchronized to a global takt in a flexible fashion. Consequently, an IT-based solution is required for global standardization and also to enable a company to stay nimble when adapting Rhythm Wheel parameters in the face of unpredictable planning dynamics.

At the local plant level, the *Rhythm Wheel Designer* has been added to existing tools such as the SAP SCM PP/DS application. This tool supports planners effectively in identifying and configuring the best *production sequence, production quantities*, and *cycle times* for their Rhythm Wheels. Later, these preconfigured plans are combined with actual pull replenishment signals during local *LEAN Rhythm Wheel Planning* runs to generate only consumption-based production orders. If the designed cycle time boundaries are violated during execution, planners can apply a short-term Factoring heuristic to move back within the designed boundaries. The Factoring functionality provides the required flexibility for "reacting to volatility" caused by short-term supply disruptions or unforeseen market events.

However, deviations between the forecast-based Rhythm Wheel design parameters and the executed, consumption-based Rhythm Wheel schedule will always remain. To monitor such deviations and overall performance, we developed the *Rhythm Wheel Monitor*. Here, we monitor cycle times, run-to-target results, capacity utilization, and IRL development. Now planners can continuously evaluate the adherence of executed plans to the optimal preconfigured Rhythm Wheel set-up. The same locally monitored data—cycle time variation and inventory developments—are passed on to the global supply chain level to allow for end-to-end supply chain synchronization. IT-empowered planning standardization and global visibility will give the new LEAN SCM Planning concept staying power rather than being only a temporary fashion in SCM, providing the benefits we outline next.

3.3 HOW LEAN SCM PLANNING DRIVES CORPORATE SUCCESS IN THE VUCA WORLD

As discussed above, traditional planning has worked well enough in predictable and stable business environments. To withstand the challenges of the VUCA world, however, a shift in the planning paradigm is required. Overall, LEAN SCM directly addresses the root causes of poor customer service and weak financial performance in a company's operations—by actively managing variability. In this section, we highlight the benefits your company will gain when changing its planning paradigm, resulting in better customer service, better asset performance, and lower inventories along the end-to-end supply chain. By changing the rules of the game in supply chain planning, your company can:

- Create a step-change in supply chain performance.
- Turn customer satisfaction into a competitive advantage.
- Let your supply chain drive financial success.

3.3.1 Creating a Step Change in Supply Chain Performance

Today, supply chains have to compete along several dimensions to successfully contribute to a company's market success. Although shareholders continually expect cost reductions all over the globe, competing on cost alone is not sufficient for virtually any supply chain. To attract new customers and keep old ones, it is absolutely necessary to offer agreed service. Similarly, only a supply chain that acts in a timely and flexible way will have the agility and resilience to withstand the turbulent times of today's VUCA world.

LEAN SCM helps your company increase its competitiveness along the four key dimensions of supply chain performance: cost, time, flexibility,

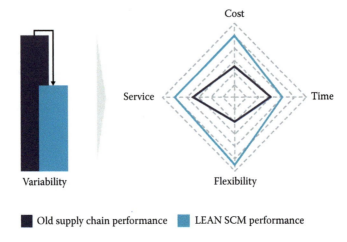

Old supply chain performance   LEAN SCM performance

FIGURE 3.11
LEAN SCM improves on the key dimensions of supply chain performance.

and service (see Figure 3.11). To this end, LEAN SCM actively addresses the key root causes of "performance waste" in a supply chain. By addressing variability, uncertainty, complexity, and ambiguity in supply chain planning and management, it raises the level of performance along all key supply chain dimensions described above. However, LEAN SCM goes beyond short-lived cosmetic corrections and ad-hoc changes. By introducing a paradigm change in supply chain planning, LEAN SCM creates a true and sustainable step change in supply chain performance.

Take for example variability, which causes so many issues in today's supply chains. Most supply chain organizations and the people therein are simply overwhelmed by frequent changes in supply and demand patterns. Existing supply chain planning processes and systems struggle to keep pace with these changes. By reducing variability along the supply chain, your company can rid itself of these issues, effectively supporting the organization and its people, and make more out of the planning systems already in place.

3.3.2 Better Service Leads to Customer Satisfaction and True Competitive Advantages

Your company's supply chain is the key interface with its customers. A well-run supply chain ensures that customers receive the right product at the right place in the right quantity and at the right time. In life science industries, ensuring a stable and reliable drug supply is more than just a business-related challenge—it is an ethical and legal obligation. Many companies thus regard best-in-class service as integral to business success.

However, high levels of customer satisfaction do not happen automatically in the VUCA world. The factors that affect customer satisfaction must be addressed proactively. As product portfolio differentiation increases and market developments become ever more dynamic and uncertain, managing variability has become a major challenge along the entire supply chain. If variability is not addressed adequately, reliability will suffer and there is a substantial risk of poor performance in key determinants of customer satisfaction, such as maintaining high service levels and avoiding stock-outs.

LEAN SCM addresses and reduces variability along the supply chain effectively. By "dampening" the variability and triggering production based on real consumption instead of unreliable forecasts, it becomes possible to "calm the supply chain" and reduce unwanted fluctuations in production and replenishment. The rewards are substantial. A 30% reduction in variability goes a long way as it directly increases service levels or reduces the risk of stock-outs in your company's supply chain by more than 60% with no additional investments needed (see Figure 3.12).

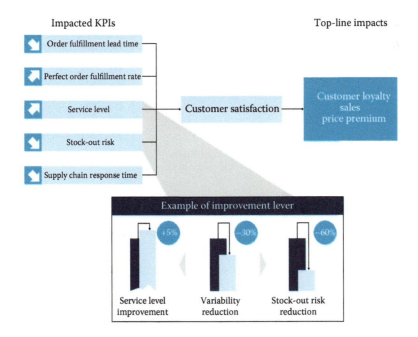

FIGURE 3.12
How LEAN SCM raises customer satisfaction.

3.3.3 World-Class Operational Supply Chain Performance Means Financial Success

Beyond raising your company's customer service levels, improving supply chain planning through LEAN SCM also contributes to better overall financial performance. In this sense, planning forms the backbone of SCM as it ensures that production assets and working capital are deployed efficiently and effectively. These are among the sustainable benefits of better planning (see Figure 3.13).

Consider again the impact of variability. Many supply chain managers worry that their supply chains need to carry higher and higher inventories to buffer rising variability along the supply chain and ensure continuity of operations. In many cases, up to 50% of all inventories can be traced back to the need for protection against such variability. And the truth is those inventories often cannot be reduced without negatively affecting the desired customer service levels.

Therefore, LEAN Supply Chain Planning is designed to address the roots of variability. LEAN SCM thereby goes beyond eliminating the obvious

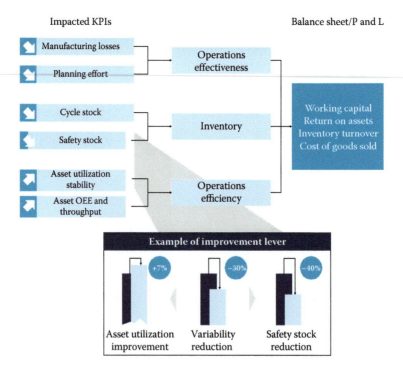

FIGURE 3.13
How LEAN SCM improves key financial indicators.

sources of waste and inefficiency in a supply chain such as overproduction or redundant operations and activities. By introducing the LEAN SCM paradigm change in planning, variability itself is reduced by active "dampening" on assets, "funneling" orders through Rhythm Wheels, and "calming the supply chain," which in turn reduces the immense cost involved in buffering variability.

For instance, reducing variability in demand signals by 30% reduces safety stocks by around 40%, based on the typical mechanics of a company's inventory management system. Similarly, more effective variability management also reduces the need for costly capacity buffers. Thus, by changing the underlying planning problem, companies can unlock new opportunities to improve financial performance and master the challenges of the VUCA world successfully.

To summarize, improving supply chain planning also raises financial performance. By improving operational efficiency, effectiveness, and inventory deployment, your company can significantly reduce both the cost of goods sold and working capital consumed. And variability is just one root cause of "performance waste" that is actively addressed by LEAN SCM. As a holistic planning and management approach to supply chains, it also provides a new way to cope with the uncertainty, complexity, and ambiguity that are inherent in today's VUCA world. Considering the often high value of inventories and production assets as well as the total costs that are directly linked to operations and the supply chain, LEAN SCM is a necessity for corporate success.

CHAPTER SUMMARY

Many companies struggle to adapt supply chain planning to the new business realities of the VUCA world. LEAN SCM provides a paradigm shift in supply chain planning; it is designed to enable planning to cope successfully with today's variability and complexity in supply chains by providing new planning concepts that are fully integrated into a company's corporate IT systems and supply chain organization.

To pave the way for a true step change in supply chain performance, companies need to overcome the multiple drawbacks of traditional MRP-based planning and its variants in ERP and APS solutions. Furthermore, they need to adapt conventional lean approaches to the new reality, so they can support global supply chain synchronization. To achieve these objectives, five building blocks of LEAN SCM are fundamental:

- The introduction of High-Mix and Breathing Rhythm Wheels makes it possible to smooth variability and limit volatility propagation upstream, two areas in which standard product wheels fail.
- Two-sided variability management allows companies to efficiently buffer both demand and supply variability in inventory instead of relying solely on very costly buffers in capacity.
- Effective alignment of inventory targets and cycle times allows companies to introduce a global takt in manufacturing and achieve true global supply chain synchronization.
- The separation of tactical and operational planning functions reduces the need for unrealistically high levels of forecast accuracy and reduces planning complexity considerably.
- The full integration of all LEAN SCM concepts into organization, processes, and corporate IT systems through dedicated planning add-ons guarantees sustainable change.

On the basis of these building blocks, companies can bring together planning paradigms that are often seen as conflicting: LEAN SCM introduces lean flow along the supply chain, but ensures high agility to respond to changes in the business environment through effective variability control mechanisms. High response capabilities toward business changes also provide greater resilience along the entire supply chain. In this way, the LEAN Supply Chain Planning paradigm combines leanness, agility, and resilience for sustainable SCM.

The Lean SCM Planning framework represents a step change in supply chain performance. Several industry cases in this book confirm that process industries can achieve substantial improvements in customer service and key financial figures. All these examples from leading chemical and pharmaceutical companies show that LEAN SCM can provide a decisive competitive advantage in process industries.

Part II

How to Design and Build LEAN SCM

4

Prepare Your Supply Chain for LEAN SCM

Just as you might think twice before building a five-story house on sandy ground, so you would not want to try to improve your company's strategic, tactical, or operational planning processes if its physical supply chain set-up was costly and inefficient. So in order to optimally mitigate the effects of variability, complexity, and uncertainty in today's VUCA world, it is worth "cleaning up" your company's supply chain before implementing LEAN SCM. The following rule of thumb underlines the corresponding impact on your company's costs: a 1% reduction in variability reduces the cost of inventory by about 1.5% and that of capacity requirements by about 1.7%, on average.

This chapter provides valuable insights into how to prepare your company's supply chain for LEAN SCM Planning improvements. Here, we introduce and explain how to use several methods and tools with which to address the following challenges:

- Ambiguous and unclear supply chain strategies
- Poorly managed variability in customer demand and complexity in the product portfolio
- Unleveled capacity along the supply chain
- Poor agility of supply chain structures and supply chain partners
- Physical manufacturing variability within a production site

Overcoming these challenges is an important enabler for implementing LEAN SCM. Note, however, that preparing your company's supply chain for LEAN SCM implementation is not a fundamental prerequisite for realizing significant benefits from LEAN SCM. But it will certainly help in exploiting its potential more fully. Therefore, we recommend preparing the supply chain properly as the first step on your LEAN SCM journey.

In Section 4.1, we focus on strategic supply chain segmentation. Segmentation makes it possible to slice and dice the supply chain into manageable pieces and to define optimal, segment-specific strategies. Section 4.2 provides a methodology for assessing and improving the end-to-end supply chain from a global top-down perspective. We show how to level capacity across supply chain stages, how to reduce portfolio complexity, how to increase supply chain agility, and how to actively reduce variability in customer demand. In Section 4.3, we introduce a bottom-up, shop-floor-oriented, analysis-based improvement tool box. We highlight ways to reduce variability in shop-floor operations and carve out stable product flows within a production site by leveled flow design.

4.1 SEGMENT AND STRATEGIZE YOUR SUPPLY CHAIN

Preparing your company's supply chain for LEAN SCM starts with strategic considerations. Most importantly, management needs to clarify the company's SCM strategy:

- Does your company need more than one strategy for its supply chains (see Section 4.1.1)?
- How can product portfolios and customers be structured to segment the supply chain (see Section 4.1.2)?
- What is the right balance between distinct strategic goals, for example, service vs. cost focus (see Section 4.1.3)?

The following sections support answering these questions in a way that will address your company's SCM issues effectively. This will help reduce complexity and make the related challenges more manageable. In other words, a clear strategic direction for managing your company's supply chains will serve as a guiding light for fundamental trade-offs and decisions such as selecting the right planning mode.

4.1.1 How Many Supply Chain Strategies Are Needed?

For many decades, it was enough to supply standard products and services to customers. However, a strong customer focus has nowadays become vitally important to marketplace success. As a consequence, companies must develop and distribute a growing number of product variants to serve the complex needs of their customers. In the chemical industry, for example, the number of individual products has grown by more than 200% over the last 15 years. This high number of product variants contributes to the mind-boggling complexity of today's VUCA world.

As a consequence, supply chains that have performed well in a less complex environment are now facing an enormous gap between their capabilities and the new challenges. Consequently, your company's supply chain needs to evolve in order to stay competitive. The refinement of the supply chain strategy provides the foundation for this evolution. In the VUCA world, a one-size-fits-all approach is no longer adequate and must be replaced by a more differentiated approach. Nowadays, the objective must be to create tailored supply chains which are guided by a range of strategies that coincide with the specific requirements of customers. In the end, such a differentiated approach should lead to higher profits by increasing revenues, reducing costs, or achieving both (see Figure 4.1).

To build tailored supply chains for a wide range of business needs, your company's supply chain should be properly segmented (see Figure 4.2). With the help of supply chain segmentation, a highly heterogeneous business can be clustered into homogeneous groups. For each of the resulting supply chain segments, an optimal supply chain strategy should then be assigned.

Effectively slicing a company's supply chain into segments involves a challenging trade-off between similarities across product characteristics,

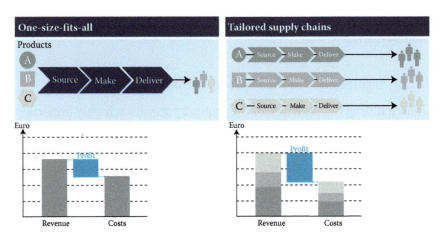

FIGURE 4.1
Tailored supply chains and strategies for increasing profits.

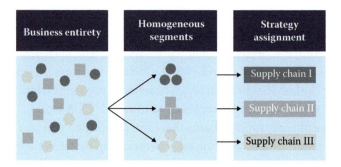

FIGURE 4.2
Segmentation to prepare for assigning tailored strategies.

customer requirements, and production environments. Such a trade-off would also affect the total number of segments required. In general, the more heterogeneous the business is, the more supply chain segments are needed. It is, however, important to avoid segmenting into too many supply chains. There is a point beyond which adding supply chain segments no longer benefits the business as a whole but merely increases the complexity of the overall strategy.

4.1.2 Structure Customers and Products to Build Supply Chain Segments

Within a supply chain segment, product and demand characteristics can be highly heterogeneous. Products and customers should therefore be clustered to master such complexity, thereby facilitating the creation of supply chain segments.

4.1.2.1 Segmentation of the Product Portfolio

Understanding the structure of the product portfolio is highly important to effectively master complexity along the supply chain. From a LEAN SCM perspective, the most important factors to consider are demand volume and demand variability. While some products are sold in high volumes for which demand is typically very stable and easy to predict, others are sold only sporadically with less predictable demand. A combination of widely spread ABC and XYZ analyses can be used to segment the product portfolio based on those two characteristics. As depicted in Figure 4.3, the ABC–XYZ matrix is clustered in three categories: runners, repeaters, and

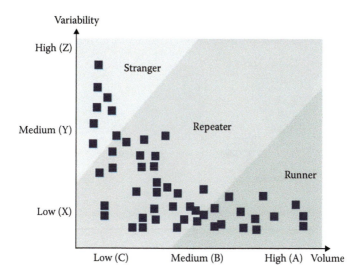

FIGURE 4.3
Segmentation of products on an ABC–XYZ matrix.

strangers. Runners are fast-moving products with high demand volumes and low demand variability, repeaters exhibit medium-demand volume with medium variability, while strangers are the remaining products with low demand volumes and high demand variability.

On the basis of the ABC–XYZ segmentation, distinct planning strategies can then be defined to optimally plan production and replenishment of runners, repeaters, and strangers. Refer to Chapter 5 for the details of optimal planning mode selection.

4.1.2.2 Segmentation of the Customer Portfolio

Customer-specific needs can vary widely. To optimally fulfill customer needs, it is essential to build homogeneous customer clusters. Clusters can, for example, be built with regard to lead time requirements, service level expectations, or cost-sensitivity. Another criterion for clustering customers can be the country or region of origin, since, for example, legal requirements can be similar. The pharmaceutical market in Japan, for instance, is extremely restrictive with regard to lead times and order fulfillment. If those standards set by Japanese authorities are not met, then companies risk losing their licenses to operate in this market. In contrast, other regions can be more tolerant. Segmenting customers by region is

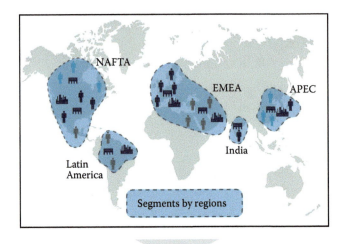

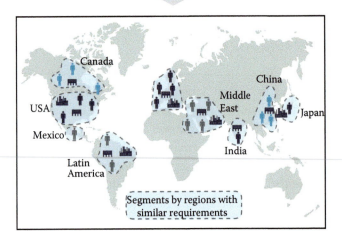

FIGURE 4.4

Segmentation of customers by region.

therefore one opportunity to build groups based on detailed common features. Figure 4.4 illustrates this division into regional segments.

No matter how customer clusters are carved out, it is important to understand that each customer segment has its own needs and should therefore be approached with a tailored strategy. The next section will explain how to assign appropriate supply chain strategies to defined segments in order to have a positive impact on profitability. We will also emphasize that the definition of supply chain strategies is important for the strategic LEAN SCM process since such strategies impact the selection of suitable LEAN supply chain modes.

4.1.3 Assigning Strategies to Defined Supply Chains

Once supply chain segments are defined, the next step is to strategize each supply chain. In general, any supply chain strategy should maximize a company's profitability. Therefore, the impacts of distinct supply chain strategies on profits need to be evaluated. On the one hand, supply chain strategies obviously affect costs on the operations side, while on the other they create customer value added that is reflected in revenues (Figure 4.5). Even though a supply chain strategy should assign a high priority to the creation of customer value added, a company needs to evaluate carefully whether the created customer value added justifies the resulting costs. In other words, it is crucial to find the right balance between these conflicting dimensions to maximize profitability.

When defining a supply chain strategy, it is important to remember that customer requirements are highly heterogeneous. Some customer groups might be price-sensitive whereas others strongly emphasize short lead times. Accordingly, supply chain strategies should be defined such that customer-specific requirements in each supply chain segment are addressed. As shown in Figure 4.6, there are four key "order winner" dimensions to be considered.

Achieving high performance along all dimensions is highly challenging and sometimes even infeasible. For instance, simultaneously offering

FIGURE 4.5
Supply chain strategies affect costs as well as revenues.

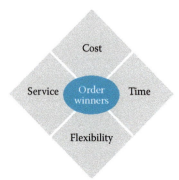

FIGURE 4.6
Customers' key requirements can be clustered into four order winner dimensions.

products at low prices (order winner: cost) while also satisfying high customer service requirements (order winner: service) is difficult to accomplish. Therefore, the trade-off between various dimensions needs to be carefully investigated. The right dimension to focus on is defined in the supply chain strategy for each segment based on customer-specific requirements (see Figure 4.7).

This enables the highest possible value added for all customers. Box 4.1 provides an example illustrating how a European diagnostics manufacturer increased profits by splitting its one-size-fits-all approach into two distinct, customer-specific supply strategies.

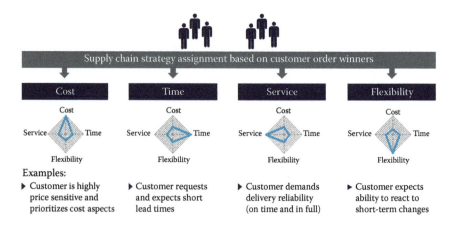

FIGURE 4.7
Supply chain strategy assignment based on customer order winners.

BOX 4.1 HOW A DIAGNOSTICS COMPANY
SURVIVED THE ECONOMIC CRISIS BY
REFINING ITS SUPPLY CHAIN STRATEGY

In the first decade of the twenty-first century, a European diagnostics company faced severe challenges due to the global economic crisis. Sales broke down in several markets, especially in Europe. Therefore, the company triggered a supply chain improvement program that aimed not only to cut costs but also to increase sales.

One of the first steps included the analysis of the company's customer portfolio and benchmarking against its competitors. The analysis revealed that a key improvement potential could be realized by adapting the supply strategy to hospitals and research institutes. For both customer segments the same supply chain strategy was applied, which was focusing on the order winner dimensions of cost and time simultaneously. Hence, the company aimed to offer its products within short lead times and at low prices for both customer segments. As a result, promised delivery times were regularly exceeded and product margins were low due to low prices and high supply chain costs (see upper portion of Table 4.1).

A key finding of the analysis was that the requirements of the two customer segments were very different. On the one hand, the research institutes did not value short delivery times but rather gave priority to low prices. On the other, hospitals requested short delivery times to ensure consistent product availability while having limited storage space. Tailored supply chain strategies were therefore designed based on the identified order winners.

A cost strategy was assigned to the research segment. In the end, a price discount of 10% was agreed on in exchange for an extension of the delivery promise by 2 weeks. This strategy made it possible to level production, which helped to cut down overtimes in production and to avoid the large number of emergency deliveries. This significantly reduced operating costs. For the hospital segment, a strategy with a focus on time was defined. By establishing a delivery priority, high product availability within half a day could be achieved. At the same time, the prices for hospitals were increased by 12% on average across the product portfolio to account for the faster service

TABLE 4.1

Profit Increase through Tailored Supply Chain Strategies

One-Size-Fits-All Approach						
	Research Institutes (Focus on Time and Value)			Hospitals (Focus on Time and Value)		
Performance Indicators	Value	Units	Total	Value	Units	Total
Revenue	$ 25	30 k	$ 750 k	$ 25	70 k	$ 1750 k
Variable cost	$ 10	30 k	$ 300	$ 10	70 k	$ 700 k
Margin		$ 450 k			$ 1050 k	
Aggregated margin				$ 1500 k		

Tailored Supply Chain Strategies						
	Research Institutes (Focus on Value)			Hospitals (Focus on Time)		
Performance Indicators	Value	Units	Total	Value	Units	Total
Revenue	$ 22,5	30 k	$ 675 k	$ 28	70 k	$ 1960 k
Variable cost	$ 9,5	30 k	$ 285 k	$ 10	70 k	$ 700 k
Margin		$ 390 k			$ 1260 k	
Aggregated margin				$ 1650 k (+10%)		

times and flexibility. However, the prices still remained below those of competitors and were accepted.

Using this new dual-strategy approach, the company was able to improve its supply chain performance remarkably, as the lower portion of Table 4.1 demonstrates. Additionally, sales representatives believe that this tailored approach will produce an additional significant increase in sales volumes as customer requirements are perfectly met. Due to this remarkable success, the diagnostics company aims to refine the supply chain strategies for other segments as well.

The differentiation of supply chain segments and subsequent strategy definitions represent an important enabler for the strategic LEAN SCM process, which aims to select the best-fit supply chain planning concepts (see Chapter 5 for details).

Summary

In this section, we explained that tailored supply chain strategies are needed to sufficiently meet customer requirements in today's VUCA world. In this context, we provided guidance for structuring your company's products and customers in order to segment the supply chain and to define appropriate supply chain strategies. Additionally, we emphasized that a focus on one or more key order winners per supply chain is an important prerequisite for defining LEAN supply chain modes (see Chapter 5 for details). To complete the preparation of your company' supply chain, we now demonstrate how to approach supply chain segments from the top down in Section 4.2 and from the bottom up in Section 4.3.

4.2 ALIGNING THE SUPPLY CHAIN FROM A TOP-DOWN PERSPECTIVE

A top-down analysis of your company's supply chain should provide a quick and pragmatic analysis of areas needing improvement before launching the implementation of the LEAN SCM paradigm. It is legitimate to ask here whether there are factors that enable LEAN SCM that should be established before starting. The answer is that there are no hard show-stoppers, but the more enablers there are, for example, supply chain structures to enable agility, the more benefits from LEAN SCM your company will enjoy as a result of the paradigm change.

This section offers insights into:

- Achieving transparency and a top-down map of the supply chain (Section 4.2.1)
- Identifying and assessing gaps that may diminish the results of LEAN SCM (Section 4.2.2)
- Applying instruments for agility and preparing a supply chain for LEAN SCM (Section 4.2.3)

4.2.1 Create End-to-End Transparency in Supply Chains

The very first step in top-down analysis is to map the supply chain. This may seem like an easy exercise—take a pen and chart it on a white board—but there are several questions that arise almost immediately:

- What is the most effective strategy for achieving transparency regarding what really happens along the supply chain?
- What information is needed in order to not get lost in the complexity of details?
- What kind of mapping is needed to achieve consensus on the facts?

This section provides an overview of a possible approach, one that has proved successful in several supply chains and industries and represents a good balance between high granularity of details and overall effectiveness. An alternative is creating a value stream map (VSM), which we present in Section 4.3. Based on our experience, we believe that creating a VSM is a very good approach to strategy at the shop-floor level but has its limitations for midsize and large-scale supply chains and is too detailed to offer a global end-to-end view of the supply chain.

The first step is to map the structure of the supply chain for a certain segment carved out in Section 4.1: vendors, manufacturing locations, stock-keeping points, customer segments, and connections between them, as shown in Figure 4.8.

Before enriching a supply chain map with details for later analysis, two important decisions need to be made:

- Selecting the representative time period for observatiovn.
- Selecting a common unit of measurement so that all stages along the supply chain are comparable.

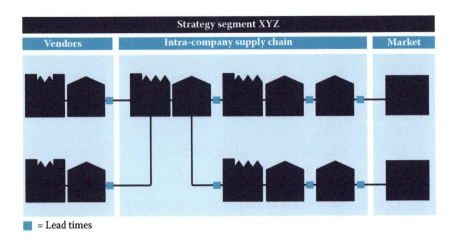

= Lead times

FIGURE 4.8
Example of a supply chain structure map.

After making these two important decisions on time period and unit of measure, it is time to map customers, stock-keeping points, vendors, and manufacturing.

4.2.1.1 Selecting a Representative Time Period for Observation

First, select the right time period over which to measure key performance indicators (KPIs). For a company with stable demand for at least a year, we recommend a period of 1–2 years for consideration (see first graph in Figure 4.9). In cases involving trends or seasonality, there are some useful "tricks": for supply or demand seasonality, measure by season (either high or low), for demand that shows a representative trend, begin by averaging out the trend (see second and third graph in Figure 4.9).

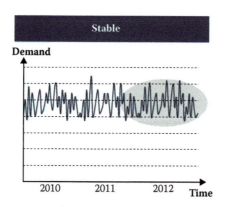

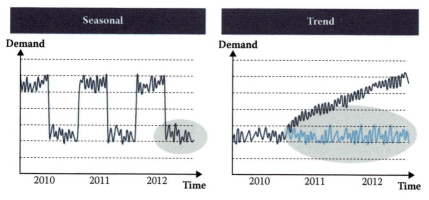

FIGURE 4.9
KPI measurement periods for various demand situations.

4.2.1.2 Selecting a Common Unit of Measure

Second, select a common unit of measurement for such KPIs as inventory, demand, and so on. Such a unit must meet an important criterion: it must be comparable—it is quite difficult to compare bulk liters, for example, with numbers of packages of finished goods.

A good choice is, of course, to use a monetary amount to measure the value of a KPI: dollars, euros, or francs. In reality, however, it is very challenging to come to consensus on real money values within an organization and then to track them end-to-end along a supply chain. That's why we recommend the following shortcut: take the biggest cost-dominant component that is an integral part on the first supply chain stage and use this unit of measurement as equivalent to the other stages. For example, in a life science industry an active pharmaceutical ingredient measured in grams at various stages may be used to unify information regarding blends, dosages, package sizes, and so on. A similar approach applies to the chemical industry, in which a mix of components could be measured and used as a portion of cost of goods sold. If none of these options is applicable, it is still a better choice to use days-of-coverage for inventory than, for example, absolute values like pieces or packs.

4.2.1.3 Mapping the Customers

After creating the high-level map of the supply chain and determining time periods and units of measurement, it is time to delve into the details, starting with the customer side.

The most important information from the demand side is average demand over the selected period as well as minimum and maximum demand values per product segment. We recommend conducting such an analysis for a reasonable number of geographical regions. It is also important to determine the customer takt, which is the time period between two orders for the same product per segment. In this context, we introduce the German word "takt," which was originally associated with music, referring to the beat or rhythm in accordance with which music is played. In other words, the takt specifies how often products are ordered. To capture takt variability, we recommend determining the minimum and maximum takt boundaries (cf. Figure 4.10).

Moving upstream, let us take a close look at stocks and suppliers—the most important factors affecting replenishment—and an even closer look

Total customer demand		Demand		Customer takt	
Runners	5000	4300	1	1	
		5800			2
Repeaters	4700	3750	10	6	
		4950			12
Strangers	1200	370	60	14	
		2150			90

Average	Minimum
	Maximum

FIGURE 4.10
Segments based on as-is characteristics of customer demand.

at production sites, which serve as capacity-relevant factors within the LEAN SCM paradigm.

4.2.1.4 Mapping Stock-Keeping Points and Suppliers

The next factors to be mapped and characterized are stock-keeping points. These include warehouses, in- and out-bound stocks at production sites, and raw material stocks filled by suppliers. The relevant KPIs to be evaluated are average inventory levels and their max and min boundaries as well as a replenishment takt. The latter as a counterpart to the customer takt represents the typical time between orders to replenish the same product upstream. The max and min boundaries of the replenishment takt finalize the mapping of a stock-keeping point. The process begins with the last stock-keeping point downstream and moves step-by-step upstream. The best methods for estimating the above-mentioned KPIs are either to take a representative product for each segment and location and then scale the number up to the whole segment volume or to take an average group of products and use that as a peer group along the entire supply chain. Remember that it is more important to see the whole picture by top-down analysis than to have very precise numbers.

Regarding the last upstream entities along the supply chain, the same methodology should be used to map vendors that supply a company with

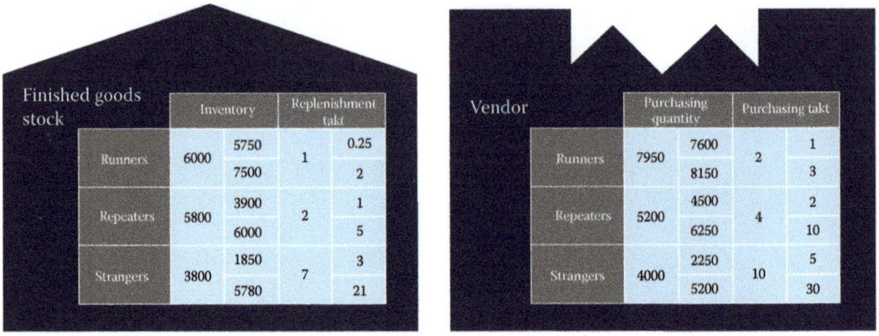

FIGURE 4.11
The characteristics of stock-keeping points and replenishment by vendors.

raw material or any further input. Measuring replenishment involves analyzing average purchase order sizes as well as min/max boundaries. Here again it is important to account for a takt, the purchasing takt—the time period between the placement of two purchase orders for the same material. At this point, the factors involved in storing raw materials, intermediate goods, and finished goods are mapped (Figure 4.11).

4.2.1.5 Mapping Production Sites

The next and final factor to consider in top-down mapping is production sites (Figure 4.12). First, to measure capacity, we suggest taking the average

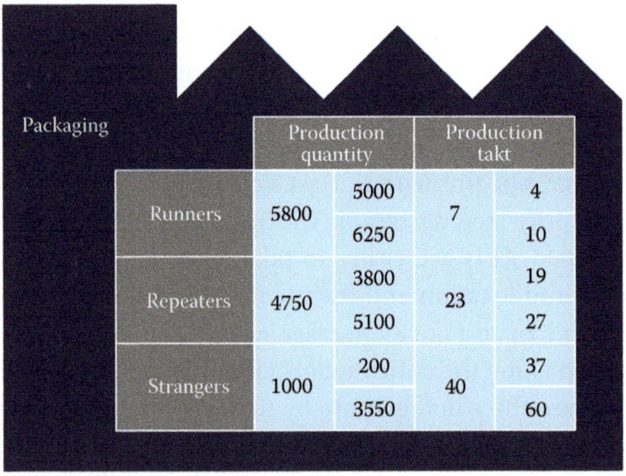

FIGURE 4.12
Characteristics of production sites, inventories, and vendors.

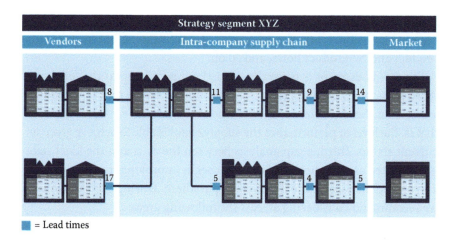

= Lead times

FIGURE 4.13
Results of mapping in a top-down analysis.

production quantity of a representative product in a given segment within the defined time period and extending it to the maximum and minimum quantities produced at this stage. Second, the production takt should be measured. As we show in subsequent chapters, the production takt is the heart of the LEAN SCM paradigm. A production takt is the typical time period between production campaigns for the same product at a given analyzed production stage.

A company with typically divergent supply chain structures should use a fixed ratio for material flows between stages. The same approach should be used for parallel or shared resources, in which case an aggregated, static ratio should suffice for top-down analysis. The result of such as-is mapping is a top-down picture of the current supply chain (Figure 4.13). At this stage, it is crucial that all stakeholders involved in the supply chain be committed to the mapped structure and figures.

4.2.2 Identify and Assess Gaps to Improve Supply Chain Synchronization

On the basis of the transparency achieved through the mapping process we have described, it is now possible to analyze the current KPIs of the supply chain. Here, we offer several examples of the most important dimensions that are crucial for identifying areas that should be targeted for improvement along the supply chain and for enabling the implementation of LEAN SCM. These indicators are not prescriptive; feel free to look at other specific indicators of your company's supply chain: shelf-life, material substitutability, shared resources and jumping bottlenecks, co-product and by-product flows, and so on.

The gap assessment is generally carried out according to the following three steps: first, the big picture is consolidated based on the top-down assessment. Second, the ideal state under the current set-up is defined, that is, the level of performance the supply chain is able to achieve given its current set-up. Third, a gap analysis between the as-is and the ideal state is conducted as an indicator of where to begin improving the supply chain and paving the way for the LEAN SCM paradigm. During gap assessment, we recommend considering at least the following areas:

- Inventories
- Capacities
- Global takt
- Global lead time

4.2.2.1 Inventories

We begin with the analysis of inventories, not as a matter of principle but because inventories are always good indicators of proper supply chain functioning and preparedness for transitioning to the LEAN SCM paradigm. From our experience, it is very helpful to visualize the current state of the supply chain based on the information collected during the top-down as-is analysis.

This visualization (see Figure 4.14) provides a comparative insight into the inventory situation at a glance. It is important to keep in mind here that inventories should be scaled in accordance with the cost-of-goods-sold ratio of the particular supply chain stage. Alternatively, universal measures such as a drug substance equivalent, as mentioned previously, can be used. At this stage, based on an early diagnosis, the following conditions should raise suspicions:

- More inventory is held by strangers than by runners.
- The inventory spread (min–max) is wider for repeaters than for strangers.
- 90% of inventory is concentrated at one stage (e.g., raw materials).

A more effective diagnosis can be achieved using a benchmark to assess the current situation. Let us call this benchmark an ideal state—which

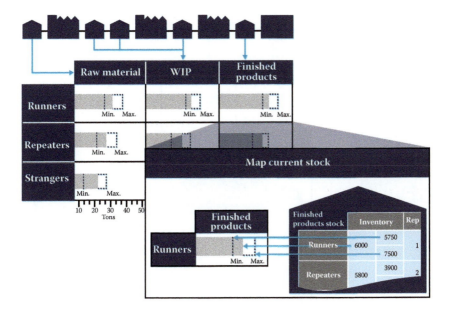

FIGURE 4.14
As-is analysis of stock.

includes the optimal sizing and allocation of inventories in a representative supply chain. To calculate an ideal stock level, we suggest using a very pragmatic approach that considers lead times, average demand, and buffers for variability. We provide a more sophisticated approach in Section 6.2. Remember that the top-down analysis we have described here is not about being precise in every detail, but about seeing the big picture (Figure 4.15). Again, it is very important here either to scale the inventories based on the cost-of-goods-sold level or to use some universal unit of measure.

Now, by comparing the current and ideal states, it is possible to derive the most critical issues to emerge from the analysis (see Figure 4.16).

4.2.2.2 Capacities

The next important point regarding the health check of the supply chain is the capacity of production assets. To create a clear view of the as-is situation, take average minimum and maximum produced quantities and compare these data with maximum available capacity (Figure 4.17). This yields a clear end-to-end picture of the range of capacity utilization along the supply chain based on the information collected during the top-down analysis.

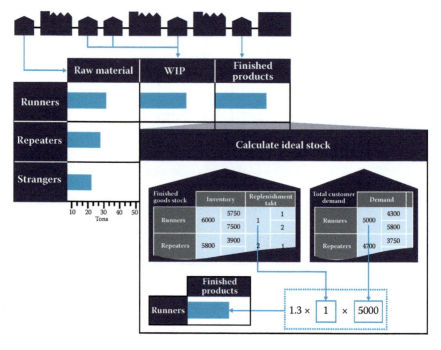

FIGURE 4.15
Ideal stock.

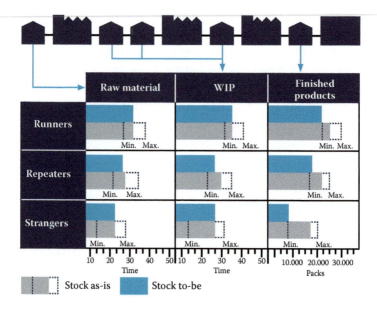

FIGURE 4.16
Comparison of as-is with ideal stock.

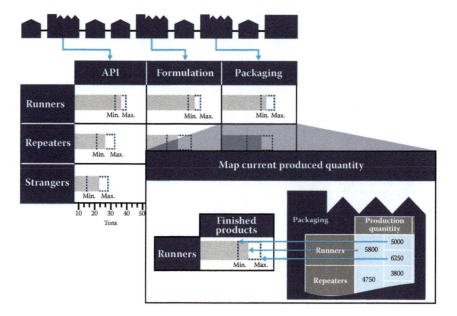

FIGURE 4.17
As-is analysis of capacity.

If several product groups run on the same resource, split the involved resources for the defined segments. In the opposite case, if there are multiple production resources at one stage, aggregate them to obtain overall logical capacity. As already discussed, the top-down analysis ultimately is not about being precise but about gaining a holistic view.

Already at this stage an early diagnosis should be forthcoming. For example, if there is a huge gap between average capacity utilization and maximum available capacity in one segment and a very small gap or even excess capacity in other segments, there is likely a misbalanced allocation between products and assets. To get an even better diagnosis once again a benchmark is needed—this time we suggest using customer demand (see Figure 4.18). This will provide a clear picture of capacity utilization that will indicate whether quantities produced equal what customers are ordering.

By comparing the current supply chain map with an ideal state analysis, it is possible to identify excessive discrepancies between minimum and maximum capacity utilization at some stages—this is an indicator of the amplification of variability (the bullwhip effect). Two potential root causes are misaligned planning approaches between supply chain stages and unleveled physical capacity along the supply chain (shortages and capacity excesses at various stages). LEAN SCM will help to eliminate the first

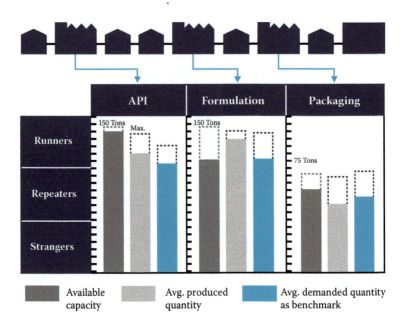

FIGURE 4.18

Comparison of as-is and to-be analysis of produced quantities.

cause; Chapter 5 provides details on the best approach to adopt. The second cause should be addressed in advance to the extent possible, before launching implementation of the new planning paradigm. Unbalanced production capacity will either reduce the end-to-end benefits of LEAN SCM or will require special workarounds that will dramatically increase the effort during implementation.

4.2.2.3 Global Takt

The issues with inventory and capacity utilization that we have been discussing are symptoms of inefficiency. Now, let us move to another important indicator that describes a root cause of such inefficiency and should be eliminated before introducing LEAN SCM to enable efficient end-to-end synchronization.

All stages and functions along the supply chain have a takt. We have described its variation between min/max boundaries in connection with the customer takt in the previous section. Comparing the takt of distinct supply chain stages provides a clear picture of the level of synchronization that has been achieved along the supply chain (see Figure 4.19).

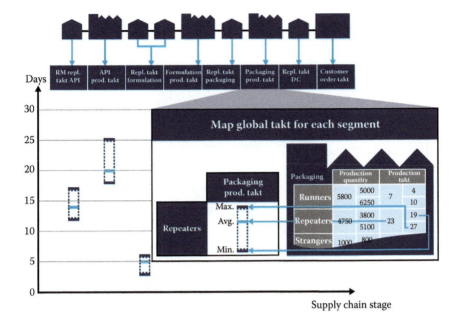

FIGURE 4.19
Map the global takt to allow comparison between the takt at various stages.

This stage of the analysis should yield an early diagnosis. A huge spread in the takt between adjacent stages can indicate one of two cases. The first case: material or products are waiting in stock, because the downstream stage has a longer takt. The second case: the production asset at a downstream stage has an idle run because the upstream stage has a longer takt.

Again, a pragmatic benchmark is needed to better understand the major gaps in the takt. A to-be state can be defined by harmonizing the average takt between stages and finding feasible boundaries for the min/max spread around it (see Figure 4.20).

The better the supply chain is prepared using an harmonized physical takt between assets, the easier it is to achieve the benefits of LEAN SCM Planning concepts that enable end-to-end takt synchronization. We provide greater detail on this point in Section 6.3.

4.2.2.4 Global Lead Time

The final characteristic or KPI that we consider here is global lead time. Global lead time connects the supply chain to customer satisfaction in two ways. First, it directly indicates the time required by the supply chain to convert raw materials into finished products delivered to the customer.

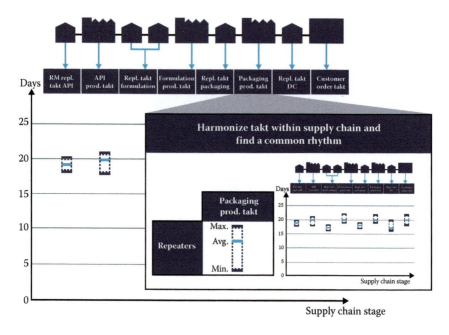

FIGURE 4.20
Define the to-be state by harmonizing the average takt.

Second, it indicates the effectiveness of the supply chain, because the time a product spends moving along the supply chain beyond the physical conversion time is time in which it waits in inventory and ties up working capital (and increases the price of the product to customers).

To analyze overall global lead time, calculate the sum of the physical lead time (e.g., actual manufacturing time for a given batch) and the average takt per stage (e.g., the representative interval until the product is processed within this stage again). The sum of the takt and the physical lead time reflects, for example, the time between two production runs of the same product plus the production time itself. Consolidating the calculated sums for all stages across the supply chain yields the global lead time per segment, as illustrated in Figure 4.21. The difference between the physical lead time and the overall global lead time contains potential inefficiencies which should be addressed before implementing LEAN SCM with its principles of short takt times and pull replenishment.

Inventories, asset capacity utilization, takt, and global end-to-end lead time per segment are the major indicators that we recommend focusing on when conducting top-down analysis. We provide further insights into KPIs related to enabling the implementation of LEAN SCM in Section 9.2.

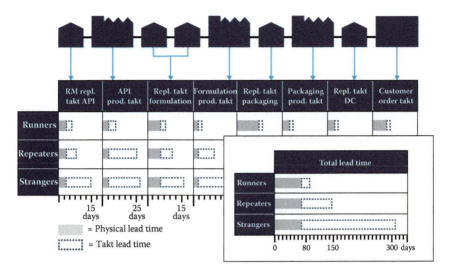

FIGURE 4.21
Mapping global lead time.

4.2.3 Adopt Three Measures for Preparing the Supply Chain

In the previous section, we demonstrated how to map and analyze the supply chain within strategically defined segments. Based on that analysis, we presented a clear picture of potential areas for improvement and factors that should be addressed to ensure a smooth introduction of LEAN SCM and realize rapid gains in efficiency as well as sustainable benefits for the supply chain through the use of the new paradigm.

Gaps in the supply chain may have any of several causes, often related either to planning and coordination issues or to one of the following areas:

- Physical structure and capacity footprint of the supply chain
- Variability caused by customers or suppliers
- Product portfolio complexity

These aspects should be addressed to the extent possible before launching the LEAN SCM journey to ensure that the network structures enable operational agility, variability is managed in the right way, and physical material flow is leveled.

The next sections present a selection of tools that can be used to support the adoption of LEAN SCM. These tools include a network redesign approach to address the physical footprint, proactive demand-shaping to address customer-induced variability, and a complexity-management approach to address the product portfolio.

4.2.3.1 Network Design: Strengthen the Foundation for Agility

Operational inefficiency is often caused by poor decisions at the design stage of a supply chain. Network design is part of the strategic planning process in a company that determines the structure and physical configuration of its supply chain. Operating an agile supply chain depends on having sufficient "strategic flexibility" to adapt network structures and relationships between all network entities quickly to changing conditions. During the strategic network design process, a company defines the number and locations of its warehouses and manufacturing plants, the allocation of capacity and technology requirements to facilities, the assignment of products to plants, and the flow of goods along the entire supply chain (Figure 4.22).

Network design decisions are among the most important supply chain decisions as their implications are significant and long lasting. These decisions involve capital-intensive investments that are highly irreversible. Furthermore, network design also provides the framework for tactical and operational supply chain processes and thus will affect the efficiency and agility of sourcing, manufacturing, and transportation significantly over

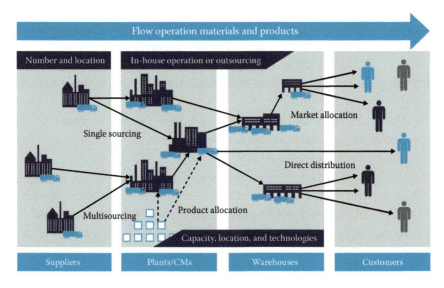

FIGURE 4.22
Key decisions in strategic network design.

a longer period. Therefore, the design of resilient and agile supply chains is widely perceived as a strategic imperative for SCM.

Defining the strategic direction of and objectives for a supply chain begins with choosing principles to guide network design. Industry benchmarks as well as the experience of top companies in process industries demonstrate the importance of the following aspects:

- **Lead time reduction** is characterized by optimizing internal throughput times and lead times to customers.
- **Simplification** reduces network complexity, for instance by concentrating products or processing steps at certain facilities and outsourcing the processing steps that add less value.
- **Leveled capacity allocation** eliminates capacity bottlenecks and helps to avoid over-investment in assets and low utilization rates.
- **Structural agility**, through flexible plants and flexible workforce models, for example, makes it possible to respond to demand fluctuations or supply chain disruptions. A flexible manufacturing footprint also forms an efficient hedge against currency and cost risks.
- **Surge capacity** at the plant and supplier levels ensures that the supply chain can provide high levels of customer service when demand increases.

Generally, the strategic network design process should run through a structured sequence of planning steps (see Figure 4.23). First, the objectives and corresponding network strategy need to be defined. This encompasses identifying all relevant targets in alignment with the overall supply chain strategy, ranking them according to their relevance, and investigating appropriate trade-offs. The second step comprises an in-depth analysis of the as-is supply chain capability using information and data from IT systems as well as from key stakeholders to identify current strengths and weaknesses and to refine objectives. Third, alternative

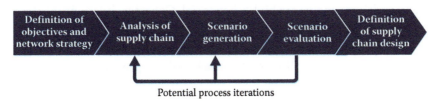

FIGURE 4.23
Strategic network design should rely on a structured process.

supply chain scenarios are developed from a business perspective. To elaborate the business scenarios, optimization tools can be used in this phase, for example, to decide on optimal facility locations or capacity levels. The developed scenarios undergo a validation and prioritization process in the fourth step. Both quantitative and qualitative factors have to be taken into account when ranking the identified scenarios according to their benefits and business feasibility. In the final step, a detailed business case and implementation plan is essential for the final selection of a given scenario.

4.2.3.2 Demand Variability: Shape it Proactively

As already stated in Chapter 1, variability, volatility, and uncertainty are of great relevance in today's challenging VUCA world. Demand variability and volatility are caused by variations in customer order placement and order quantities. Uncertainty arises from the lack of predictability regarding demand and is increased by variability. Both dimensions make it difficult for supply chain planning to match demand with supply.

Much effort has been put into managing increased variability and uncertainty. Companies are continuously searching for new concepts to improve forecasting accuracy, for example, in order to accurately predict variability in customer demand. Less attention has been given to concepts that are focused on eliminating the root causes. One possibility is proactively managing and coordinating customer demand. Instead of managing varying and unpredictable customer order placements and order quantities, customer order behaviors can be influenced by setting the right incentives, which can be financed by correlating cost savings. When offered financial compensation, some customers might be willing to enter into certain agreements, such as, for example, committing to ordering products every second week to achieve a stable customer takt, or ordering in a predefined lot size. In a lot size-dependent production environment, such agreements are highly beneficial. Consider situations in which customer orders are slightly below or above a company's optimal production lot size. By passing on some portion of a company's cost savings, its customers might be willing to align their order behavior to help it manage such variability. Appropriately implemented demand shaping reduces demand variability and uncertainty. This enables a company to reduce its safety stock and supply chain planning effort and improves overall supply chain performance (see also Figure 4.24).

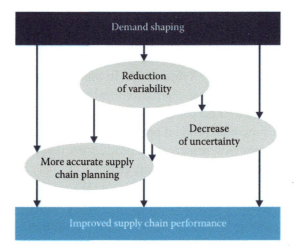

FIGURE 4.24
Demand shaping improves supply chain performance.

Generally, not all customers are willing to change their order behavior due to process-related restrictions or inflexible organizational structures. However, proactively shaping at least some customer demand will have an impact on supply chain performance. As stated in Chapter 1, a minor reduction in variability significantly improves the level of service and reduces the risk of stock-outs along the supply chain. What is more, customers capable of changing their order behavior could pay lower prices. The higher price for customers whose demand cannot be shaped at least partly compensates for the costs of the remaining variability, uncertainty, and higher supply chain planning effort.

To fully exploit the benefits of demand shaping, it can be extended to other supply chain stages, for example, to suppliers. Suppliers will likely also appreciate more reliable order behavior and are willing to offer a discount for stable and takted order behavior.

4.2.3.3 Portfolio Complexity: Get Rid of Unprofitability

Facing accelerating competition in recent years, many firms across industries have found it necessary to truly embrace their customers: companies in process industries have widened their product portfolios to address new markets with even shorter life cycles, offered fully customized products to satisfy unique customer needs, and reshuffled their supply chain set-ups to ensure timely delivery of short-notice orders. Frequently, limited cost transparency has nebulized the financial benefits (or harm) of these actions. On top of that, many external factors in today's VUCA world challenge companies, as we discussed in Chapter 1: volatility or uncertainty also affects the entire value chain. To make things even more difficult, some industries such as pharmaceuticals are tightly regulated and adherence to regulations across multiple countries can also lead to additional spread in a regional product portfolio. The result of all these factors has become very obvious: unnecessary product portfolio complexity and corresponding complexity in the supply chain.

Some examples of such complexity affecting the value chain include:

- **Planning and overhead**: Nontransparent coordination across divisions and functions and unclear lines of accountability exacerbate a responsive execution of orders.
- **Sourcing and procurement**: Proliferation of materials and suppliers decreases economies of purchasing scale and constitutes higher transaction and administrative costs.
- **Manufacturing**: More manufacturing assets are required, while operating efficiency declines, due to shorter lead times, frequent changeovers, lower utilization rates, and higher levels of scrap and rework.
- **Delivery and logistics**: Warehousing and administrative costs increase, whereas flexibility and customer responsiveness are hampered.
- **Sales and marketing**: Overly complex, historically based pricing and trade terms with limited links to performance might result in unnecessary risks if a smaller customer with better characteristics is acquired by a larger one.

Frequently, complexity is deemed to be inherently negative. Indeed, if complexity is not managed, it can harm business performance in a multiplicity of functions within a company. Nevertheless, complexity can be a value-adding factor as well. A trade-off between a product's profitability and its strategic relevance should be considered to assess the value contribution of products in a company's portfolio.

Traditional profitability calculations analyze profitability as a gross margin reduced by a share of the overall costs allocated to a given product. These costs include strategic sourcing costs, R&D overhead, general sales and administration share, and so on (see Figure 4.25).

For LEAN SCM purposes, we recommend extending traditional complexity management considerations to address the costs of variability in

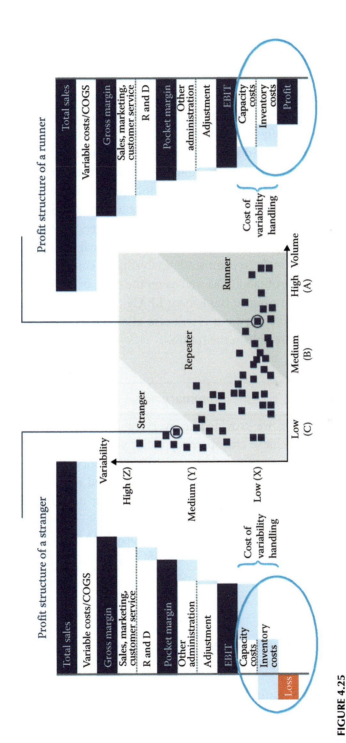

FIGURE 4.25

Costs of variability handling in complexity management.

handling a given product. These costs include buffering along the supply chain by adjusting asset capacities and inventories to address supply and demand fluctuation.

The majority of low-volume products have smaller contribution margins, because there is less potential to achieve economies of scale. At the same time, such low-volume products exhibit more pronounced variability characteristics—they usually belong to the strangers segment (cf. Figure 4.3). And as variability rises, so do the costs involved in managing such variability by either inventory or capacity adjustments along the supply chain. A rule of thumb applies here: to manage a 1% increase in variability requires a 1.5% increase in inventory and a 1.7% increase in capacity. Furthermore, advanced planning and optimization methods, or even individual workarounds, are often required to manage the supply of such products.

We therefore strongly recommend analyzing a company's product portfolio based on profitability as well as on the costs of variability management before beginning to implement LEAN SCM. This will not only increase overall profitability, but it will also allow for as much planning standardization as possible in the future.

Summary

Within this section we provided an overview of top-down mapping methodology to achieve transparency along your company's supply chain, explaining our approach to analyzing inventory, capacity, and global takt as well as lead times. We then introduced several tools for optimizing the results of such analyses, including network design, active demand shaping, and complexity management. These tools will help your company establish end-to-end supply chain management to prepare it for introducing the new LEAN SCM paradigm.

4.3 ALIGNING THE SUPPLY CHAIN FROM A BOTTOM-UP PERSPECTIVE

On the basis of the results of top-down analysis and the resulting overall improvements, we now focus on the local facilitation of LEAN SCM at particular stages along the supply chain. The bottom-up analysis presented over the next several sections will support your company not only by preparing local sites along the supply chain for LEAN SCM, but also by laying a foundation for the integration of global end-to-end supply chain planning and local production planning.

In the context of LEAN SCM, the traditional motivation of value-stream analysis is extended. The detailed analysis of selected processes at the site level is used not only to identify potential areas of improvement for classical operational excellence initiatives but also to gain in-depth knowledge of existing product flow paths at the plant level as a basis for an effective planning. Leveled and synchronized material flows as well as adequate allocation of products to manufacturing lines are essential for the selection of optimal production and replenishment modes, as we show later in Chapter 5. Having a deep understanding of process flows within a plant is a crucial prerequisite for fully exploiting the potential of LEAN SCM.

Key questions that need to be answered during bottom-up analysis include:

- How long are lead and processing times?
- Where in the stream do waiting times occur?
- Where are (excess) inventories accumulated?
- Where are bottlenecks (e.g., capacity shortages)?
- How well are the manufacturing takt and customer takt balanced?

In the remainder of this section, we present an approach to VSM and design which is extended to cover the needs of LEAN SCM. We explain in step-by-step fashion how a VSM is created, how the results can be interpreted, how the performance of a value stream can be improved, and which actions should be taken to prepare the ground for LEAN SCM.

4.3.1 Gain Transparency into Local Value Streams

Before conducting a detailed value-stream analysis, we recommend creating transparency and ensuring a shared understanding of the various value streams at the site level. Therefore, pre-analyses, namely product family analysis and flow-path mapping, are conducted up front. This helps to bring the big picture into focus and set priorities for the detailed analysis to follow.

4.3.1.1 Product Family Analysis

Product family analysis is the first step in every value stream analysis and the key to all further activities as it separates the entire value stream

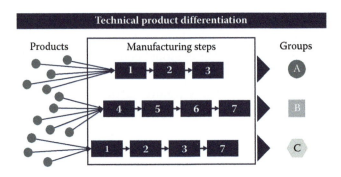

FIGURE 4.26
Product differentiation based on technical characteristics.

into manageable sub-streams. The full product range is broken down into groups of products that share a significant part of the value stream and can therefore be managed together.

Although the grouping of products can follow several approaches, we suggest focusing primarily on similarities in the production process or resource requirements, as illustrated in Figure 4.26. Technical restrictions need to be incorporated to obtain an exact picture of potential limitations that have to be considered later on.

In addition to technical product differentiation, LEAN SCM includes factors that are relevant to planning, focusing on demand volumes and demand variability as well as order patterns. Transparency in this regard is needed for leveled flow design, which is described in Section 4.3. Furthermore, it is the basis for the assignment of LEAN production and replenishment modes as well as their parameterization, as discussed in Chapters 5 and 6.

4.3.1.2 Flow Path Mapping

On the basis of a product family analysis, it is possible to map the flow paths of product streams in a plant. Each flow path represents one group of products that shares a distinct part of the network. In process industries, complexity typically increases toward the end of the value chain due to the high number of variants that are specified and packaged for different markets, customers, or countries.

In this context, the flow path map is quite helpful for both scoping and communication. It creates transparency and visualizes the relevant material flows in a plant. The flow path can be shared and used to bring

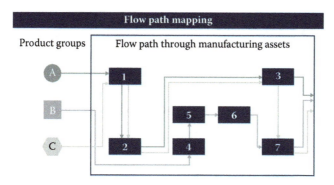

FIGURE 4.27

Flow path mapping creates transparency along value streams.

people operating in separate plant operations on the same page to have fact-based rather than emotional discussions (see Figure 4.27). In other words, a flow path map helps to establish a common language for discussing processes, which is also valuable with regard to change management: only transparency and clear communication enable management to overcome organizational barriers under the slogan "We've always done it that way."

Experience shows that it can be useful to further specify the scope of the value-stream analysis to bundle a company's limited resources efficiently. The refined scope can cover the key value streams of a network or focus on known issues, for example, streams with exceptionally high stock levels or long lead times. A Pareto analysis regarding sales volumes in the product portfolio can be conducted as an additional means of prioritizing value streams (see Figure 4.28).

Having obtained a high-level picture and shared understanding of the existing value streams within a plant, the next step is to conduct a more detailed analysis of the characteristics and performance of those value streams. We explain how to do this in the following section.

4.3.2 Analyze Value Streams to Prepare the Shop Floor for LEAN SCM

Focused value stream analysis at the site level is conducted for streams that have been identified as being in scope during pre-analysis. Typically such an analysis covers every process step from the receipt of raw materials to the shipment of finished goods.

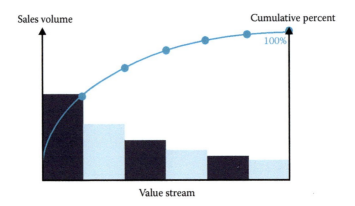

FIGURE 4.28
Pareto chart to set priorities for bottom-up analysis.

As in supply chain analysis, here both the information and material flows are analyzed in order to provide a holistic picture of the main process stages of the value streams within a plant.

4.3.2.1 Current State Map

The first step in value stream analysis is drawing a map of the current process. Typically, such a value-stream map consists of four elements: a high-level image of the value chain, a detailed depiction of the process steps, data boxes with key metrics for each process step, and the overall timeline of the value stream. Figure 4.29 illustrates the schematic layout for such an analysis.

In the upper section of the map, the key value stream stages are depicted, such as a supplier delivering the required raw material, a manufacturing site producing finished goods, and a customer demanding and receiving goods. All sub-processes that are mapped in the following analysis can be associated with these high-level stages of the value chain so that the big picture of the value stream is recognizable at any time. Now let us look at the notation in a nutshell.

Below the main stages of the value chain, the corresponding process steps are listed. All inventories of raw materials, work-in-process (WIP), or finished goods that exist between process steps are indicated by triangles. Both physical as well as information flows between value stream stages or process steps are indicated by arrows.

Below each process step, a corresponding data box is added. Here, all parameter values of relevant metrics that are measured during the detailed

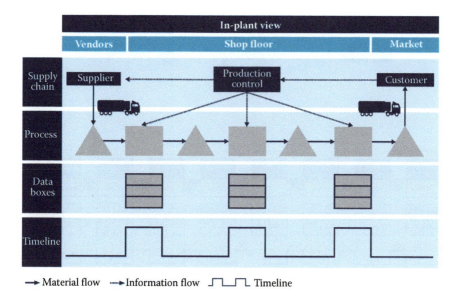

→ Material flow ⋯▶ Information flow ⌐⌐⌐ Timeline

FIGURE 4.29
Schematic layout of a VSM.

analysis are listed. A detailed description of the most important KPIs for value stream analysis is provided in the following section.

Last but not least, the timeline is depicted as a square wave below the value stream. It shows the process lead times and especially visualizes value-added and nonvalue-added time. Typically, the upper segments of the square wave depict the nonvalue-added time, while the lower ones stand for the value-added time. The timeline is a first indicator of potential points of improvement along the value stream. The ratio of value-added and non-value-added time reveals where waste or imbalances regarding capacities and takt are present. Figure 4.30 provides an example of a complete VSM.

4.3.2.2 Key Shop-Floor Performance Metrics That Matter for LEAN SCM

The key metrics for detailed value stream analysis at the site level are quite similar to those considered during top-down analysis. From a plant perspective, shop-floor KPIs, such as overall equipment effectiveness (OEE), are also of interest and thus are included in the analysis. Bear in mind that a VSM should provide only as much detail as is actually required. The process flow and potential process bottlenecks need

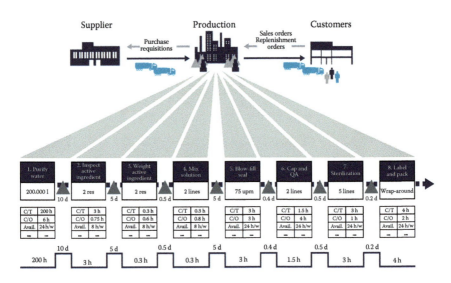

FIGURE 4.30
Completed VSM.

to be visible, and the parameters should provide enough information to enable both a feasible planning design as well as root-cause analyses for the elimination of waste.

Typically, the following performance metrics are analyzed. Nevertheless, as a general principle, the focus of value stream analysis and the selection of relevant performance metrics should be adapted to a company's specific business context. From a LEAN SCM perspective, however, some metrics, such as lead times or production rates, transportation times, changeover times, resource utilization, and current production mode, are essential for the selection and configuration of production and replenishment modes later on and should definitely be included in the analysis.

4.3.2.3 Lead Time

Lead time is defined as the time it takes one unit or batch to complete a process step. Consequently, the lead time does not only necessarily refer to the actual processing time of a production step but also covers transportation times, waiting times, quality control, changeover time, order lead times, or generally all time-consuming activities that are performed within one process step.

Experience shows that long lead times usually imply high levels of WIP, a shop floor crammed with material, rather nervous scheduling due to frequent rush orders, and fluctuating capacity utilization.

4.3.2.4 Value-Added Time

Value-added time is the proportion of the lead time that is actually used for transforming products into something a customer is willing to pay for. Waiting or transportation, for example, does not transform raw materials into finished goods and thus is regarded as nonvalue added. It should be kept in mind here that value always needs to be defined from the perspective of the end customer.

The ratio of value-added time to nonvalue-added time indicates how well the value stream is configured. The higher is the fraction of value-added time, the better is the performance of the value stream. A low ratio indicates long waiting times and uncoordinated processes and generally is a good reason to look more closely at additional KPIs to identify potential improvement areas more precisely. Experience shows that in process industries the fraction of value-added time is often less than 5%.

4.3.2.5 Changeover Time

Changeover time is defined as the time that is needed to change from one product type to another in a production process. In process industries, the dependence of changeover times on the production sequence needs to be considered. Production sequencing is a powerful lever for freeing capacity and reducing the overall changeover effort.

4.3.2.6 Takt

In every manufacturing or service system, requirements are ultimately generated by customer demand. In manufacturing, a customer takt is the "rate of customer demand" or the "pace of customer demand" and measures the total customer demand over a certain time, expressed either as a time factor or as a rate.

Hence, the customer takt defines the rate at which every task needs to be performed (by analogy to the global takt explained in the previous section), that is, it indicates the maximum amount of time which is available within one process step to produce one unit of output. In other words, the manufacturing takt at every stage needs to be shorter than the customer takt in order to permanently satisfy customer demand. If for a given process step the manufacturing takt is only slightly shorter than the customer takt—manufacturing produces a given output only slightly faster than customers demand requires—this process step incurs a high risk of

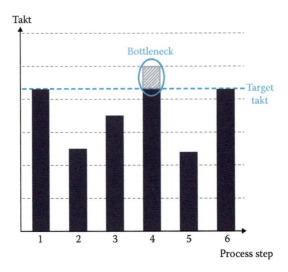

FIGURE 4.31
Takt of different process steps at the shop-floor level.

shortages or stock-outs in cases of demand fluctuation. If, on the other hand, the manufacturing takt is dramatically faster than the customer takt, excess inventories can accumulate since the production rate is higher than the consumption rate.

Another reason for using the term "takt" in the context of process design is that it evokes the vision of a conductor, leading and coordinating an orchestra. To achieve a harmonic interplay within the value chain, it is—as is the case with an orchestra—not sufficient to focus on the outcome of only one system or workstation. Well-tuned interaction between all stages is essential to maintaining a smooth value stream.

If analysis shows that the manufacturing takt of consecutive process steps differ widely, the process as a whole is not optimally configured (see Figure 4.31). Generally, the process step with the longest takt time is the capacity bottleneck and usually the place where excess inventories accumulate and waiting times for subsequent steps are generated.

4.3.2.7 Inventory

The volume of inventory held at various stages of the value stream indicates where material is accumulating due to potential capacity bottlenecks. Both the average inventory level and the coverage in days of supply should be measured and noted below the stock-keeping points on the VSM.

Surges in material flow can easily be recognized by excess inventories, which generally can indicate process steps that are not aligned sufficiently, suggesting that one process step lacks capacity compared with others.

4.3.2.8 Utilization

Utilization is the fraction of time during a process step in which it is not idle. In other words, a utilization indicator provides information about how much of the available capacity of a process step is actually used and indicates where in the value stream bottlenecks and excess capacity are present.

4.3.2.9 Overall Equipment Effectiveness

OEE breaks the performance of a manufacturing step into three separate but measurable components: availability × performance × quality. Each component points to another aspect of the process that can be a reason for suboptimal performance and thus a target for improvement actions. It is unlikely that any manufacturing process can run at 100% OEE. Practice shows, however, that an OEE of 20–30% is not uncommon.

Availability separates unplanned downtimes from theoretical operating times, representing the percentage of the scheduled amount of time that can be actually used for production. To determine whether changeover time should be considered in light of availability is usually a company-specific decision. If changeover times are included, production is motivated to reduce the overall changeover time in order to improve availability and OEE—which is not generally bad. The downside is that increasing batch sizes in order to reduce the overall number of set-ups also helps to reduce the overall changeover time and improve OEE. This contradicts the principles of lean manufacturing, as an increase in production lot sizes automatically involves an increase in inventories. Experience shows that conflicts of interest between production and SCM are intensified if—as in most companies—production performance is measured by OEE while SCM performance is measured by inventory levels.

Manufacturing performance represents the speed at which production actually runs in relation to its designed speed, accounting for losses of every kind such as reduced output rates due to suboptimal component quality.

Manufacturing quality stands for the fraction of good units, that is, the proportion of output that is produced within quality specifications in relation to total output.

4.3.2.10 Transportation Data

Regarding the transportation steps in the value stream, frequencies of delivery, transport lot sizes, and transportation lead times should be noted on the VSM.

4.3.2.11 Information Flows

Since information processing within a value stream is often quite time consuming, a VSM also visualizes the key information flows, including both customers and suppliers, and shows the main data processing processes. The frequency of data generation and delays in data processing steps are noted in the data boxes below these processing steps.

4.3.2.12 Planning Mode

From a LEAN SCM perspective, it is essential to have a complete picture of the distinct production modes that are used within the production network. Therefore, the current production mode, such as a Rhythm Wheel, is noted for each process. Production modes are discussed in detail in Chapter 5.

Although the elements that are noted in the data boxes focus on distinct aspects of the whole process, they often point in the same direction, providing a clear picture of the problem areas that prevent a value stream from flowing smoothly. Generally, imbalances in the value chain must be understood as waste. To achieve sustainable improvement, these imbalances have to be eliminated. Leveled flow design is one way to increase the performance of a value stream by creating balanced and stable material flows.

4.3.3 Aim for Leveled Flow Design

Flow without interruptions has traditionally been considered a crucial element of lean manufacturing. This holds true for LEAN SCM as well: leveled production flow is one of the most important enabler applying LEAN SCM.

Leveled flow design aims at achieving the optimal fit between product and resource portfolios and a smooth flow of product streams with stable resource utilization. Accordingly, three main tasks can be distinguished:

- Ensuring stable and efficient production processes
- Allocating products to manufacturing lines
- Focused leveling of product flows

4.3.3.1 Ensuring Stable and Efficient Processes

Stable and reliable processes are the backbone of successful LEAN Planning. Repetitive production modes such as the Rhythm Wheel depend on reliable manufacturing assets. If, for example, changeover times or production rates fluctuate widely, it is difficult to achieve an optimal Rhythm Wheel design (see Chapter 6). Although LEAN Planning methods can function in an environment that is not perfectly stable, the full benefits will be realized only if the production processes are reliable and operate at a constant speed and output quality.

To improve both the stability and efficiency of manufacturing operations, well-known lean tools, which have their origin in the Toyota Production System, can be used. Various lean methods have been successfully applied for decades and are still widely used today by practitioners across all industries. Some of the most important tools are Single Minute Exchange of Die (SMED) and Root Cause Analysis with tools such as the Ishikawa (Fishbone) Diagram or Five Why, Poka Yoke, Total Productive Maintenance (TPM), and 5S.

Some of the mentioned tools, however, need to be modified to meet the specific requirements of process industries. When applying SMED to reduce changeover times, for example, it should be understood that, at process industry sites, such as pharmaceutical and food processing plants, cleaning often occupies a significantly larger proportion of the overall changeover time than set-up tasks in terms of mechanical or electrical modifications to equipment, which represent the bulk of the work involved in assembly plant set-ups. What is true for SMED applies to the other tools as well: to fully leverage the waste reduction potentials within a value stream both industry- and company-specific characteristics should be considered rather than blindly following textbook approaches.

4.3.3.2 Allocating Products to Manufacturing Lines

From a production point of view, it would be ideal to have a mix of high-volume products with constant demand on one resource to achieve flawless flow in the value stream. It is every planner's dream to have dedicated machines for a high-volume mix, enabling high resource utilization on a constant basis while at the same time maintaining low inventory levels and minimal planning effort. Unfortunately, reality paints a different picture. For process industries, a high-mix environment with a rat's tail of low-volume products on every line is typical. This makes shared resources literally inevitable.

As a consequence, complexity management from a planning perspective is required. Complexity management is based on a reasonable clustering of products and the adequate allocation of those clusters to production lines.

Typically, product allocation is conducted in three steps. First, both product and resource portfolios are analyzed and then products are allocated to resources.

4.3.3.3 Product Portfolio Analysis

Understanding the structure of a product portfolio is the first step toward achieving feasible and efficient product allocation. Various characteristics of a product portfolio can be decisive for matching products and resources:

- Volume and value of demand
- Variability of demand in terms of quantity and time
- Product families
- Product formats
- Quality specifications

Typically, demand volume and variability are the major criteria. Here the same methodology that was recommended for supply chain segmentation (Section 4.1.2) is applied to the local product portfolio. Figure 4.32 illustrates the results of product portfolio analysis.

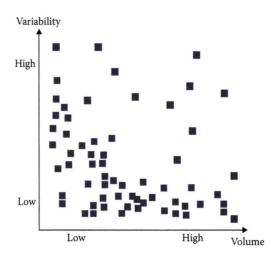

FIGURE 4.32
Results of portfolio analysis at the shop-floor level.

To simplify this process, the ABC–XYZ matrix is additionally clustered in three categories: runners, repeaters, and strangers.

4.3.3.4 Resource Portfolio Analysis

Experience shows that a company's resource portfolio is typically heterogeneous, such that the available lines for a given process step are not identical in construction. As a consequence, the resources exhibit varying characteristics or strengths and weaknesses that are relevant to the allocation of products, such as:

- Line speed
- Changeover flexibility
- Output yield

It is crucial to consider differing machine capabilities in planning and utilize technical strengths as much as possible. Figure 4.33 shows how a resource portfolio can be clustered to generate transparency for the allocation of products. In this example, line speed and changeover flexibility are considered to obtain a differentiated picture of the resource portfolio, which in this case consists of three lines.

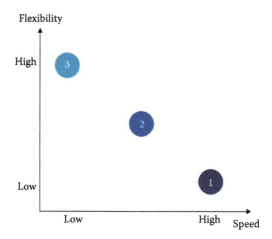

FIGURE 4.33
Segmentation of the resource portfolio.

4.3.3.5 Allocation of Products to Resources

Unfortunately, there is no one-size-fits-all solution for the allocation of products to resources. The allocation strategy needs to be defined according to the characteristics of individual items in the product and resource portfolios. If, for example, a highly mixed product portfolio has to be matched with a homogeneous resource portfolio, distinct allocation strategies can be derived from the characteristics of the product portfolio. Figure 4.34 shows possible allocation strategies based on demand characteristics ranging from an equal distribution of runners, repeaters, and strangers on all lines to more or less dedicated lines that are reserved for one of these groups.

If the resource portfolio is heterogeneous, for example the resources differ significantly in terms of speed and flexibility, this should also be incorporated in the allocation. In this case, slow resources with quick set-ups should be reserved for products with high variability and low volumes while fast resources with slow set-ups should preferably be used for products with low variability and high volumes. Figure 4.35 shows how the product and resource portfolios can be laid one above the other.

Naturally, different decision criteria lead to different allocations. An allocation according to volume and variability as described above differs from an allocation based strictly on product families. The decision as to which criteria should be considered for matching products and resources needs to be made on a case-by-case basis, depending on where the greatest potentials lie. If a product family is heterogeneous regarding demand

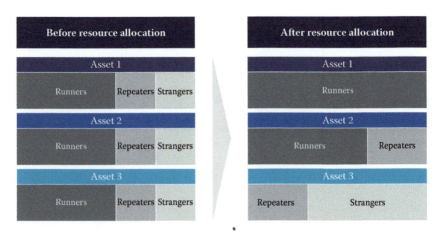

FIGURE 4.34

Allocation strategy for homogeneous resource and heterogeneous product portfolios.

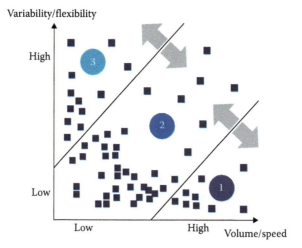

FIGURE 4.35
Allocation strategy for heterogeneous product and resource portfolios.

volumes and variability, for example, it might not be beneficial to keep it together on one resource. On the other hand, changeovers within the same product family usually require less effort, which favors not splitting the product family into separate resources.

4.3.3.6 Focused Leveling of Product Flows

Imagine what happens on a motorway when there is lots of traffic and a road is narrowed from three lanes to two: if the capacity of the two lanes is not sufficient for the number of cars driving through, the result will be either slowed traffic flow or in the worst case a full-blown traffic jam. If the motorway is extended to three lanes again after a couple of kilometers, traffic will again flow smoothly (see Figure 4.36). In this case, the two-lane stretch of the motorway is the bottleneck, failing to provide the required capacity and preventing the stream from flowing smoothly.

The traffic example can be easily transferred to manufacturing and service operations. If a process step lacks capacity, two things will happen: before the bottleneck process, WIP material will be jammed; downstream of the bottleneck processes will have to wait until enough input material has been provided. To smooth the material flow and level the production at a downstream stage of the supply chain, a steady supply of material needs to be ensured. Ideally, this is achieved by reducing the takt time of the

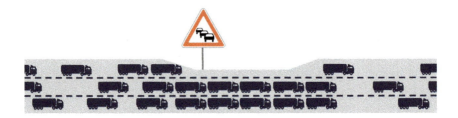

FIGURE 4.36
Traffic jam before motorway bottleneck.

bottleneck process (see Figure 4.37). Several factors should be addressed to improve the manufacturing takt:

- Changeover time
- Processing time
- Learning effects (applying repetitive production patterns)
- Machine reliability
- Inspection time
- Number of shifts

If it is not possible to achieve full balancing of the takt, for example in batch manufacturing operations, mismatches between two process stages can be buffered by installing controlled work in process stocks. Bear in mind that the buffer stock is sized based not on demand patterns but on the time imbalance between the two processes.

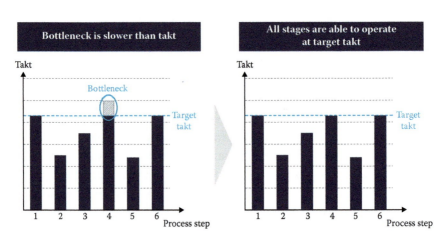

FIGURE 4.37
Improvement of takt at the shop-floor level.

The production processes themselves are, however, not the only levers in terms of leveling material flow and smoothing capacity utilization that can be adjusted. Chapter 5 discusses a range of strategic alternatives regarding production planning modes. Repetitive or cyclic production modes, for example, make use of the fact that the sequence in which products are manufactured does not need to match the sequence in which they are purchased by customers. Production quantities are kept as constant as possible and triggered regularly according to a predefined takt or "rhythm" that enables a smooth and leveled material flow.

In general, the leveled flow exercise allows for the reallocation of products within technological and legal constraints. However, product-allocation mixes on the lines sometimes remain. This may imply that all product groups (runners, repeaters, and strangers) should be managed on a single line. In such cases, extended planning approaches are necessary. Within the Breathing Rhythm Wheel concept, for example, producing every product at every cycle (the EPEC-rule) will not be sufficient. Instead, a High-Mix Rhythm Wheel must be applied. Further details regarding the High-Mix Rhythm Wheel are provided in Chapter 5.

Summary

In this section, we provided a methodology for changing your company's supply chain from a bottom-up perspective to prepare local sites for LEAN SCM. A pre-analysis builds a baseline while focused value stream analysis provides a holistic picture of the main process stages of value streams within a plant. Leveled flow design concludes the section, supporting the achievement of optimal fit between the product and resource portfolios and smooth flow of product streams with stable resource utilization.

CHAPTER SUMMARY

In this chapter, we presented important enablers for exploiting the full potential of LEAN SCM. We introduced a range of tools and approaches to support your company's efforts to change planning conditions in a way that mitigates variability, complexity, and uncertainty at various stages along the supply chain.

In Section 4.1, we stressed the importance of developing a clear understanding of strategy. We recommended using supply chain segmentation. We explained how segmentation based on customer order winners provides a customer-oriented, value-added perspective.

In Section 4.2, we provided a top-down mapping methodology to achieve transparency regarding inventory, capacity, global takt, and lead times. Furthermore, we showed how gaps can be identified and assessed in order to detect areas of potential improvement along your company's supply chain. To improve and prepare your supply chain for LEAN SCM, we explained network design, active demand shaping, and complexity management.

In Section 4.3, we focused on preparation for LEAN SCM at particular local stages of the supply chain. First, a pre-analysis encompasses product family and flow-path mapping. A focused value stream analysis should be performed in the second step. This includes creating a current state map and KPI analysis. Third, leveled flow design should be aimed for by ensuring stable and efficient processes, allocating products efficiently, and allowing for a smooth flow.

You should now be able to prepare your company's supply chain to adopt the LEAN SCM paradigm. Doing so will establish an effective physical supply chain set-up which represents a solid foundation for LEAN Planning. In the next chapter, we offer detailed insights into strategic LEAN Planning configurations, which define how planning and coordination are managed along the value chain.

5

Strategic LEAN Supply Chain Planning Configuration

The strategic LEAN Supply Chain Planning process defines how planning and coordination are managed along the value chain. Within the range of possible approaches, not all ensure timely and efficient supply of goods. In process industries, supply chain practitioners are increasingly dissatisfied with how their planning approaches perform. What about your company? Is it satisfied with its current approach to planning?

Experience shows that implementing the LEAN SCM principles introduced in Chapter 2 enables a company to maximize the potential of planning, especially in the process industry environment. These principles are anchored in LEAN SCM Planning concepts, namely supply chain modes. To conduct detailed analysis and appropriate mode selection according to a company's specific needs, supply chain modes should be divided into two categories: replenishment modes and production modes. The replenishment mode determines in which quantity and at what time products are ordered from other supply chain stages. The production mode defines in which sequence, time, and quantity the products are finally scheduled at each stage (see Figure 5.1). The key challenge within the strategic LEAN Supply Chain Planning configuration is to select the best-suited replenishment and production modes for a company's supply chains.

This chapter answers three principal questions:

- What are LEAN alternatives to traditional planning concepts?
- What are the benefits of LEAN Planning concepts for the process industry?
- How does a company select appropriate LEAN modes for its supply chains?

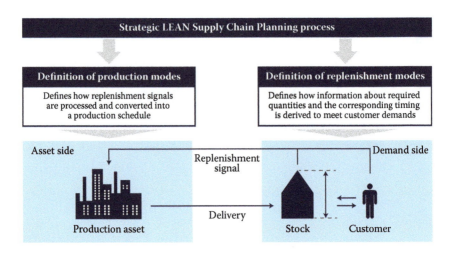

FIGURE 5.1

Production and replenishment modes are defined within the strategic LEAN Supply Chain Planning process.

To provide corresponding answers, the chapter is structured as follows. In Section 5.1, we introduce LEAN replenishment modes. In addition to forecast-based push replenishment, consumption-based pull replenishment modes are introduced.

In Section 5.2, we provide detailed insights into selected LEAN production modes. We describe kanban and the innovative Rhythm Wheel concept in detail. The section focuses on the Rhythm Wheel concept as it addresses the specific needs of process industries. Besides explaining the benefits of this production mode, the section provides insights into precisely how the Rhythm Wheel concept works.

Section 5.3 is dedicated to supply chain mode selection. Within this section, we provide a structured approach to identifying the best-suited replenishment and production modes for your company's supply chain. The decision support provided will help companies tackle the challenge of selecting appropriate supply chain modes.

Owing to constantly changing business environments, production and replenishment modes must be reviewed and adjusted regularly to sustainably ensure the competitiveness of the respective supply chain. The strategic renewal process that we introduce in Section 5.4 provides excellent guidance for understanding how such a structured process should look.

5.1 WHAT TO PRODUCE: REPLENISHMENT MODES

Choosing suitable replenishment modes is an integral part of the strategic LEAN Supply Chain Planning process. Only through appropriate mode selection can a company assure timely and efficient supply of goods along the value chain. The replenishment mode defines at which time and in which quantity orders are created along the supply chain. We refer to this information as replenishment signals, which are then transmitted accordingly to the production assets.

In this section, we introduce two general classes of replenishment modes: push replenishment and pull replenishment, depending on the way in which production is triggered. With push replenishment, production is triggered by a forecast-based plan. With pull replenishment, it is triggered only by real consumption. Since the use of consumption pull is one of the major LEAN SCM principles, this chapter focuses on these replenishment modes. Figure 5.2 provides an overview of all replenishment modes discussed, and represents the underlying structure of the chapter.

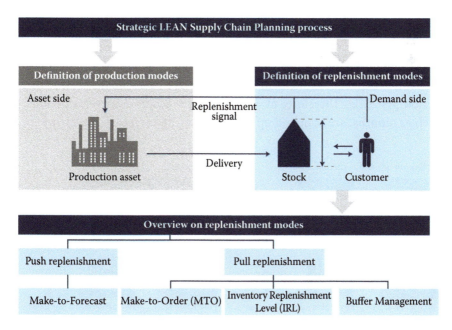

FIGURE 5.2
Overview of relevant replenishment modes.

5.1.1 Sell What You Make: Forecast-Based Push Replenishment

In push replenishment modes, production schedules are derived on the basis of forecasted demand. Here, an attempt is undertaken to predict future demand accurately enough to match produced quantities with anticipated demand. Once such a forecast is made, production orders are "pushed" into the production system, in the hope that demand occurs as expected.

How well push replenishment works depends critically on forecasting accuracy. If forecasting error is high, push-managed supply chains typically lack in performance due to poor customer service, excess inventories, and heightened planning nervousness from short-term production plan adjustments. That is why companies across all industries have drawn so much attention to opportunities that improve forecasting accuracy. Improved supply chain coordination (see Figure 5.3) and the application of structured forecasting methods have proved to be very effective in this context.

Despite ongoing enhancement of forecasting capabilities, most companies have yet to meet the challenge of accurately predicting future demand for most of the products in their portfolios. Therefore, forecasting inaccuracy remains an obstacle for push-managed supply chains seeking to realize their full performance potential.

Under LEAN SCM, aggregated forecasting can be used, since it is more reliable due to statistical pooling effects. While it is, for example, possible to estimate total aggregated demand within a 3-month period, it is not realistic to expect accurate demand forecasts for every single week of this period per SKU. Therefore, to achieve higher accuracy, several periods, such as months, are summed up for an aggregated forecast. Nevertheless,

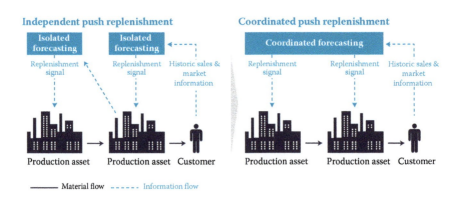

FIGURE 5.3

Comparison between independent and coordinated push replenishment.

for the more volatile products in a company's portfolio, even an aggregated forecast may not be accurate enough to reliably trigger production. In this case, LEAN SCM proposes to configure the replenishment parameters with the help of the aggregated forecast alone; replenishment and production are then triggered only by real consumption. We therefore discuss alternative LEAN pull replenishment modes in the next section.

5.1.2 Make What You Sell: Consumption-Based Pull Replenishment

In this section, we introduce LEAN replenishment modes that provide an alternative to push replenishment by following the LEAN SCM principle of using consumption pull. The replenishment trigger and thus the final production schedule is based on actual consumption, which in turn is indicated either by customer orders or by decreasing stock levels.

When implementing consumption pull, a company's production processes respond to the voice of the customer and produce only quantities that are really sold. High service levels as well as low inventories can be achieved in this way. Furthermore, consumption pull ensures that only reliable information about customer demand is passed along the supply chain (see Figure 5.4). It follows that constant re-scheduling activities can be avoided, which significantly reduces overall supply chain nervousness.

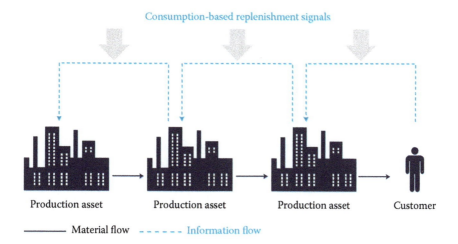

FIGURE 5.4
Pull replenishment propagates information about real consumption through the supply chain.

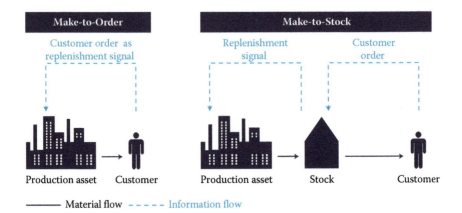

FIGURE 5.5

Comparison of MTO and MTS in the context of pull replenishment.

Pull replenishment modes can be further distinguished into two classes, Make-to-Order (MTO) and Make-to-Stock (MTS) (see Figure 5.5). In MTO replenishment, orders from customers function directly as replenishment signals for production. On the other hand, in MTS replenishment, a placed order is delivered from stock. The corresponding removal of goods triggers a replenishment signal indicating the need for replenishment. Note that MTS is not identical to push replenishment. MTS just means that stock is held for a certain product. The stock can be replenished with either mode, push or pull replenishment.

In the following sections, we introduce three distinct modes that achieve pull replenishment. In addition to MTO, we describe two MTS replenishment modes—IRL replenishment and Buffer Management.

5.1.2.1 Make-to-Order

The MTO replenishment mode is a straightforward form of pull replenishment. In this scenario, customers place orders to be replenished by production assets. The customer order serves directly as a replenishment signal. Once an order is placed, the production asset schedules the order such that the customer's due date is met. After production is complete, the ordered quantities are delivered to the customer.

The major benefit of the MTO replenishment mode is that there is no need to hold inventories because orders are not produced on stock in advance but only when required by the customer. This is especially ben-eficial if:

- Inventory holding costs are high
- Shelf lives are short
- Demand is sporadic
- Products are characterized by a high degree of diversification or customization

MTO's potential to achieve notable inventory reductions should make it highly attractive to supply chain practitioners. However, MTO replen-ishment is not possible in every case. To carry out MTO replenishment successfully, two major prerequisites must be met: adherence to customer lead time expectations and sufficient capacity buffer.

The first prerequisite concerns customer expectations concerning the delivery times of ordered products. Customers must be willing to wait until the production of the ordered quantities has finished and the goods have been delivered. Hence, the order lead time may not exceed the cus-tomer lead time expectation (see Figure 5.6).

If customer lead time expectations are long enough to cover the order lead time of more than one supply chain stage, it is beneficial to move the "MTO boundary" further upstream along the supply chain. By doing so, customer orders trigger production at a supply chain stage that is further upstream. The resulting production output is then processed stepwise at

FIGURE 5.6
Adherence to customer lead time is a prerequisite for MTO replenishment.

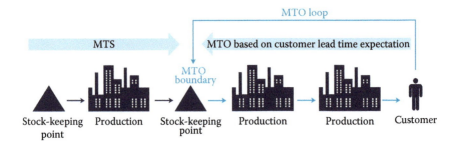

FIGURE 5.7
The MTO boundary depends on the customer lead time expectation.

downstream supply chain stages until the final products are delivered to the customer. In this way, significant inventory savings can be achieved within the so-called "MTO loop," as depicted in Figure 5.7.

In cases in which the order lead time exceeds the maximum lead time that is still accepted by the customer, the MTO boundary cannot be moved further upstream. At this point, stocks are required to decouple supply from demand. The replenishment of these stocks must now be based on MTS replenishment modes. Bear in mind that MTS can be based on either pull or push replenishment.

The second prerequisite concerns the availability of sufficient capacity. If, for example, many products are ordered in the same period, not all orders can be produced in that period due to capacity restrictions. Therefore, a capacity buffer (of resources and equipment) is required to manage order peaks. However, high capacity buffers imply low average resource utilization, which should typically be avoided in capital-intensive process industries.

If these two prerequisites of sufficient customer lead time expectation and capacity buffer cannot be met, products cannot be made-to-order. Instead, they have to be made-to-stock. In the next section, we introduce two MTS pull replenishment modes: the IRL concept and Buffer Management.

5.1.2.2 Inventory Replenishment Level (IRL)

IRL replenishment is another mode that involves consumption pull, one of the major LEAN Supply Chain Planning principles. In IRL replenishment, stocks are used to decouple production from demand and to effectively buffer variability. These stocks are often called "supermarket stocks." Be aware that even though the quantities produced are MTS, replenishment is pull-driven insofar as the replenishment signal is triggered by actual consumption, not by forecasted demand.

The IRL represents the target inventory level while the replenishment interval determines how often replenishment signals are sent to the appropriate production asset. Once these parameters are defined, replenishment signals can be triggered automatically based on the following logic. Every time inventory is reviewed, current inventory (CI) levels are compared against the defined IRL. Whenever the CI falls below the IRL, a replenishment signal is sent to the production asset to indicate the need for replenishment. The quantity to be replenished is derived dynamically by the difference between the IRL and the CI. As demand typically fluctuates over time, replenishment quantities typically vary, too, as Figure 5.8 illustrates. Once production has taken place and delivery occurs, the inventory level increases.

To manage replenishment in the IRL replenishment mode, only two basic supply chain parameters need to be defined: the IRL and the replenishment interval. To calculate those, a range of factors must be taken into account, such as demand volume and variability, supply variability, and production and transportation times. These factors will be described in detail in Section 6.2.

The benefits of IRL replenishment are the use of consumption pull, efficiently buffering demand variability in supermarket stocks, and managing replenishment within only two basic parameters. However, we now

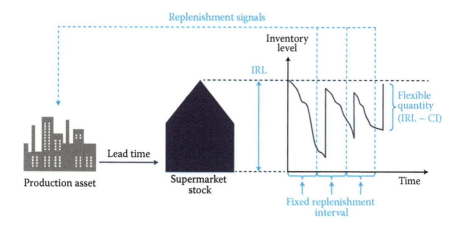

FIGURE 5.8
IRL replenishment with fixed replenishment intervals and flexible replenishment quantities.

consider another promising pull replenishment mode that is often applied in process industries: Buffer Management.

5.1.2.3 Buffer Management

Buffer Management is another efficient way to implement pull replenishment using supermarket stock that decouples production from demand. Much like IRL replenishment, Buffer Management requires appropriate management of only a few parameters: the reorder point (ROP) as well as the minimum and maximum inventory levels of supermarket stock.

When using Buffer Management, the objective is to maintain the inventory level between defined minimum and maximum boundaries. Hence, replenishment must be triggered whenever inventory levels are in danger of falling out of this range. Because final replenishment requires a certain lead time, the application of an ROP serves as an efficient measure for ensuring that the inventory stays within the defined range. When the inventory falls below that defined ROP, a replenishment signal is created. In practice, the quantity that is replenished is typically fixed (see Figure 5.9).

Under Buffer Management, stock is often divided into three zones: green, yellow, and red. Such visualization simplifies understanding the concept and supports monitoring and alerting. Typically the green zone is the area between the maximum inventory boundary and the reorder point. As long as the inventory level remains within this zone, no replenishment orders

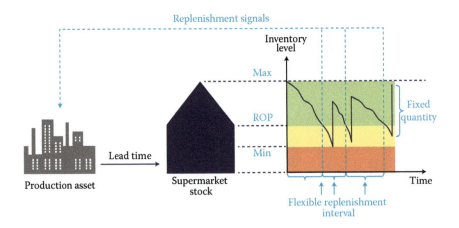

FIGURE 5.9
Buffer Management with flexible replenishment intervals and fixed replenishment quantities.

LEAN pull replenishment modes	Supermarket stock	Buffering variability in inventories	Replenishment quantity	Replenishment interval
MTO	No	No	Dynamic	Dynamic
IRL	Yes	Yes	Dynamic	Fixed
Buffer Management	Yes	Yes	Fixed	Dynamic

FIGURE 5.10
Key characteristics of pull replenishment modes.

are placed. The yellow zone is the area between the reorder point and the minimum stock level. As soon as the yellow zone is entered, a replenishment signal is triggered. Inventory should be monitored with greater attention in the yellow zone. The red zone is the area between the minimum stock level and the zero line. Only in exceptional circumstances, such as demand peaks or major supply events, should inventory be allowed to drop into this area.

As in IRL replenishment, a range of factors must be taken into account to derive the optimal values for the described replenishment parameters when applying Buffer Management. For detailed information about deriving optimal sizes of replenishment parameters, we again recommend reading Section 6.2.

The benefits accruing from using Buffer Management as a replenishment mode are very similar to those associated with IRL replenishment. Implementing pull helps to overcome the drawbacks of push replenishment; demand variability is efficiently buffered in stocks; and rule-based replenishment management with only three parameters makes for a low-effort mode in practice.

For a comparative overview, Figure 5.10 provides a summary of all previously introduced pull replenishment modes and their key characteristics.

Summary

The selection of appropriate replenishment modes for the supply chain is a key challenge within the strategic LEAN Supply Chain Planning process. In this section, we introduced the replenishment modes that are relevant within the LEAN SCM framework, while the following section provides insights concerning relevant production modes.

We have shown that replenishment modes differ mainly with regard to how replenishment is triggered. This relates to both timing and quantity as well as to whether the replenishment signal is triggered based on forecast (push) or real consumption (pull). Because it is so difficult to make accurate forecasts, push replenishment often yields unsatisfactory results, such as supply chain nervousness, low service levels for some products, and excess inventories for others. Considering these drawbacks, implementing consumption-driven replenishment as a major LEAN SCM principle represents a promising alternative. MTO, IRL, and Buffer Management have been introduced as pull replenishment modes that follow this principle.

All of the above-mentioned replenishment modes are worth considering as options for practical use. To choose appropriate replenishment modes for your supply chains, company-specific factors need to be considered explicitly. In Section 5.3, we provide a detailed approach that will help your company find the best-suited replenishment modes for its supply chains.

5.2 HOW TO PRODUCE: PRODUCTION MODES

In addition to choosing suitable replenishment modes, appropriate production modes must also be determined, which represents the second major decision in the strategic LEAN Supply Chain Planning process (see Figure 5.11). Selecting a production mode means deciding how replenishment signals are processed and transformed into final production orders with precise production quantities and production times. In other words, the production mode defines how the production schedule is derived based on received replenishment signals.

In the course of this section, we introduce several state-of-the-art LEAN production modes. We pay special attention to the Rhythm Wheel concept as it embodies core LEAN SCM principles such as the use of repetitive production patterns and leveling of production and resource utilization. Applying the Rhythm Wheel concept appears highly beneficial, especially in process industries.

5.2.1 Kanban and Its Advancements for Process Industries

As lean approaches have proved able to overcome the obstacles involved in traditional production planning and scheduling, they are increasingly catching

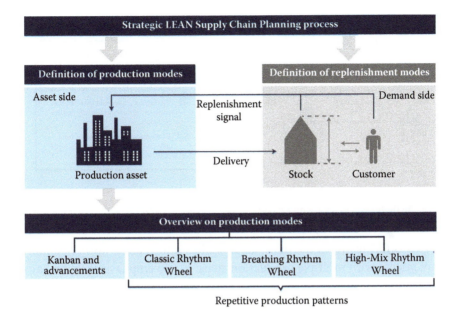

FIGURE 5.11
Overview of relevant production modes.

the attention of supply chain practitioners. These alternative approaches not only simplify planning but also lead to more efficient planning results. One well-known lean approach in the area of production planning and scheduling is known as the kanban system. The origin of kanban dates back to the late 1940s, when the automotive manufacturer Toyota developed the system to improve its production efficiency. Since then, the kanban system has spread from Japan across the world and is widely recognized as an integral part of the just-in-time (JIT) philosophy. Under a kanban system, detailed planning and scheduling are not a centralized task. Once the appropriate number of kanbans (specific cards that are used on the shop floor) has been determined, the creation of a production schedule is taken over by self-regulating and independent control systems. A key characteristic of a kanban system is that it implements pull replenishment, under which production reacts to real consumption. More precisely, kanban systems are typically based on supermarket pull. When demand occurs, it is satisfied with goods from stock which are then replenished by production. Figure 5.12 summarizes how production is scheduled under a kanban system.

Since production consumes prematerial that needs to be replenished as well, the preceding production step can also be managed with kanbans.

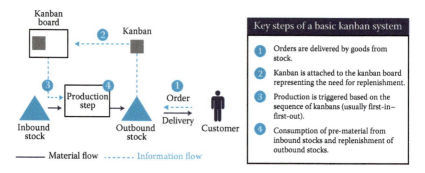

FIGURE 5.12
Basic methodology of a kanban system.

Figure 5.13 illustrates the material as well as information flows in a kanban system with multiple production steps. It also emphasizes that a kanban system may consist of several production control systems working independently.

An enhancement of the kanban approach is constant work-in-process (ConWIP). The basic idea is similar, but in ConWIP the kanban control system embraces not merely one but several production steps. In this way, the individual production and control systems no longer work independently; instead, the various production steps are coupled. The total number of kanbans can be reduced in this case, which means a reduction of work in process.

Another form of the traditional kanban is the "advanced kanban." So far, the kanban systems described here schedule production orders on a first-in–first-out (FIFO) basis, following the sequence of their arrival. As production in the process industries is typically characterized by very time-consuming and sequence-dependent changeovers, this scheduling

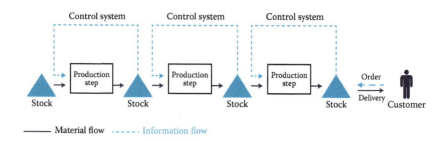

FIGURE 5.13
Kanban system with multiple production stages.

rule leaves room for improvement. Box 5.1 offers a practical example in which a big pharmaceutical manufacturer adjusts the kanban approach to cover its industry-specific needs. The basic idea behind the adjustment is that production orders are batched to reach production quantities that are economically efficient.

BOX 5.1   ADVANCED KANBAN SYSTEM OF A LEADING PHARMACEUTICAL COMPANY

In the context of a major lean manufacturing initiative, a leading pharmaceutical manufacturer carried out pull replenishment at a key production site in Europe by implementing a kanban system. The kanban system manages the replenishment of stock for bulk material, that is, medicinal tablets. In the industry-specific pharmaceutical production environment, the kanban system in use differs from the basic approach in two major respects:

1. Consolidation of production batches to campaigns due to high changeover effort
2. Sequencing of production campaigns due to the high sequence dependency of changeovers

These modifications are operationalized with the help of a kanban board, which is depicted schematically in Figure 5.14.

Whenever a drum of tablets is consumed by the packing unit, the attached kanban (card) is removed from the drum and attached to the wait section of the kanban board. As soon as a certain predefined number of kanbans is reached, they constitute a campaign and are passed on to the work section. In the work section, the planner allocates campaigns to various production resources and determines the production sequence.

Accordingly, the kanbans are moved on to the corresponding position in the in-process section, where they trigger production. As each process step—granulation, compressing, coating, and quality control—is completed, the corresponding kanban moves forward on the board. After completion of the production process, the kanban is finally attached to the finished drum of tablets.

FIGURE 5.14
Schematic picture of a kanban board.

Also, the inventory section of the kanban board shows the stock levels of each product. This visibility enables the planner to react appropriately whenever a product approaches a stock-out situation. In the case of a potential stock-out, the planner is allowed to attach a red emergency kanban to the board, which indicates high priority for production.

By realizing consumption pull and significantly reducing planning complexity, kanban and its enhancements appear to offer a promising alternative to traditional production planning and scheduling. However, one might wonder if there is still room for improvement regarding applicability to process industries. Indeed, there is, as we will show in the next section.

5.2.2 Product Wheels and Rhythm Wheels for Cyclic Production Planning

In this section, we introduce an innovative production planning and scheduling approach that considers key characteristics of process industries—the so-called Rhythm Wheel or product wheel approach. This approach both satisfies the wish for a LEAN Planning process and ensures efficient scheduling. Industry experts such as Peter L. King and Raymond C. Floyd have already introduced "cyclic scheduling" and "product wheel" applications in process industries. But the general product wheel approach is rather suitable for large volume products with relatively stable sales. Therefore, specific Rhythm Wheel approaches have been developed for high-mix product portfolios and volatile environments.

5.2.2.1 General Idea of Rhythm Wheel Planning and Scheduling: What You Need to Know ...

The general idea of the Rhythm Wheel concept is fairly simple, and like-wise brilliant: Use a pre-configured optimal production schedule and repeat it over and over again. As shown in Figure 5.15, such a repeating schedule can be visualized by a wheel on which each segment stands for a product being produced at an asset. Segment size represents the time required to complete the planned production runs. In a metaphorical sense, the wheel rotates over the course of time at a certain rhythm, which explains the name "Rhythm Wheel." Translating a schedule that is created by a Rhythm Wheel into a commonly used Gantt chart, a repetitive production cycle with a corresponding cycle time is created (see Figure 5.15).

Other than the optimal production sequence, only a few production parameters have to be determined to pre-configure a Rhythm Wheel. According to a major LEAN SCM demand principle, an aggregated forecast is used for this pre-configuration (see Figure 5.16). Chapter 6 provides an overview of the parameters to be chosen and detailed insights into how to derive their optimal values.

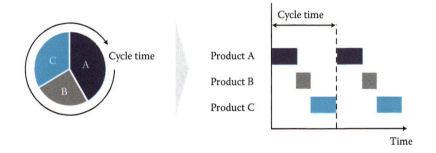

FIGURE 5.15
Illustration of the Rhythm Wheel as a repetitive production mode.

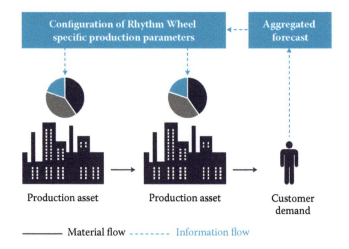

FIGURE 5.16
An aggregated forecast is used to configure the Rhythm Wheel.

It is important to understand why supply chain champions in the process industries have already implemented the Rhythm Wheel concept and why others are sure to follow in the near future. We therefore discuss the major benefits of the Rhythm Wheel approach in the next section.

5.2.2.2 Key Benefits: Why Your Company Should Put the Rhythm Wheel on Its Agenda ...

By following several LEAN SCM principles, such as the use of repetitive production patterns and leveling production and utilization, the Rhythm Wheel approach makes it possible to enjoy major benefits, especially for companies in process industries. The key benefits of the Rhythm Wheel concept cover four major areas, as shown in Figure 5.17.

5.2.2.2.1 Optimal Production Sequence

A key characteristic of the Rhythm Wheel approach is that production runs are continuously scheduled in the optimal changeover sequence in terms of the costs and time that are required for changeovers between consecutive products. The clue is that the optimal changeover production sequence has to be determined just once, namely in the design phase of the Rhythm Wheel. Once this is done, the production process follows the designed sequence from then on. Optimizing the repeating sequence

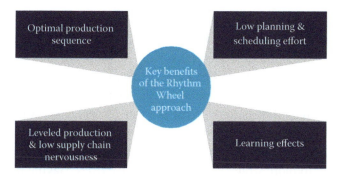

FIGURE 5.17
Key benefits of the Rhythm Wheel approach.

in which products are produced once and permanently sticking to it significantly improves production efficiency. Particularly in process industries, in which production is characterized by highly sequence-dependent changeover times and costs, valuable capacity can be freed up because less time is spent on changeovers. This freed-up capacity enables greater production flexibility and increased production volumes on an asset, or it simply allows for a reduction in overtime, fewer shifts, or even disinvestment of production assets.

5.2.2.2.2 Leveled Production and Low Supply Chain Nervousness

A key characteristic of the Rhythm Wheel concept is that an optimized production schedule is created and then continuously repeated. This allows for stable and leveled capacity utilization which is highly beneficial in process industries. Furthermore, the constant production takt significantly increases the transparency and predictability of production. This effect has positive impacts on both local operations and the entire supply chain. In local operations, this enables efficient coordination of shop floor activities as well as better alignment with purchasing activities and logistics. Stable production additionally reduces supply chain nervousness and facilitates production planning and scheduling at upstream and downstream supply chain stages. Moreover, customers benefit because predictable production schedules lead to reliable delivery dates that can be confirmed. All these benefits emphasize that stable production patterns, which are facilitated by the Rhythm Wheel concept, represent major improvements compared with the traditional scheduling approach in which frequent re-scheduling activities are common.

5.2.2.2.3 *Low Planning and Scheduling Effort*

Another major benefit of the Rhythm Wheel approach is reduced planning effort. Once the optimal sequence has been determined, only a few tactical production parameters need to be determined to efficiently manage production (see Chapter 6 for more details). With the tactical parameters defined, the repetitive pattern of the Rhythm Wheel concept minimizes the required planning effort for local production planners. This enables local planners to shift their focus from pure scheduling tasks, including firefighting, to more strategically important activities. Having more time available, planners can, for example, foster operational excellence initiatives, which again can have a significant impact on a company's production efficiency. Furthermore, the simplicity of the Rhythm Wheel concept facilitates the interpretation of planning results. As a consequence, planners as well as employees on the shop floor show a very high degree of acceptance concerning the concept and actively promote its application.

5.2.2.2.4 *Learning Effects*

A key advantage of the Rhythm Wheel concept concerns the repetitiveness of production, which generates learning effects. When process activities are repeated over and over in the same sequence, the people managing those processes need less time to execute them and can achieve higher quality and more reliable results with the accumulated experience. Especially on the shop floor, companies see the impact of this approach very quickly. Improvements such as faster changeovers, reduced scrap rates, and higher average production speeds lead to significantly increased production efficiency. In this way, the implementation of the Rhythm Wheel concept supports and even promotes a continuous improvement process that is assigned high priority within lean manufacturing initiatives. This aspect highlights another decisive difference between the Rhythm Wheel approach and either traditional production planning and scheduling or kanban, in which the potential benefits of repetitive production patterns are never realized.

5.2.3 How to Manage Variability with Different Rhythm Wheel Types

In light of the benefits of the Rhythm Wheel approach, we expect this LEAN Planning concept to shape the future of production planning and

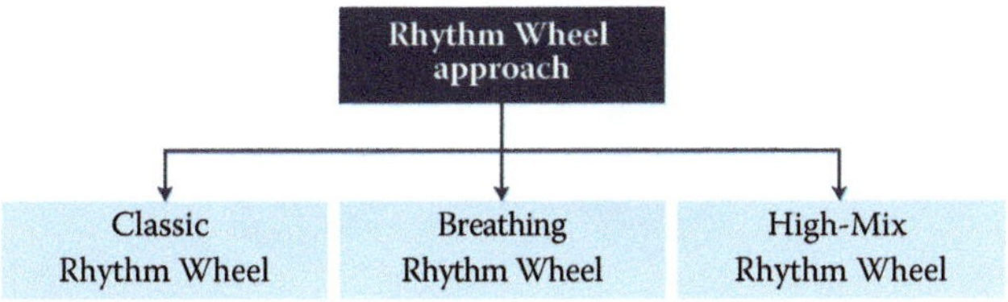

FIGURE 5.18
Various Rhythm Wheel types ensure the broad applicability of the Rhythm Wheel concept.

scheduling in process industries. Hence, companies would do well to place this topic on their strategic agendas.

However, since one size generally does not fit all, the Rhythm Wheel approach needs to reflect that companies pursue a range of strategies, face varied business environments, run several production processes, and so on. Therefore, several Rhythm Wheel types are required to fully exploit the potential of this approach, especially for the effective management of variability. Within the general Rhythm Wheel approach, three types can be distinguished, which will be explained in the following sections (see Figure 5.18).

5.2.3.1 The Classic Rhythm Wheel

The key characteristic of the Classic Rhythm Wheel design is that, in addition to involving a pre-defined fixed production sequence, constant production quantities are produced in every Rhythm Wheel cycle. The result is a repeating schedule with a pattern characterized by fixed times and fixed quantities. So once the production sequence and quantities have been determined, the Classic Rhythm Wheel design makes it easy to derive reliable start and end times for production runs within a given Rhythm Wheel cycle. Figure 5.19 illustrates a Classic Rhythm Wheel with a production cycle of five days. With this type of Rhythm Wheel, planners can predict that the production of a certain product always starts on the same day of the week.

With a Classic Rhythm Wheel, production does not react to demand variations. Instead, production quantities are leveled over time, which we call production leveling, one of the major LEAN SCM principles. Hence, production quantities are shifted from periods of high demand to periods of low demand to maintain high, stable capacity utilization. By leveling

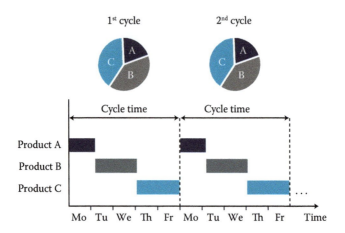

FIGURE 5.19
Production pattern with a Classic Rhythm Wheel.

production, production assets can be utilized as fully as possible, since no excess capacity is needed to cover peaks of production quantities. Especially for companies in process industries, which typically have very capital-intensive production assets, high, stable utilization represents a major benefit. This makes the Rhythm Wheel approach especially suitable for the process industry.

Because the business environment changes constantly, the designed cycle of the Classic Rhythm Wheel needs to be reviewed regularly. Relevant changes in the environment such as changes in the demand pattern are highly relevant in this context. It is therefore advisable to establish a structured process in order to manage adjustment decisions and the corresponding implementation of changes in the configuration efficiently. We describe how this process should look in Section 6.4, where we cover the tactical renewal process.

Using a Rhythm Wheel with fixed quantities is a powerful approach that can be applied broadly in a practical context to achieve all the above-mentioned benefits. However, the strength of a highly stable and predictable production schedule based on fixed production quantities comes at a cost: low flexibility for reacting to actual demand variations. If these variations are medium or even high in magnitude, demand peaks require undesirably high safety stock levels. Therefore, with medium-to-high demand variability, the Classic Rhythm Wheel reaches its limits of applicability. But is this the end of the Rhythm Wheel story? Not at all! As we show in the next section, there is an alternative type of Rhythm Wheel that makes it possible to retain the benefits of the Classic Rhythm Wheel on the one hand, while adding flexibility for more effective reactions to demand variability on the other.

5.2.3.2 Breathing Rhythm Wheel

The Breathing Rhythm Wheel differs from the Classic Rhythm Wheel insofar as it allows flexibility in production quantities. This in turn enables production quantities to be dynamically adjusted to occurring demand and thus embodies one of the major LEAN SCM demand principles— using consumption pull. Hence, the Breathing Rhythm Wheel is perfectly suited to be combined with pull replenishment modes, in which production is triggered on the basis of real consumption (see Chapter 7 for details). Whenever demand occurs, the Breathing Rhythm Wheel schedules the corresponding production quantities at the appropriate production asset. Due to the dynamic adjustment of production quantities to actual demand, the application of the Breathing Rhythm Wheel concept has two major impacts as compared with the Classic Rhythm Wheel:

1. Deviation between designed and actual Rhythm Wheel cycles
2. Need for cycle time boundaries to level production quantities

5.2.3.2.1 Deviation between Designed and Actual Rhythm Wheel Cycles

Under the Breathing Rhythm Wheel, production quantities are allowed to be dynamically adjusted to real demand, so actual production quantities may vary over time and thus deviate from the designed quantities. As a consequence, production times and thus the rhythm cycle may vary as well. As illustrated in Figure 5.20, some cycles may be a bit longer or shorter than others. However, with several products on the Rhythm Wheel, the variations tend to balance each other out, so that the Rhythm Wheel cycle typically remains more or less constant.

Here not only the production quantities but also the pre-defined production sequence may deviate from those included in the design. In cases of high demand variability, there might be cycles with no demand for a given product. Strictly following a pull replenishment model would mean that such a product would be skipped in these cases, that is, it would not be produced within such a cycle. However, even though the actual sequence may deviate from the designed sequence, in most cases the sequence

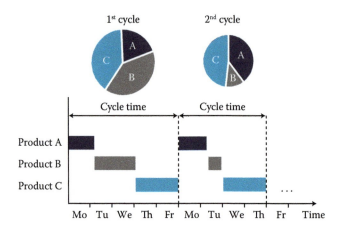

FIGURE 5.20
Production pattern with a Breathing Rhythm Wheel.

remains optimal. Imagine, for example, the tablet-pressing process in the pharmaceutical industry, in which the optimal sequence is typically achieved by producing tablets from low to high concentration of an active ingredient; let us assume the levels of concentration are 10, 20, 50, and 100 mg. If the process skips pressing the 20 mg tablet, the sequence for the other tablets remains optimal. Therefore, the deviation of the actual rhythm cycle from the designed cycle does not significantly impact the efficiency of the production schedule that is created by the Breathing Rhythm Wheel approach.

5.2.3.2.2 Need for Cycle Time Boundaries to Level Production Quantities

One of the major LEAN SCM principles postulates the importance of leveling production and utilization. To realize this principle appropriately within the Breathing Rhythm Wheel approach, Rhythm Wheel cycles should be prevented from fluctuating widely. In practice, this can be achieved by the use of minimum and maximum cycle boundaries. If, for example, high demand in a period indicates that the Rhythm Wheel cycle is about to run beyond that range, production quantities are cut off while in succeeding cycles the produced quantities slightly exceed actual consumption to refill the stocks. This ensures smooth production and stable resource utilization, since production quantities remain more or less constant over time. The application of the Breathing Rhythm Wheel concept in conjunction with an efficient method for achieving production leveling makes it possible to find the optimal balance between flexibility and the benefits of repetitive production patterns (see Box 5.2). There are several possible approaches to effectively attaining production leveling via minimum and maximum cycle boundaries. We discuss these so-called "factoring" approaches in Section 6.1.

So far we have introduced the Classic and the Breathing Rhythm Wheels. We have demonstrated that both concepts are very well-suited for production planning and scheduling in process industries; the Classic

---

BOX 5.2 THE BREATHING RHYTHM WHEEL IN PHARMACEUTICAL MULTISTAGE BULK PRODUCTION

A major European pharmaceutical manufacturer decided to implement Rhythm Wheel-based planning in its multilevel bulk operations. A pilot was set up in one production unit in order to gain first experiences with the concept and to evaluate the benefits. In the specific bulk production unit, the active pharmaceutical ingredient (API) is transformed into tablets in three major steps: first, the API and other ingredients are mixed and granulated. Second, the granulate is pressed into the form of the tablet (compacting). Third, the tablet is coated to obtain the desired dissolving characteristics.

In the specific processing unit, the bottleneck operation was the granulation step. The Rhythm Wheel was designed to achieve the best changeover sequence and continuous high utilization on this operation. Once the Rhythm Wheel schedule was created, the other processing steps were adjusted to it. In this case, there was a forward push of orders from granulation to compacting, and from compacting to coating (see Figure 5.21).

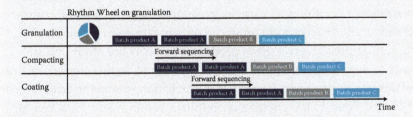

FIGURE 5.21
Application of Breathing Rhythm Wheel in multistage bulk production.

The results of the pilot implementation were so intriguing regarding changeover and capacity improvement that the pharmaceutical manufacturer started to roll out the Rhythm Wheel concept in bulk planning in other production units as well. Since the first processing step (granulation) was not always the bottleneck operation, the approach needed slight adjustment. The Rhythm Wheel still created the optimal production schedule for the bottleneck operation, but now, operations further upstream were scheduled backwards, whereas operations downstream of the bottleneck were pushed forward.

Rhythm Wheel works best in an environment of low demand variability while the Breathing Rhythm Wheel can also manage higher demand variations. As we show, the High-Mix Rhythm Wheel even broadens the applicability of the Rhythm Wheel concept since it perfectly suits the needs of a company that features heterogeneous product portfolios that are produced at one asset.

5.2.3.3 High-Mix Rhythm Wheel

Under both previously described Rhythm Wheel concepts, Rhythm Wheels are designed such that every product occurs once within a cycle. In the case of a heterogeneous product portfolio on a Rhythm Wheel-managed asset, this is not necessarily the optimal design. The High-Mix Rhythm Wheel closes this gap and broadens the applicability of the Rhythm Wheel approach to heterogeneous product portfolios.

In Section 4.3, we emphasized that the application of leveled flow design is a powerful means of avoiding the commitment of undesirable high-mix product portfolios to a single production asset. In this way, runners, repeaters, and strangers are produced on separate lines. However, allocating similar products to a given production asset is not always possible in practice, as several constraints must be considered. In this context, technical limitations or validation aspects in particular prevent the achievement of leveled flow. As a consequence, resource commitment for a heterogeneous product portfolio possibly remains in place, which means that both low-volume and high-volume products share the same resource.

What happens if either the Classic or the Breathing Rhythm Wheel is applied to heterogeneous product portfolios? According to the design

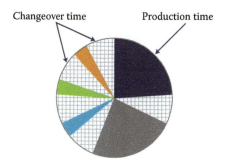

FIGURE 5.22

Applying the Classic or Breathing Rhythm Wheels to a heterogeneous product mix.

entailed in both approaches, every product occurs once within a Rhythm Wheel cycle. Hence, in all cycles the production resource needs to be set up independently for every product, whether the quantity to be produced is large or small. As changeovers in process, industries are typically very time consuming and costly, it is inefficient to set up a resource for small quantities in every cycle. Figure 5.22 illustrates how changeover times would be relatively high compared with the actual production time needed for low-volume products.

The key design improvement idea for producing every product only once per cycle is rather intuitive: defining varying production rhythms for products sharing the same resource. Instead of producing low-volume products every cycle, they are scheduled only once over a certain number of cycles.

By using varying production rhythms, changeover times for low-volume products can be reduced, which frees up valuable capacity. This additional capacity can either be saved by reducing overtime within an asset or it can be used in other ways. One promising alternative to using such freed-up capacity is reducing the designed Rhythm Wheel cycle and setting up the asset for high-volume products more frequently. In this way the cycle time for high-volume products can be reduced, which makes it possible to lower inventory levels for these products. Note that, while inventories of high-volume products are reduced, higher inventories are needed for low-volume products since their cycle time increases. To achieve optimal production rhythms, the trade-off between inventories of low-volume products and those of high-volume products should be considered. Figure 5.23 illustrates this trade-off with the help of inventory charts for a high-volume and a low-volume product.

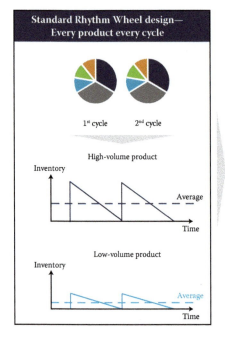
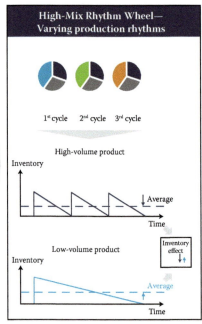

FIGURE 5.23

Impact of the High-Mix Rhythm Wheel on inventories in cases involving heterogeneous product portfolios.

We provide additional details about the concrete configuration of High-Mix Rhythm Wheels and about the determination of optimal production rhythms in Chapter 6. In that chapter, we also offer insights into the implementation of production rhythms.

For a comprehensive overview, see Figure 5.24 as it provides a summary of all previously introduced Rhythm Wheel types and their key characteristics.

Rhythm Wheel type	Optimal production sequence	Production leveling	Rhythm Wheel cycle time	Production quantities	Production rhythm
Classic RW	Yes	Yes	Fixed	Fixed	Every product, every cycle
Breathing RW	Yes	Yes	Dynamic	Dynamic	Every product, every cycle
High-Mix RW	Yes	Yes	Dynamic	Dynamic	Different production rhythms

FIGURE 5.24

Key characteristics of Rhythm Wheel types.

Summary

In this section, we addressed the fact that, in addition to selecting suitable replenishment modes, companies must also select appropriate production modes as they define how production is planned along the supply chain. Several production modes were introduced in this context.

By realizing consumption pull as one of the major LEAN SCM principles and significantly reducing planning complexity, kanban was introduced as a promising alternative to traditional production planning and scheduling. However, some improvement potentials with regard to the applicability to process industries were identified.

The Rhythm Wheel concept realizes major LEAN SCM principles and thus seizes remaining improvement potentials. The Rhythm Wheel follows the idea of predefining an optimal production schedule and repeating it over time. Applying the Rhythm Wheel concept and thus implementing repetitive production yields major benefits for process industries such as low supply chain nervousness and optimal production sequences. To apply the Rhythm Wheel concept to a wide range of business environments, we introduced three types of Rhythm Wheel to effectively manage variability: the Classic, Breathing, and High-Mix Rhythm Wheels.

Your company should expect to find that at least one of the production modes we have introduced here is worth considering. It should identify the production mode that is best suited to its operations. In the next section, we provide methodological support that will help your company to define the right production modes for its supply chain.

5.3 SUPPLY CHAIN MODE SELECTION: COMBINING PRODUCTION AND REPLENISHMENT MODES

In previous sections, we have introduced replenishment and production modes that are relevant in the context of LEAN SCM. What, however, is the best supply chain mode set-up for companies in process industries? It is obvious that there is no single correct answer to this question. There are simply too many company-specific factors that impact the suitability of a given supply chain mode, such as supply chain strategy, demand patterns,

FIGURE 5.25
Approach to supply chain mode selection.

and supply chain characteristics. To identify the best-suited supply chain mode set-up for your company, we recommend following a structured approach that incorporates four major steps, as depicted in Figure 5.25.

In the following section, we provide insights into the four major steps of the recommended approach, which provides your company with valuable decision support to help it identify the best supply chain mode set-up.

5.3.1 Define the Configuration Scope of the Supply Chain Segment

The first step in following the recommended approach to selecting the best-suited supply chain modes entails defining the parts of the value chain that fall within the scope of the supply chain mode selection. Three key aspects should be considered when defining the scope in the process of selecting the best-suited supply chain modes: determining the control span, adopting an end-to-end approach, and subdividing the supply chain into more manageable components.

To define the overall scope, the control span concerning a company's supply chain must first be clarified (see Figure 5.26). Companies operating the complex supply chains that are characteristic of process industries generally do not control the entire supply chain from the production of raw material to the delivery of finished goods to the end customer. In most cases, one or several production steps are outsourced or the distribution of finished goods to customers is taken over by external partners. Hence, supply chain stages that are out of a company's control can be excluded straightaway from the scope of the supply chain for purposes of mode selection.

Second, an end-to-end approach should be adopted. The selection of a supply chain mode at one stage impacts conditions in other parts of the supply chain. Assume, for instance, that the Rhythm Wheel concept is implemented at one of the production assets in a company's supply chain. More stable and predictable production results in smoothed order

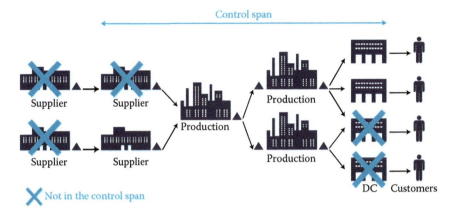

FIGURE 5.26
Definition of a company's control span.

patterns. As a consequence, upstream supply chain stages see more stable demand patterns which, as we will see, have a significant impact on supply chain mode selection at that stage. Therefore, when seeking the best production and replenishment modes for a supply chain, we advise adopting an end-to-end approach.

Owing to the size and complexity of global supply chains in process industries, it can be useful to slice and dice the supply chain into more manageable sub-supply chains in a third step. Supply chain mode selection for sub-supply chains can be pursued either in parallel or successively. When sub-supply chains are sliced, some simple guidelines should be followed. These guidelines suggest that subdividing a supply chain is especially sensible where material flows do not interact with other products, product areas are very different from one another, or a range of management responsibilities is involved (Figure 5.27).

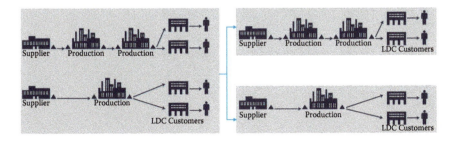

FIGURE 5.27
Subdividing the supply chain due to the independence of material flows.

5.3.2 Analyze Key Impact Dimensions of Mode Selection

Once the scoping phase is completed, an in-depth analysis with respect to the key factors influencing supply chain mode selection is launched. These factors can be grouped into three dimensions: strategy, demand, and supply-related factors. These three dimensions form the LEAN SCM triangle for supply chain mode selection (see Figure 5.28). Analyzing these dimensions provides valuable insights and enables solid preassessment of the suitability of one or another supply chain mode set-up.

Insights concerning the analysis of key impact dimensions fit the following structure:

- Replenishment mode evaluation
- Production mode evaluation

5.3.2.1 Replenishment Mode Evaluation

In Section 5.1, we introduced several replenishment modes, including push replenishment and several pull replenishment modes—MTO, IRL replenishment, and Buffer Management. On the basis of an analysis that covers the three dimensions of the LEAN SCM triangle, the following sections provide valuable insights into choosing the best-suited replenishment modes for your company's supply chain.

5.3.2.1.1 Strategy Analysis

In Section 4.1, we discussed the need to define a company's supply chain strategy such that the optimal balance between the four key order winner dimensions is achieved: cost, time, flexibility, and service. With such a strategy in place, clear guidance is provided for designing and managing

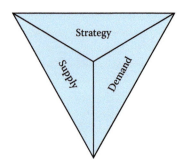

FIGURE 5.28
LEAN SCM triangle for supply chain mode selection.

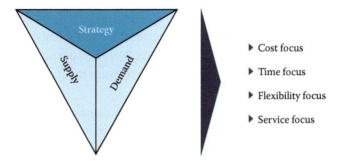

FIGURE 5.29
Strategy-related factors for replenishment mode selection.

the supply chain. As a consequence, supply chain strategy must be taken into account during the process of selecting replenishment and production modes. We address how to consider strategies with respect to replenishment mode selection in this section (Figure 5.29).

The cobweb diagram in Figure 5.30 represents the defined balance between relevant target areas within a supply chain strategy. The diagram indicates that in the depicted case the supply chain needs to focus on service and flexibility to match customer needs. Such a strategy is certainly reasonable and worth considering, especially in environments such as the pharmaceutical industry with lifesaving drugs. The consequences of stock-outs are dramatic. Therefore, a focus on service and flexibility is essential for such supply chains. In such cases, pull replenishment has considerable promise for success. Triggering production based on consumption ensures the required flexibility for reacting to demand variations and thus makes it possible to maintain the required service levels.

FIGURE 5.30
Supply chain strategy and its impact on replenishment mode selection.

5.3.2.1.2 Demand Analysis

Demand patterns are certainly key determinants in replenishment mode selection and thus analyzing such patterns is very important. Figure 5.31 lists demand-related factors that are highly relevant to identifying suitable replenishment modes.

As experience shows, it makes sense to begin demand analysis by investigating demand volume and variability. To do so, it is useful to plot relevant products on a two-dimensional matrix. Typically, the *y*-axis represents demand variability as measured by the coefficient of variation, which is defined as the standard deviation in demand divided by mean volume per period. The *x*-axis displays mean volume per product per period. Figure 5.32 shows a product portfolio that contains a mix of high-volume/low-variability products on the one hand and low-volume/high-variability products on the other. Figure 5.32 also includes recommendations as to which LEAN replenishment modes are most promising for various parts of the product portfolio.

High-volume products with low demand variability are usually characterized by a high degree of forecast accuracy. In this case, push replenishment—triggering replenishment based on forecasting—is a promising option since produced quantities are likely to meet actual demand with sufficient accuracy. Experience shows that in the other segments that are characterized by lower demand volumes and/or higher demand variability, forecasting error increases significantly, which reduces the suitability of push replenishment. Therefore, pull replenishment modes are recommended in these segments as production there is triggered based not on forecasts but on actual demand.

In cases of high demand variability, products are characterized by rather low volumes. For these products, MTO is typically applied as the

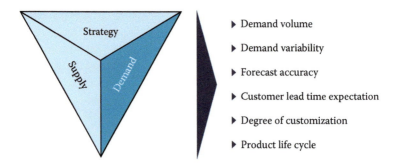

FIGURE 5.31
Key demand-related factors for replenishment mode selection.

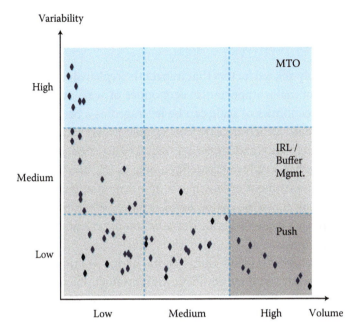

FIGURE 5.32
Volume/variability analysis for replenishment mode selection.

replenishment mode, for two reasons. First, other pull replenishment modes, such as IRL and Buffer Management, use supermarket stock, which means that inventories are held throughout the entire year, although sales volume is rather low. Second, the products in this segment are typically not basic commodities but rather are characterized by a high degree of customization. Therefore, it is often not possible to keep these products in stock.

Another demand-related factor that is worth considering is product life cycle, which helps in predicting how products will develop with respect to key demand-related factors such as volume and variability. In the new product launch phase, product demand volumes are typically low with high demand variability, which indicates that MTO applies. With increasing product maturity, however, product volumes tend to increase while demand variability should decrease, which calls for realigning the replenishment mode.

5.3.2.1.3 Supply Analysis

In addition to strategic and demand-related factors, supply-related factors also play an important role in a company's selection of replenishment modes. Figure 5.33 provides an overview of key supply-related factors that should be taken into account when choosing a replenishment mode for a given product.

Among supply-related factors that impact the replenishment mode decision, two are of major importance: economies of scale and lead time. In this context, economies of scale can be leveraged as a benefit of batching production orders, which occurs, for example, when usual or extraordinary changeover costs and times are required to changeover from one product to another. Lead time is defined here as the time that is required to fulfill a replenishment order, that is, the time that passes from order receipt until delivery to the customer takes place. The implication of these two supply-related dimensions for the selection of replenishment modes is illustrated in Figure 5.34.

If economies of scale are low, due to short changeover times for example, small-batch production does not incur high production costs. With small batches, production can react quickly to occurring demand, which favors a pull replenishment mode that triggers production based on actual consumption. On the other hand, large economies of scale favor push production as forecasted demand from several periods can be batched and consolidated into a single production run, which implies high saving potentials.

The second supply-related factor depicted in Figure 5.34 is lead time. For products with long lead times, production is not able to react quickly to real occurring demand. In these cases, production based on actual consumption is not recommended. Instead, push replenishment should be applied. On the other hand, short lead times enable production to react accurately to occurring demand, which favors pull replenishment.

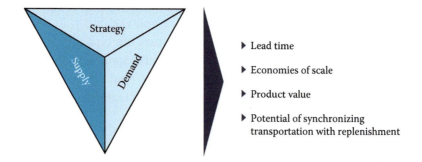

FIGURE 5.33
Key supply-related factors for replenishment mode selection.

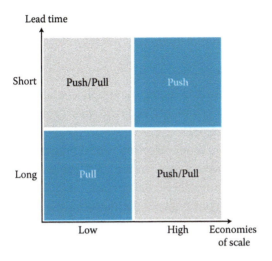

FIGURE 5.34
Impact of major supply-related factors on replenishment mode selection.

If a company decides to implement a pull replenishment mode for a set of products within its product portfolio, the following two supply-related factors support the decision as to which of the potential modes to select: product value and the potential for synchronizing transportation with replenishment. High product value favors MTO replenishment while low product value favors IRL replenishment or Buffer Management, because in this case supermarket stock does not incur high inventory costs. The potential for synchronizing transportation with replenishment is relevant to the choice between IRL replenishment and Buffer Management. However, such a choice requires a careful tradeoff. On the one hand, Buffer Management uses fixed replenishment quantities, which allow for alignment between transportation lot sizes and quantities to be replenished. So, for example, the full truck load capacity can be continuously exploited, which reduces logistics costs. On the other hand, IRL replenishment enables replenishment to be carried out at fixed intervals, which facilitates transportation planning and thus should have a positive impact on cost as well. To evaluate the final impact on cost, a company's specific situation must be considered.

5.3.2.2 Production Mode Evaluation

The second step in the suggested approach covers the analysis of key factors that impact production mode selection. Again, the three dimensions of the LEAN SCM triangle need to be analyzed to choose among potential production modes.

5.3.2.2.1 Strategy Analysis

When selecting a production mode, the supply chain strategy that defines the balance between the four key order winner dimensions needs to be considered (see Figure 5.35).

Regarding production mode selection, we highlight two areas: cost and time. If a company's operations strategy focuses on cost and time (see Figure 5.36), the Rhythm Wheel concept is an excellent approach. By leveling production and continuously maintaining the optimal production sequence, utilization remains constant at a high level, enabling a company to reduce production costs. Moreover, repetitive production patterns bring learning effects that additionally increase production efficiency. Focusing on time also favors the Rhythm Wheel concept, as continuous production in the optimal production sequence helps to reduce cycle times and thus delivery time to the customer.

But which of the Rhythm Wheel concepts should be chosen? The answer is closely linked to a concrete weighting of the four target areas. If a

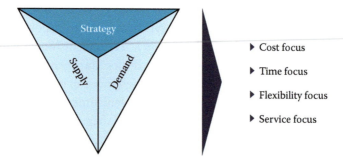

FIGURE 5.35
Strategy-related factors for production mode selection.

FIGURE 5.36
An example of an operations strategy and its impact on replenishment mode selection.

company's operations strategy also draws attention to the service dimension and flexibility, the Breathing Rhythm Wheel represents a promising production mode. The Breathing Rhythm Wheel makes it possible to dynamically adjust production quantities which provide the flexibility needed to respond quickly to actual consumption and ensure high service. If such a weighting indicates a strong focus on cost, the Classic Rhythm Wheel is best suited. By continuously outputting a product at the same time and quantity, production and changeover costs can be minimized. Yet the High-Mix Rhythm Wheel can also be a good alternative in such a case as it accommodates optimal production rhythms. The High-Mix Rhythm Wheel has a positive impact on the time dimension as well. Since not all products are produced in every cycle, average cycle time decreases, thereby reducing the time needed to react to customer demand.

5.3.2.2.2 Demand Analysis

It is extremely important to analyze demand-related factors when selecting a suitable production mode. Figure 5.37 displays key factors that should influence the selection of production modes.

Despite numerous influencing factors, here we suggest again that demand analysis begin with volume and variability, so we have plotted a product portfolio of an asset on a two-dimensional matrix. Figure 5.38 shows examples of product portfolios and the respective recommended production modes.

The product portfolio depicted in the first chart of Figure 5.38 is characterized by products with high volume and stable demand. If such a portfolio is assigned to a production asset, a Classic Rhythm Wheel is highly beneficial as in such a case the full benefits of the Rhythm Wheel concept can be applied. With repetitive production patterns in place, the optimal

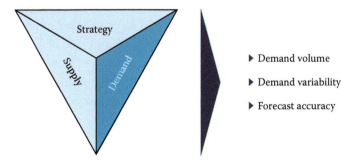

FIGURE 5.37
Key demand-related factors for production mode selection.

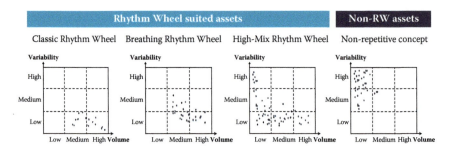

FIGURE 5.38
Volume/variability analysis for production mode selection.

production sequence is followed continuously, production is leveled and highly predictable over time, and learning effects increase overall production efficiency. Since volumes are very stable and are typically characterized by high forecast accuracy, the required service level can be achieved and maintained at minimum levels of inventories as well.

If the product portfolio at an asset looks like the one depicted in the second matrix of Figure 5.38, greater flexibility is required to enable the asset to react effectively to real demand. The Breathing Rhythm Wheel is an excellent design for such product portfolios since it allows for dynamic production quantities while maintaining the benefits of the Rhythm Wheel concept.

If a heterogeneous portfolio must be produced on one production asset as shown in the third matrix of Figure 5.38, the High-Mix Rhythm Wheel is recommended as the production mode. In contrast to other Rhythm Wheel concepts, the High-Mix Rhythm Wheel allows for various production rhythms for products being produced on an asset, that is, low-volume products are not produced in every cycle. This avoids frequent changeovers for small production quantities, which frees up valuable capacity.

If the product portfolio consists only of a large number of low-volume products with high variability (as on the right-hand side of Figure 5.38), the Rhythm Wheel approach is not the perfect fit for the asset. The benefits of a leveled and repetitive production sequence would be lost due to many sporadic production runs. In this case, a nonrepetitive production mode would be the better pick.

5.3.2.2.3 Supply Analysis

In addition to considering a company's strategic and demand-related aspects, supply-related factors must also be taken into account to properly

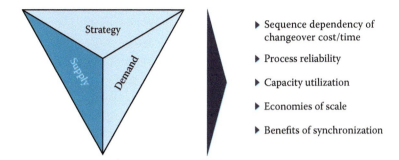

FIGURE 5.39
Key supply-related factors for production mode selection.

evaluate production modes. Figure 5.39 provides an overview of key supply-related factors that should influence the selection of appropriate production modes.

Again, we highlight especially important factors in this context: sequence dependency of changeover costs and times and the reliability of the production process. The sequence dependency of changeover costs and times represents the degree to which a suboptimal production sequence impacts such KPIs as capacity utilization or total production cost. A reliable production process implies that machine breakdowns are rare, scrap rates are low, and the quality of finished goods is constantly at a high level. In this case, the production schedule can be executed as planned. We illustrate the implications of these two supply-related dimensions for the selection of production modes in Figure 5.40.

Since the Rhythm Wheel concept enables an asset to produce continuously in the optimal sequence, its application is very beneficial in environments characterized by highly sequence-dependent changeover costs. Even if such sequence dependency is not present, however, the considerable advantages of the Rhythm Wheel concept favor this production mode. However, the Rhythm Wheel concept is not a silver bullet for all production environments. In an environment in which the reliability of the production process is low, for example, when production assets break down regularly, it is impossible to fully exploit the full potential of the Rhythm Wheel concept. There are two reasons for this. First, the Rhythm Wheel cycle is likely to fluctuate widely, which precludes constant production intervals and the realization of corresponding benefits. Second, in such an environment, frequent switching between alternate lines is likely

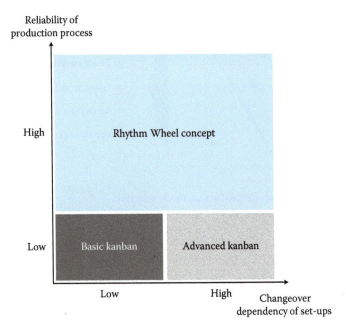

FIGURE 5.40

Impact of major supply-related factors on production mode selection.

required to avoid stock-outs. Such a scenario cannot be effectively managed with the Rhythm Wheel concept.

In cases of low reliability of the production process, kanban is a suitable alternative to the Rhythm Wheel. Using a kanban system, production orders are dynamically allocated to substitutive production assets, providing the required flexibility in such an environment. However, the basic kanban concept follows a FIFO scheduling rule. If the sequence dependency of changeovers is high, the advanced kanban mode needs to be applied, as described in the practical example offered in Section 5.2.1.

Some additional supply-related factors and their impact on production mode selection are summarized in Figure 5.41.

The second step in our recommended approach to selecting supply chain modes is now complete concerning the analysis of various dimensions of the LEAN SCM triangle. Box 5.3 presents an example of a company in the food processing industry which conducted such an analysis to identify the best-suited production modes for a European site.

Supply-related factor	Kanban	Classic RW	Breathing RW	High-Mix RW	Reasoning
High current capacity utilization	−	+	+	+ +	Leveled production, optimal production sequence and optimal production rhythms relieve capacity utilization
High economies of scale	−	+	+	+ +	Optimal production lot-sizes help to seize high economies of scale
High benefits of synchronization	−	+ +	+	−	Fixed production patterns allow for best possible supply chain synchronization

FIGURE 5.41

Impact of additional supply-related factors on production mode selection.

BOX 5.3   EVALUATION OF THE RHYTHM WHEEL CONCEPT FOR A COMPANY IN THE FOOD PROCESSING INDUSTRY

To stabilize and level production, a yoghurt manufacturer evaluated the suitability of the Rhythm Wheel concept for a major European site. The scope of the evaluation included four packaging lines which covered a broad product range. The product portfolio covered multiple flavors, several cup sizes, and various labels for delivery to several European markets. To investigate the suitability of the Rhythm Wheel, the manufacturer conducted an analysis with regard to key supply- and demand-related factors.

Figure 5.42 shows the results of the analysis for the first filling line. Besides the analysis concerning demand volume and variability (on the left-hand side), several other factors were investigated (right-hand side).

Based on its detailed analysis, the company was able to make a carefully reasoned decision regarding the four lines. Finally, the Rhythm Wheel concept was implemented on three of the four filling lines. As filling line 3 was aged and characterized by very low process reliability, scheduling remained a manual task on that line.

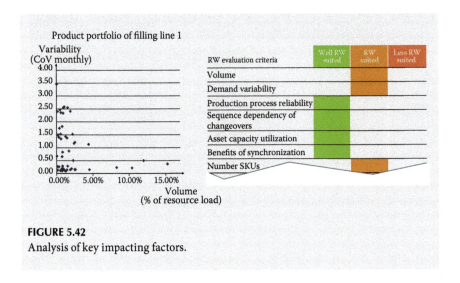

FIGURE 5.42
Analysis of key impacting factors.

Your company should now be able to make the first evaluations regarding which of the replenishment and production modes are suitable for its supply chain. However, we recommend two additional steps to make the best decision regarding the final supply chain mode set-up.

5.3.3 Select the Appropriate Supply Chain Modes

Once the analysis we have just described is completed, the third step in the recommended approach tackles the challenge of consolidating the findings of the analyses and deriving the most promising replenishment and production modes within the defined scope of the supply chain. This is not an easy task, since the findings of the analysis with respect to the three dimensions may be inconsistent or even contradictory. For instance, demand analysis might favor applying a Classic Rhythm Wheel due to stable demand with high volumes, while supply analysis favors kanban because process reliability is rather low. In such cases, a company must carefully weigh the potential benefits and drawbacks of each approach. In the end, however, consolidating the findings typically leads to several alternative supply chain mode set-up scenarios, each of which appears beneficial. Here, we emphasize two additional aspects that should be considered when potential supply chain mode set-ups are pre-selected:

1. Interaction between replenishment and production modes
2. Supply chain interdependencies

5.3.3.1 Interaction between Replenishment and Production Modes

We recommend giving explicit consideration to potential interaction between production mode selection and replenishment mode selection. This is important since some combinations of replenishment and production modes are not feasible for all companies, while others are simply less effective or efficient than the optimal combinations. Figure 5.43 depicts the LEAN SCM compass for supply chain mode selection, providing a guide to understanding the degree of compatibility

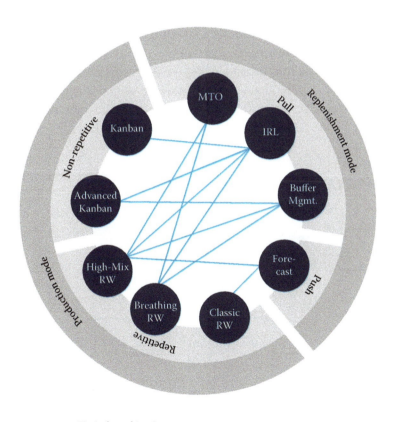

————— Typical combinations

FIGURE 5.43
LEAN SCM compass for supply chain mode selection.

between replenishment and production modes by displaying typical combinations.

5.3.3.2 Supply Chain Interdependencies

Another aspect beyond interaction between replenishment and production modes at a single stage that should be considered is that of interdependencies between replenishment and production modes along the entire supply chain. This is an issue because the definition of a supply chain mode at one stage impacts supply chain mode selection elsewhere along the supply chain, especially at upstream stages. If the Rhythm Wheel concept, for instance, is assigned to a company's production asset within its supply chain, production patterns will stabilize, which allows for smoothed order patterns. Upstream supply chain stages are in that case likely to face more stable demand, which needs to be considered in the supply chain mode set-up.

To capture these interdependencies, we suggest running a backwards assignment of supply chain modes, starting at the downstream supply chain stage closest to the external customer and moving stepwise upstream. Figure 5.44 visualizes the recommended backwards assignment

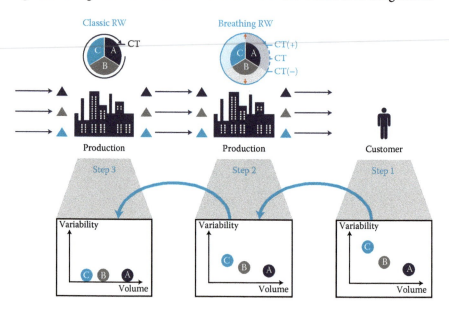

FIGURE 5.44

Backwards assignment of supply chain modes and its implications.

Alternatives	Production stage 3	Production stage 2	Production stage 1
Supply chain mode set-up 1	FC & Classic RW	IRL & Breathing RW	IRL & Breathing RW
Supply chain mode set-up 2	IRL & Breathing RW	IRL & High-Mix RW	MTO & High-Mix RW
Supply chain mode set-up 3	IRL & Breathing RW	IRL & Breathing RW	Buffer Management & High-Mix RW

FIGURE 5.45
Examples illustrating pre-selection of supply chain modes.

and illustrates its impact on relevant demand-related factors for supply chain mode selection.

Using such an iterative assignment of supply chain modes and with the help of the LEAN SCM compass, the consolidation of findings from step two of the suggested approach makes pre-selection of beneficial supply chain mode set-ups possible. Figure 5.45 illustrates several promising scenarios for supply chain mode set-up. In the fourth step of the approach, we suggest applying simulation-based validation to identify the best supply chain mode set-up for your company.

5.3.4 Evaluate Your Decision Quantitatively

Our review of the previous steps of the suggested approach has shown how promising supply chain mode set-ups can be derived. From this point, however, making the final selection of appropriate supply chain modes is difficult, due mainly to the complex interdependencies that typically operate within global supply chain networks. Predicting the impacts of one or the other promising supply chain mode set-up is hardly possible, especially when findings from in-depth analyses are contradictory.

To resolve this uncertainty, we recommend complementing the qualitative results with a quantitative analysis that predicts the impacts of pre-selected supply chain modes on supply chain performance. In this context, a simulation-based analysis provides the required quantitative results. The corresponding benefits of supply chain simulation are depicted in Figure 5.46.

To conduct a simulation-based analysis, a simulation model must be built first, incorporating all relevant supply chain processes. In particular, production capacities, throughput rates, stock-keeping points, transportation patterns, and lead times are typical variables that are modeled to create an appropriate picture of real-world conditions along a supply chain. Since such a supply chain simulation aims at providing quantitative decision support, KPIs must be defined, depending on the underlying supply chain strategy and its corresponding objectives. Such KPIs as customer service levels, inventories, changeover cost and times, capacity utilization, transportation cost, and total production cost are typically used to evaluate overall performance.

Once the model has been built, the pre-selected supply chain mode set-ups can be evaluated to identify the best choice for your company's supply chain. Figure 5.47 displays screenshots that show the end-to-end simulation of a pharmaceutical supply chain which supported the decision on supply chain mode set-up.

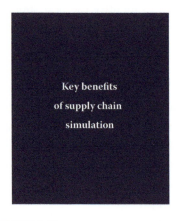

FIGURE 5.46
Benefits of supply chain simulation for supply chain mode selection.

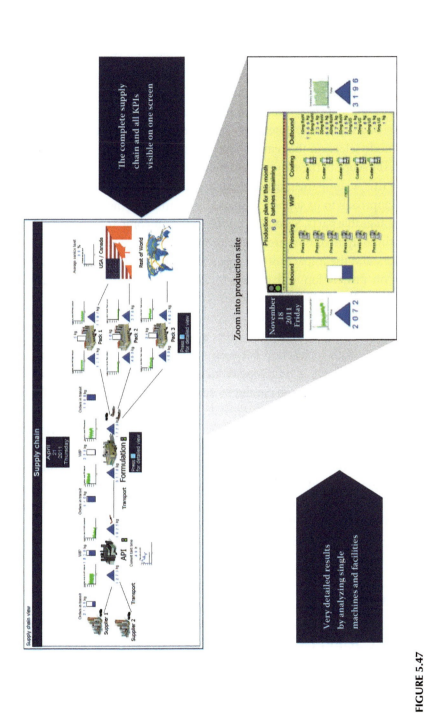

FIGURE 5.47
Supply chain simulation to validate pre-selected supply chain mode set-ups.

Summary

In this section, we have emphasized that the right supply chain mode set-up decision strongly depends on company-specific factors. Therefore, we cannot offer a general recommendation in favor of certain replenishment and production modes. To support your company's decision-making process, we have provided general advice and guidelines throughout the chapter and introduced an approach to supply chain mode selection that entails four major steps.

First, the scope of the supply chain mode selection must be defined. To achieve this, a company must determine its control span, adopt an end-to-end approach to its supply chain, and slice and dice the supply chain into manageable parts. Second, it is vital that your company conduct an in-depth analysis of the three dimensions of the LEAN SCM triangle to understand the impacts of strategy and supply- and demand-related factors on production and replenishment mode selection. The third step of our recommended approach is consolidating the findings of the in-depth analyses to make a pre-selection of the most promising supply chain mode set-ups. During this step, a company should also explicitly consider both interaction between replenishment and production modes and supply chain interdependencies. The fourth and final step of the selection approach is to conduct a simulation-based analysis of the pre-selected modes. Simulation provides quantitative results enabling a company to make a fact-based comparison of pre-selected scenarios and allows for a valid decision regarding a supply chain mode set-up.

Armed with the insights we have provided in this section, your company will be prepared to solve the challenge of finding the right supply chain mode set-up for its supply chain. When the company accomplishes this end, it has taken an important step toward realizing the full potential of LEAN SCM.

5.4 THE STRATEGIC RENEWAL PROCESS TO CONFIGURE AGILE SUPPLY CHAINS

Nothing is more constant than change! The Greek philosopher Heraclitus realized more than 2500 years ago that this is a universal axiom. Of course, this principle also holds true in supply chains. Conditions are continuously

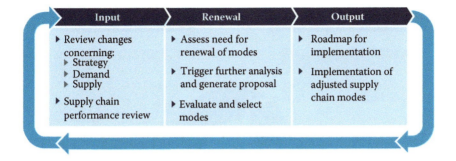

FIGURE 5.48
High-level structure of the strategic renewal process.

changing, externally and internally. Changes in network structure due to internal growth or mergers and acquisitions, expansion into new markets, changing demand patterns, and so on have a significant impact on the requirements that a supply chain has to meet.

Within the LEAN SCM framework, we have emphasized the importance of selecting the right supply chain modes to ensure that a company's requirements are met. In the face of constant change, therefore, reviewing production and replenishment modes is essential to ensuring the sustainable competitiveness of a supply chain. We therefore highly recommend conducting such a review within a structured process. The strategic renewal process that we describe in this section provides guidance for designing such a structured process.

Figure 5.48 provides an overview of the high-level structure of the strategic renewal process. It should be triggered on a regular basis within a range of 1–2 years, or when structural changes in the supply chain or business environment occur. In the course of this section, we provide further insights into the underlying phases and briefly discuss important factors within this process.

5.4.1 What Information Base Is Needed on Strategic Level?

The first phase of the strategic renewal process is the input phase, during which the collection, consolidation, and validation of data and information that support the evaluation of current supply chain modes are addressed. Hence, this phase builds the basis for decisions that will be made in the renewal phase regarding whether or not to adjust current

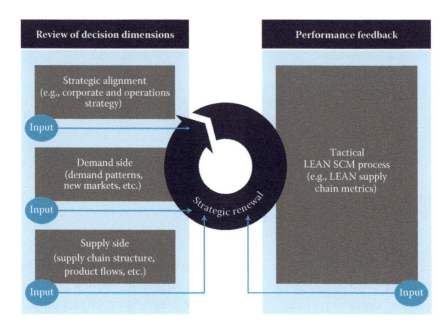

FIGURE 5.49
Input dimensions for the strategic renewal process.

supply chain modes. An overview of required input dimensions for the strategic renewal process is depicted in Figure 5.49.

5.4.1.1 Strategic Input

First, input concerning the strategic portfolio and go-to-market approach is required, which is especially relevant if significant changes have occurred. Such changes are, for example, a re-positioning of brands or a whole product group to strengthen market position, or a change in the product portfolio due to complexity management initiatives or product launches. These factors must be considered in supply chain mode set-up. The required information is typically gathered at the corporate level, which implies the involvement of the supply chain board.

5.4.1.2 Demand Input

A second decision dimension, review of the demand side, is highly relevant, since demand patterns are typically very dynamic over time. Product volumes, for instance, are impacted by a range of factors such as demand trends, the expiration of patent protection, increased market penetration, product maturity, and so on. Moreover, both volume and demand variability are constantly in flux. Changing order patterns, entries into volatile markets, or demand-shaping initiatives that aim to reduce demand variability are major root causes of these dynamic changes. Typically, sales and commercials have the closest connection to markets and customers. Therefore, they are key sources of information on demand-related inputs.

5.4.1.3 Supply Input

A third area of information required for the strategic renewal process is that of changes concerning the supply side. Highly relevant in this context are changes in the supply chain network, such as the opening or closing of production assets or distribution centers, and the adjustment of product flows, as in the re-allocation of a product family from a site in Asia to one in the United States or the other way around. These changes have a huge impact on the product portfolios to be produced on a company's assets, on lead times, resource utilization, and other factors. Since these factors are key determinants of the selection of appropriate supply chain modes, they should be considered within the strategic renewal process. A key source of the required information is the supply chain board, since it is typically involved in decisions pertaining to changes related to the supply side.

5.4.1.4 Performance Feedback

Performance feedback constitutes yet another important input that is required for the strategic renewal process. Such feedback reflects past performance and therefore serves as important feedback relating to the current supply chain mode set-up. If supply chain performance falls below expectations concerning such KPIs as service levels, inventories, or production costs, one potential root cause is that a company has chosen unsuitable supply chain modes. The tactical level of the LEAN SCM process is an important source of information pertaining to past supply chain performance that is complemented by figures from the controlling department (see Chapter 9 for details). This consolidated data represents valuable input to inform decision making within the strategic renewal process.

5.4.2 Establish Sustainable Renewal of Supply Chain Modes

Once all required input information has been gathered, the core phase of the strategic renewal process begins. The renewal phase is designed to help

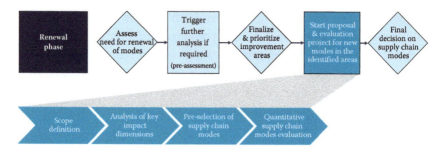

FIGURE 5.50
Main steps of the core phase within the strategic renewal process.

a company define and renew production and replenishment modes based on data gathered during the input phase. Hence, in the renewal phase, the decision is made whether to confirm or adjust current supply chain modes. Figure 5.50 summarizes the process steps that are required within the core phase of the strategic renewal process.

The first step within the core phase of the strategic renewal process involves a high-level assessment of the potential need for a renewal of current supply chain modes. If the assessment reveals that factors relevant to supply chain mode selection have not changed significantly and performance is within the targeted range, current supply chain modes can be considered appropriately chosen and remain in place. However, due to the dynamics of global and complex supply chains in the process industry, conducting a detailed analysis is often worth the effort.

Once the decision is made to trigger further analysis, it is useful to identify and prioritize the areas that stand to benefit the most from a realignment of supply chain modes. A detailed analysis of past performance helps to identify the potential benefits and thus supports the determination of areas on which to focus. If, for instance, the analysis reveals that the main root cause of an unsatisfying performance of a particular brand is an inappropriately chosen supply chain mode, high priority should be assigned to reviewing this brand in detail.

Once the focus areas for further investigation are defined, in-depth analysis is required that aims at evaluating and proposing the best supply chain mode set-up. We recommend conducting such an analysis within the structured approach that we have outlined in Section 5.3. The outcome of this approach is a fact-based recommendation regarding supply chain modes which supports the final decision on the final supply chain mode set-up. Especially if large adjustments of the current supply chain mode set-up are suggested, typically the supply chain board, the supply chain planner, and other responsible persons within the supply chain excellence center are involved in this decision.

5.4.3 Ensure Supply Chain Agility through Regular Mode Renewal

Once the renewal phase is completed, the output phase of the strategic renewal process is launched to ensure operational agility of the supply chain. The output phase is designed to prepare and conduct the implementation of the agreed-upon production and replenishment modes along the supply chain. One important element of the output phase refers to the drawing up of an implementation roadmap for integrating supply chain modes into operations. On this roadmap, the ultimate goal, the key responsibilities, and a timeline aligned with corresponding tasks, measures, and milestones are defined. Such a roadmap is essential because it helps to efficiently manage and coordinate all involved resources and provides an essential medium for communicating the targeted changes within the organization. On the basis of such a roadmap, a project is typically triggered which takes care of the actual implementation of adjusted replenishment and production modes in the supply chain.

Figure 5.51 completes the picture of the strategic renewal process by summarizing all relevant steps in a process diagram.

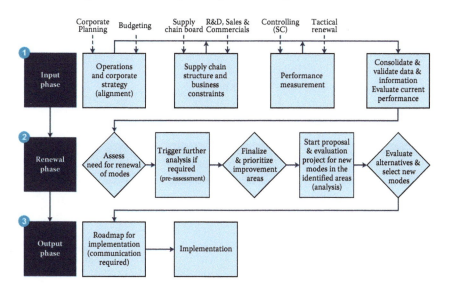

FIGURE 5.51

Focus areas for the implementation of new supply chain modes.

5.4.4 Who Is Involved to Enable Governance for Supply Chain Agility?

So far, we have emphasized how the steps and tasks involved in the strategic renewal process should look. To pursue the process steps efficiently, it is important to define clear roles within an organization for the people who will assume responsibility for the defined tasks. In this section, we describe the key roles that are relevant to the strategic renewal process. An overview is provided in Figure 5.52.

5.4.4.1 Supply Chain Excellence Center

Conducting the strategic renewal process and especially driving the implementation of new supply chain modes requires considerable supply chain expertise. So the corresponding responsibilities should fall predominantly to supply chain experts. These experts may come not only from a dedicated department, but they can also be found in other functional areas. Together these SCM experts form the supply chain excellence center. During the strategic renewal process, the supply chain excellence center should assume a leading role and accept accountability for the data-gathering process as well as the analysis and definition of strategic alternatives for supply chain mode selection. The role of the supply chain excellence center, given its consolidation of the relevant expertise, is especially important during the implementation of new supply chain modes.

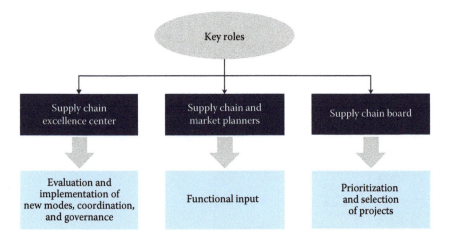

FIGURE 5.52
Key roles within the strategic renewal process.

5.4.4.2 Supply Chain and Market Planners

The demand and supply sides of a supply chain represent key input dimensions of supply chain mode selection. Hence, information gathered from these dimensions is essential to making an appropriate choice concerning supply chain modes. With his or her in-depth knowledge concerning the supply side, the supply chain planner assumes the key responsibility for providing the required data from this area. The market planner on the other hand represents a link between the company and the demand side and thus needs to share his or her insights concerning markets, countries, customers, and so on.

5.4.4.3 Supply Chain Board

The selection of supply chain modes through the strategic renewal process has a significant impact on production and replenishment planning and thus on the entire supply chain. Therefore, the supply chain board needs to be involved in the overall strategic renewal process. Key activities cover the release of the required budget, final approval of changes concerning supply chain modes, and supervision of corresponding implementation projects.

This brief overview of roles that are relevant to the strategic renewal process will be completed by additional insights into organizational aspects in Chapter 8.

Summary

Within this section, we have underlined the need to align the strategic supply chain mode set-up with a constantly changing business environment. The strategic renewal process provides a framework for managing the alignment of production and replenishment modes in a structured way. The strategic renewal process plays out in three phases: input, renewal, and output. During the input phase, all relevant information is collected to evaluate the match between current modes and underlying business requirements. If there is a need to adjust current supply chain modes, an in-depth analysis is conducted during the renewal phase to identify the best supply chain mode set-up. We recommend that this analysis follows the structured approach outlined in Section 5.3. During the output phase, the selected supply chain modes are implemented in the organization following a clearly defined roadmap. Furthermore, we have introduced four roles that are most relevant to the strategic renewal process to ensure its successful completion: the supply chain excellence center, the market planner, the supply chain planner, and the supply chain board.

CHAPTER SUMMARY

Over the course of this chapter, we have argued that the application of LEAN SCM principles to Planning enables a company to achieve a high level of improvement, especially in process industries. To familiarize your company with the impacts that result from applying LEAN SCM principles, we have first divided LEAN Planning into two conceptual parts: the replenishment mode and the production mode.

In Section 5.1, we introduced the replenishment modes that are relevant within the LEAN SCM framework. We described in detail both push and pull replenishment. With push replenishment production is triggered based on forecasts, whereas with pull replenishment it is triggered by real consumption only. In this context, we characterized IRL replenishment, Buffer Management, and MTO as promising pull replenishment modes.

In Section 5.2, we provided an overview of potential LEAN production modes. Apart from kanban, the benefits of the Rhythm Wheel concept underlined its suitability for application in process industries. To reflect the various business environments and to effectively manage variability, distinct types of Rhythm Wheels were defined, underscoring the broad applicability of the Rhythm Wheel concept.

How, though, can your company find the best-suited production and replenishment modes for its supply chain? Section 5.3 was dedicated to answering this question. We recommend following a structured approach consisting of four major steps to identify best-suited replenishment and production modes.

In Section 5.4, we emphasized the importance of regularly reviewing supply chain modes due to constantly changing business environments. The strategic renewal process that we introduced in this section provides excellent guidance on how to conduct these reviews in a structured way.

Readers should now understand how LEAN SCM concepts work, what their benefits are, and how to select appropriate replenishment and production modes for the supply chain. In the next chapter, we offer detailed insights into the proper configuration of the relevant parameters of the suggested LEAN supply chain modes to ensure efficient supply chain management.

6

Tactical LEAN Supply Chain Planning Parameterization

In the previous chapter, we introduced production and replenishment modes for LEAN SCM. These modes determine how production and replenishment are coordinated along a supply chain. The approach we presented entails selecting optimal production and replenishment modes in light of an operations strategy as well as demand- and supply-related factors. In this chapter, we show how to determine the parameters that drive the production and replenishment modes, and how to synchronize those along the supply chain.

In LEAN SCM, the supply chain is preconfigured by supply chain parameters. Operational planning and execution then adhere to this preconfiguration such that manual intervention is reduced to a minimum and the focus can turn to handling exceptions. In this way, planning and scheduling is simplified, more efficient, and transparent. Following the LEAN SCM concept, a company needs to determine only two types of supply chain parameters: production and replenishment parameters (see Figure 6.1). Production parameters determine when and in what quantity a product is produced on an asset. Replenishment parameters signal when and in what quantity a product needs to be replenished.

This chapter demonstrates how to ...

- Determine the production parameters for Rhythm Wheels.
- Right-size the LEAN stock replenishment parameters.
- Synchronize production and replenishment parameters along the supply chain.

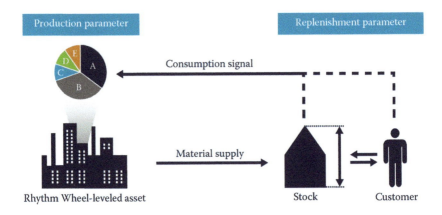

FIGURE 6.1
Production and replenishment parameters are determined to control the supply chain.

In Section 6.1, we analyze the parameter settings for the three Rhythm Wheel types: the Classic Rhythm Wheel, the Breathing Rhythm Wheel, and the High-Mix Rhythm Wheel. We explain how each type is designed and how the parameters are set for execution.

We look at the parameters for LEAN replenishment modes in Section 6.2. First, a general inventory classification is introduced. With this classification, it is possible to adopt a standardized approach to determine appropriate inventory levels along the entire supply chain. We then focus on parameter setting for the Inventory Replenishment Level (IRL) and Buffer Management replenishment modes.

In Section 6.3, we focus on the synchronization of the production and replenishment parameters along the supply chain. First, supply chain parameters are synchronized to a global takt in order to achieve a steady and leveled material and information flow. In this way, the bullwhip effect is prevented and inventory levels can be reduced. Then, we focus on the synchronization of replenishment parameters to most efficiently buffer variability in the supply chain.

For efficient and robust execution, it is important for the supply chain parameters to be always up to date. Therefore, parameters are regularly reviewed and appropriately set in the tactical renewal process, which is introduced in Section 6.4. The tactical renewal process is the framework in which production parameters (6.1), replenishment parameters (6.2), and parameter synchronization (6.3) are determined. It is a clearly structured sequence of tasks with clearly defined organizational roles and responsibilities.

6.1 SETTING UP THE PARAMETERS FOR LEAN PRODUCTION MODES

The various available production modes fit a range of asset configurations and business strategies in the manufacturing environment. Selecting the right production mode depends on demand conditions, supply chain characteristics, and operations strategy (see Section 5.3 for the right selection of production modes). In this section, we focus on parameter configuration of the repetitive LEAN production modes, namely the three Rhythm Wheel types (see Figure 6.2).

The Rhythm Wheel is among the LEAN tools that are shaping the future of planning and scheduling on modern multiproduct production assets in process industries. The Rhythm Wheel is designed to produce in a repetitive sequence in short cycle times without sacrificing customer service. In this way, it sustainably reduces the cost of goods sold, inventory holding costs, and capacity costs, because:

- Products are always scheduled in the optimal sequence. This leads to the shortest possible changeover times and frees up capacity, which allows for greater flexibility.
- The repetitive production sequence generates learning effects, which in turn further reduce changeover times and improve the yield of the production process.
- The fixed production interval, namely the Rhythm Wheel cycle time, leads to stable and leveled capacity utilization. This in turn enables more effective supply planning at up- and downstream stages and reduces supply chain nervousness.

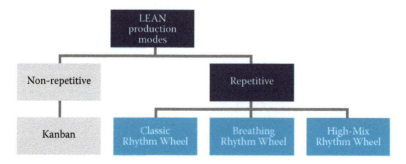

FIGURE 6.2

Parameterization of repetitive LEAN production modes.

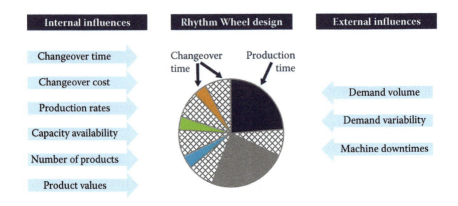

FIGURE 6.3
Internal and external influences on Rhythm Wheel design.

- The Rhythm Wheel minimizes scheduling effort and reduces fire-fighting. This frees up capacity for the planner, who can focus on other important tasks.

To manage the Rhythm Wheel properly, it is necessary to consider and understand all the internal and external influences on Rhythm Wheel design, as shown in Figure 6.3. Only by determining the right parameters in light of these influencing factors can your company take full advantage of the benefits provided by the Rhythm Wheel concept.

Before determining the right parameters, however, it is important to understand that there are three phases within the Rhythm Wheel concept (see Figure 6.4). The first is the Rhythm Wheel design phase. In this phase, the Rhythm Wheel is configured and all relevant parameters for execution are determined. The second phase is the operational scheduling of the previously designed Rhythm Wheel. In this phase, actual orders hit the preconfigured Rhythm Wheel and create a production schedule. The third

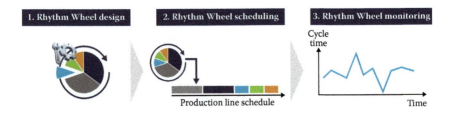

FIGURE 6.4
The Rhythm Wheel concept is separated into design, scheduling, and monitoring phases.

Rhythm Wheel type	Sequence	Cycle time	Min./max. boundaries	Factoring	Production rhythms
Classic Rhythm Wheel	✔	✔			
Breathing Rhythm Wheel	✔	✔	✔	✔	
High-Mix Rhythm Wheel	✔	✔	✔	✔	✔

FIGURE 6.5
Each Rhythm Wheel type requires a particular set of parameters.

phase is the monitoring phase. In this phase, the quality of the previously designed Rhythm Wheel is evaluated and used as an input for the next design phase—with continuous improvement. This section explains the first phase—the design of the Rhythm Wheel. We describe the scheduling and monitoring phases in Chapter 7.

In the Rhythm Wheel design phase, all relevant parameters for the three Rhythm Wheel types are determined. Figure 6.5 provides an overview of the required parameters, as not every Rhythm Wheel type requires the full set of parameters. When using the Classic Rhythm Wheel, a production sequence is chosen to minimize total changeover cost. A cycle time is defined, which is the duration of changeovers and production time required for one complete Rhythm Wheel cycle. When using the Breathing Rhythm Wheel, the so-called cycle time boundaries come into play, which prevent the cycle time from reacting nervously to demand and supply variability. Factoring specifies rules that determine how to adhere to these cycle boundaries. When applying the High-Mix Rhythm Wheel, additional production rhythms are defined, which determine how often a product is produced, such as every cycle or only after a certain number of cycles (Figure 6.5).

6.1.1 Classic Rhythm Wheel Design to Enable Flow in Stable Environments

The Classic Rhythm Wheel creates a repeating schedule that is steady over the course of time. The production schedule is characterized by a maximum degree of stability and predictability. The key characteristic of the Classic Rhythm Wheel is the use of fixed production quantities that are constantly produced in a fixed sequence in every Rhythm Wheel cycle. In other words, once the production cycle with an optimal sequence and optimal quantities is designed, production follows this schedule exactly, and repeatedly, for a defined time horizon. In this section, we analyze the two major steps involved in designing the Classic Rhythm Wheel:

- Defining the optimal production sequence
- Determining the right cycle time

6.1.1.1 Production Sequence

Building the production sequence is the first step in designing the optimal Rhythm Wheel. It is essential that the sequence be chosen to minimize changeover times. A nonoptimal sequence is wasteful, because more time than necessary is consumed by changeovers. The quality of the changeover sequence has a major impact on the cycle time. The longer the changeover time, the longer is the Rhythm Wheel cycle, and the higher are the inventory levels.

In process industries, an optimized production sequence is especially important due to the high sequence dependency of changeover times and costs. Some examples for the chemical, pharmaceutical, and food industries can be found in Box 6.1.

BOX 6.1   AN OPTIMIZED CHANGEOVER SEQUENCE IS ESPECIALLY IMPORTANT IN THE PROCESS INDUSTRIES

Changeover activities in process industries may be highly time consuming. In particular, extensive cleaning efforts between product changeovers may be required to prevent contamination by left-overs from a previous production run.

In food production processes, for example, it is very important to consumer safety that products remain uncontaminated. Cross-contact between allergens such as nuts and milk products needs to be avoided at any price. Therefore, time-consuming sterilization between changeovers is essential.

In the pharmaceutical production process, cross-contamination between active pharmaceutical ingredients must be prevented. Allowing one active ingredient to end up in a tablet for which it was not intended due to insufficient cleaning is irresponsible. Therefore, tablets on a nondedicated production line should be grouped by the same active ingredient in the production sequence. Within each group, the sequence should then be established from low-to-high dosages of the active ingredient.

In addition to the need to address contamination, there are other major differences between the changeover activities in process industries compared with those in discrete industries. Continuous production processes in the chemical industry, for instance, can make it very costly to stop a machine for a changeover. Stopping a melting plastic production process would block the tubes and pipes with the material, requiring all the equipment to be exchanged. Therefore, transitions from one to the next product are performed: molten plastic, for example, would be produced, in order, at 5%, 10%, and then 15% concentrations of a certain additive. Naturally, some material is lost during such a transition until it is back in spec. Nevertheless, transition losses can be reduced to a minimum when adhering to an optimized production sequence.

A company seeking to capture the full potential of LEAN production processes must identify the optimal production sequence to achieve a successful Rhythm Wheel design. In some cases, the optimal sequence can be found without advanced algorithms. An example of an easily optimized production sequence from the chemical pigments industry is production in a color sequence from bright colors to dark colors (see Figure 6.6). Little

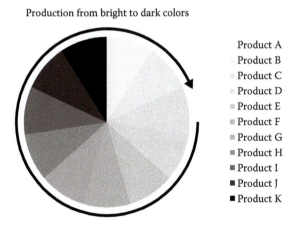

FIGURE 6.6
Products on a Rhythm Wheel are scheduled in the optimal changeover sequence.

Changeover matrix

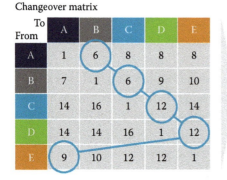

▶ Optimized changeover sequence:
A, B, C, D, E

▶ Changeover duration for one cycle:
6 + 6 + 12 + 12 + 9 = 45 h

FIGURE 6.7
Defining the optimal production sequence based on the changeover matrix.

effort is involved in changing over from a bright color to a darker color, while a changeover in the opposite direction requires intensive cleaning.

A pragmatic way to find an optimal sequence is to concentrate on product families. Major changeovers are necessary when switching between product families while changeovers within product families are often fast and cheap. Determining sequences based on product families is especially helpful when detailed changeover information is not available for every product.

If the optimal changeover sequence cannot be found by experience or simple rules, a more sophisticated approach is necessary. This usually requires a changeover matrix containing all possible changeover times between products. The optimal sequence is then derived by finding the shortest total transition time between products (see Figure 6.7). In cases with many products, software based on special algorithms is required to find the optimal changeover sequence.

6.1.1.2 Cycle Time

The cycle time is the second very important aspect when designing the Rhythm Wheel. It defines the time required to complete one production cycle from the first to the last product in the sequence. In this way, the cycle time specifies the time interval between two production runs of the same product and can therefore be denoted as the production takt. Generally, there are two primary drivers of cycle time:

- Optimality is determined from a cost perspective.
- Feasibility is determined from a capacity perspective.

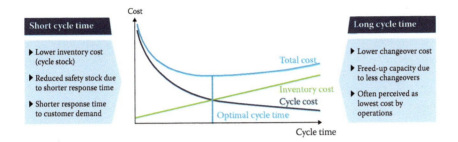

FIGURE 6.8
The optimal cycle time is determined as a trade-off between inventory and cycle costs.

6.1.1.2.1 Cost Perspective

From the cost perspective, the cycle time directly influences lot sizes and inventory levels. Obviously, lot sizes decrease as the Rhythm Wheel cycle becomes shorter. In light of the required inventory levels, shorter cycles, and thus smaller lot sizes, are beneficial as they reduce average inventory. However, smaller lot sizes imply more frequent changeovers and thus higher cycle costs. From an inventory perspective, it is generally desirable to achieve a short Rhythm Wheel cycle. From a cycle cost perspective, a long cycle time is beneficial. The optimal Rhythm Wheel cycle time balances the trade-off between cycle and inventory costs and minimizes total production cost (see Figure 6.8).

6.1.1.2.2 Capacity Perspective

From the capacity perspective, the feasibility of the designed cycle time needs to be ensured as well. Shorter cycle times require greater capacity than long cycles due to more frequent changeovers. Consequently, available capacity is the restrictive factor for the shortest possible cycle length. The following four-step approach helps in calculating the shortest possible cycle time and assuring feasibility by considering available capacity (see also Figure 6.9):

1. Determination of available production time
   First, determine the available time for production. In this step, shift patterns, maintenance, and downtime are considered.
2. Determination of required production time
   Second, calculate the required production time for meeting expected demand. Here, production rates and demand volumes are taken into account.

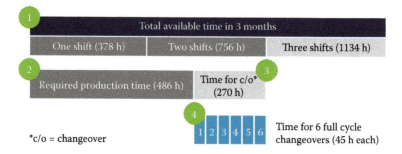

FIGURE 6.9
The four steps involved in calculating a feasible cycle time (food packaging example).

3. Capacity check of required and available production time
   Third, check capacity and the possible need for additional shifts and then derive the net available time for changeovers.
4. Calculation of cycle time
   Fourth, calculate the number of cycles and the cycle time based on the changeover time for one full cycle.

The example in Box 6.2 illustrates how a feasible cycle time was calculated for a food manufacturer's packaging line. The objective was to find the shortest feasible cycle time that utilizes capacity optimally and simultaneously allows for quick response times. This example illustrates the basic calculations involved in finding a feasible Rhythm Wheel cycle time.

**BOX 6.2   CALCULATING THE FEASIBLE CYCLE
TIME FOR A FOOD PACKAGING LINE**

After implementing the Rhythm Wheel concept at a food manufacturer's packaging site, the company needed to determine the cycle time for the next 3 months. (Note that the numbers in this example have been slightly modified for reasons of simplicity; see also Figure 7.8.)

1. Determination of available production time
   The next 3 months contained 90 calendar days, but only 54 working days. The shift plan for this time had been fixed at two shifts per day, 8 h per shift. There was a 30-min break; moreover, based on experience, 30 min of downtime and maintenance per shift

were allocated. The overall available time for production over the next 3 months was therefore 54 days × 2 shifts/day × (8−0.5−0.5) h/shift = 756 h.

2. Determination of required production time

The production rate of the packaging machine equaled 500 units/min. The packaging line therefore had a production capacity of 500 units/min × 60 min/h = 30,000 units/h. With a total customer demand forecast of 14,580,000 units over the next 3 months, the required production time to fulfill customer demand was therefore estimated to take 14,580,000 units/30,000 units/h = 486 h.

3. Capacity check of required to available production time

The time share used for production, without considering changeovers, was 486 h/756 h = 64%. This left enough time for performing changeovers (36%) and meant that a third shift would not necessarily be required. The total available time for performing changeovers with two shifts was 756,486 h = 270 h.

4. Calculation of cycle time

Over the next 3 months, 270 h could be used to perform changeovers on the packaging line. A complete changeover cycle for all products on the line in the optimized sequence required 45 h per cycle. The maximum number of cycles in the next 3 months was calculated as 270 h/45 h/cycle = 6 cycles. This meant that six full Rhythm Wheel cycles could be completed over the next 3 months. The shortest feasible cycle time was then calculated as 90 days/6 cycles = 15 calendar days per cycle, or 54 working days/6 cycles = 9 working days per cycle. The production lot sizes for every product could then easily be derived by multiplying the cycle time by the average daily demand.

Lean manufacturing greatly facilitates achieving a short Rhythm Wheel cycle time, since it improves the two main drivers of cycle length. First, production speed and production reliability are increased by lean manufacturing tools. As a higher production rate is achieved, the Rhythm Wheel cycle becomes shorter since the same demand can be produced in less time. Second, a reduction of changeover times leads to shorter cycle time. This can be achieved by applying SMED methodology. In Box 6.3, a case study describes how changeover times at a pharmaceutical packaging site were reduced by over 50%. This enabled the company to cut the Rhythm Wheel cycle time in half and to accomplish high inventory savings.

BOX 6.3 CHANGEOVER TIME REDUCTION OF 54% ON A PHARMACEUTICAL PACKAGING LINE WITH SMED

A mid-size pharmaceutical company wanted to reduce changeover times in its packaging operations. Long changeover times lead to long Rhythm Wheel cycles, which should be decreased in order to gain greater flexibility and reduce inventory levels.

A SMED initiative was launched with the goal of reducing changeover times by around one-third. The first step was to separate external and internal changeover times. External changeover times are caused by actions that can be performed before or after the packaging line is stopped for a changeover. Internal changeover times require a stoppage of the line. The second step was to analyze which internal changeover times could be converted into external changeover times, such that the packaging line was not required to stop. In a third step, measures to reduce the remaining internal changeover time were identified.

Then, for each of the measures, the effort and potential changeover reduction time were estimated and plotted on a matrix (see left-hand side of Figure 6.10). Fortunately, most of the identified measures

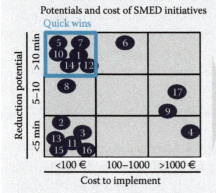 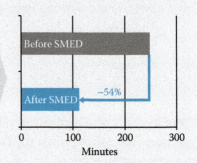

FIGURE 6.10
Prioritizing measures to reduce changeover times.

could be implemented without significant effort (see quick wins in top-left corner). One measure, for example, was to use clips instead of screws; another was to be prepared properly before a changeover.

The results of the SMED initiative were overwhelming. Changeover times were reduced by 54% by implementing only the quick wins. This made it possible to halve the Rhythm Wheel cycle time. In addition, employees were thoroughly involved from the analysis phase through implementation, which created high motivational and learning effects. SMED could then be applied by employees on other packaging lines as well.

The definition of sequence and cycle time completes the Classic Rhythm Wheel design. It is well-suited to products with high, stable demand volumes. If demand becomes more variable, the Classic Rhythm Wheel might not be the best choice due to its fixed character. The Breathing Rhythm Wheel is well-suited to manage demand variability.

6.1.2 Breathing Rhythm Wheel Design to Manage Higher Demand Variability

To achieve stable capacity utilization, cycle time and production quantities have to be leveled over time. While the Classic Rhythm Wheel is quite suitable for products with low demand variability, the Breathing Rhythm Wheel has been developed to cope with more variable products. The main difference between the Breathing Rhythm Wheel and the Classic Rhythm Wheel is that production quantities are not fixed in every cycle, but determined dynamically according to actual demand. This allows for greater flexibility in cases of medium-to-high demand variability.

When adjusting production quantities to meet customer demand, the duration of a Rhythm Wheel cycle will vary over time. Cycles with high demand lead to bigger lot sizes and therefore longer cycles. Periods with low demand lead to smaller lot sizes and shorter cycles. This means that by flexibly reacting to customer demand, the cycle time can fluctuate during execution (see Figure 6.11).

As demand becomes more variable, the Rhythm Wheel cycle time may start to fluctuate more than desired. This is where one very important LEAN paradigm comes into play: buffering variability in inventories,

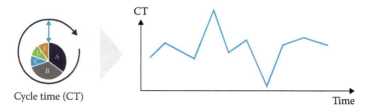

Cycle time (CT)

FIGURE 6.11
The cycle time can fluctuate due to changing production quantities in each cycle.

not on the production asset. This requires limiting the fluctuations and adhering to the preconfigured cycle time as closely as possible. Wide fluctuations in demand are buffered in inventories to achieve stable capacity utilization and smooth product flow.

To address these problems, we introduce two fundamental parameters of the Breathing Rhythm Wheel: the minimum and maximum cycle time boundaries. Min–max cycle boundaries maintain the cycle time within a lower and upper limit and prevent the cycle from fluctuating too widely. This enables a company's supply chain planner to preserve the reliability of the Rhythm Wheel cycle time. In the next sections, we explain:

1. The concept of minimum and maximum cycle time boundaries.
2. Factoring approaches to adhere to the cycle time boundaries.

6.1.2.1 Minimum and Maximum Cycle Time Boundaries

Cycle time boundaries allow for only a certain degree of variability in cycle time and production quantities. If, for instance, demand is far higher than expected, some part of it is shifted to the next cycle. It is important not to exceed the maximum cycle boundary, since huge lot sizes block the production asset for a long time, which might result in stock-outs for other products. If on the other hand, demand is far lower than expected, the asset is kept idle for some time to avoid falling below the minimum cycle boundary (see Figure 6.12). Operators of the asset must simply wait until demand rises while carrying out other tasks, such as maintenance.

Setting upper and lower cycle time boundaries ensures that production volumes are neither too big nor too small. Certain deviations from the optimal production quantity are allowed, but those deviations do not run out of control. Therefore, the upper and lower boundaries should not be too far apart, because too much variability disrupts the regular Rhythm

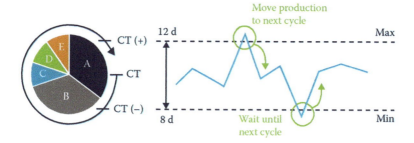

FIGURE 6.12

Factoring keeps the cycle time between the minimum and maximum cycle boundaries.

Wheel schedule and the cycle time takt. On the other hand, cycle boundaries should allow for a certain degree of flexibility in order to enable the asset to react to demand and supply variations.

The optimal design of cycle time boundaries depends mainly on the objectives set by the operations strategy. If constant and high asset utilization is the priority, the cycle time boundaries should be set very close to each other. Variability is then buffered mainly in inventories, and asset efficiency is very high. If on the other hand the operations strategy sets the focus on responsively reacting to customer demand, the cycle time boundaries need to be set further apart. Greater variability is then buffered on the asset instead of inventories, which reduces asset efficiency (see Figure 6.13).

From a global supply chain perspective as well, the reliability of the Rhythm Wheel cycle time is very important to upstream and downstream production stages. If the cycle time fluctuates too widely, upstream stages cannot deliver input material on a JIT basis; downstream stages might lack input material

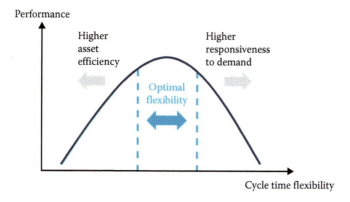

FIGURE 6.13

Optimal cycle time flexibility is a trade-off between responsiveness and efficiency.

for their production processes. These problems result in continuous adjustment of the production plans at the affected stages. These adjustments cause, in turn, further rescheduling of production plans at other supply chain stages as well, which creates unnecessary nervousness along the entire supply chain. From a supply chain perspective, it is therefore desirable to set the cycle time boundaries rather close to each other to achieve stable takt.

After having set the optimal cycle boundary parameters, care must be taken to ensure that the cycle time stays between the min/max boundaries. This is called factoring.

6.1.2.2 Factoring Types

Factoring is applied to prevent cycles from becoming either too long or too short; ensuring that the cycle time is kept within the previously defined min/max boundaries. This is achieved by rationing production quantities in cycles with high demand and moving them to cycles with relatively low demand, and vice versa. Factoring can therefore be understood as a methodology for leveling production quantities, maintaining a predefined production takt, and accomplishing a balanced material flow along the supply chain.

To achieve leveled production, several types of factoring can be applied. Depending on the characteristics of a given product portfolio, the production assets, and the strategic objectives for operations, one factoring type may perform better than another. Here, we introduce four proven factoring types: cutoff factoring, proportional factoring, rolling factoring, and IRL factoring (see Figure 6.14).

6.1.2.2.1 Cutoff Factoring

Cutoff factoring is the simplest of all factoring types. If the Rhythm Wheel cycle becomes too long and thus violates the maximum cycle boundary, the products at the end of the cycle are "cut off" and will not be produced in

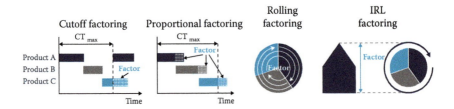

FIGURE 6.14
Several factoring types can be applied to maintain cycle time boundaries.

this cycle. They have to wait for the next cycle to be produced. This factoring mode inherits a disadvantage for products at the end of the cycle, as there is no guarantee they will be produced. Thus, more safety stock has to be held for those products to maintain the agreed-upon service levels. However, this factoring type is suitable if the strategic importance of products varies significantly. Strategically important product groups and groups with a high-profit margin should be scheduled at the beginning of the cycle. Less important product groups can then be scheduled at the end of the cycle as potential service level violations and lost sales might not be as significant for them.

6.1.2.2.2 Proportional Factoring

When using Proportional Factoring, all production quantities are factored proportionally if the maximum cycle time is exceeded. This means that all production quantities for all products are reduced by a certain percentage, such that the maximum cycle time is maintained. By using Proportional Factoring, the position of a product in the Rhythm Wheel sequence does not determine its importance since products at the end of the sequence are favored equally. Proportional Factoring is therefore well-suited to products with similar margins and strategic importance.

6.1.2.2.3 Rolling Factoring

Rolling Factoring tracks the time interval between two production runs of the same product. The time between two runs may not exceed the maximum cycle time. If it is exceeded, the production quantity is decreased. By ensuring that every individual product obeys the maximum cycle time, the maximum cycle time of the overall cycle is obeyed as well. This approach allows for considerable flexibility, as unused production time for a product with low demand can be consumed by a product with high demand in the same cycle. In this way, capacity is shifted to products that require replenishment the most.

6.1.2.2.4 IRL Factoring

IRL Factoring, which is a very elegant factoring mode, is applicable if the Rhythm Wheel is combined with an IRL replenishment mode. In an IRL replenishment mode, the replenishment quantity for a product is determined as the difference between the current inventory level and the target inventory level. If the cycle time exceeds its predefined maximum, IRL Factoring decreases the target stock level. In this way, the difference between current inventory and the target level narrows and this in turn reduces the production quantity. Note that the IRL is decreased only for the current cycle and will be restored to its original value for the next cycle. The significant advantage of this factoring mode is that production is controlled with only one parameter, namely the IRL. By increasing or decreasing the IRL, production quantities can be increased or decreased accordingly. An example of IRL Factoring in operational planning can be found in Chapter 7.

6.1.3 High-Mix Rhythm Wheel Design to Manage Diverse Product Portfolios

With Classic and Breathing Rhythm Wheels, every product on the wheel is produced in every cycle. The biggest difference with the High-Mix Rhythm Wheel is that not every product is produced in every cycle. Depending on the volume and the value diversity of the product portfolio, there is an advantage in assigning varying production rhythms to products, which means that some products have to wait a few cycles before being produced again. A production rhythm of three cycles means, for example, that a given product is scheduled only every third cycle.

In fact, this implies a difference between the asset takt and the product takt; so far, we only had one takt for products and asset. With Classic and Breathing Rhythm Wheels, the asset takt equals the product takt, since every product is produced during every cycle. In the High-Mix Rhythm Wheel concept, the asset takt and the product takt can differ (see Figure 6.15). This can significantly improve overall performance, especially if low-volume and high-volume products are produced on the same asset.

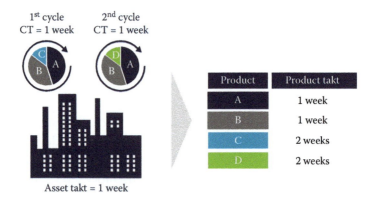

FIGURE 6.15
The High-Mix Rhythm Wheel leads to differences between asset and product takt.

In this section, we explain:

1. How to determine production rhythms
2. How to implement production rhythms
3. How to balance cycles when applying production rhythms

6.1.3.1 Determination of Production Rhythms

In many cases, demand patterns across products differ significantly. In this case, it is advantageous to set varying production rhythms, which means not producing every product in every cycle. Intuitively, high-volume products should be produced in every cycle to keep inventory levels low. Low-volume products on the other hand should not be produced in every cycle, because it is not worth making a changeover for only a very small production quantity. Therefore, it is best to produce low-volume products in larger batches that cover demand for several cycles. The effect of saving changeover times is depicted in Figure 6.16.

The more frequently a product is produced, the smaller is its lot size and therefore its average inventory level. If, for example, a product is produced in every cycle—giving it a production rhythm of one—then the lot size and inventory need to cover demand for one cycle. If the product is produced less often, for example, with a production rhythm of three cycles, then the lot size and the inventory have to cover demand for three cycles. It is important to determine the right production rhythm for every product, as this can significantly improve the performance of the Rhythm Wheel. A case study of a shampoo bottling line illustrates how assigning optimal production rhythms can reduce cycle time by around 70% and reduce inventory by over 40% (see Box 6.4).

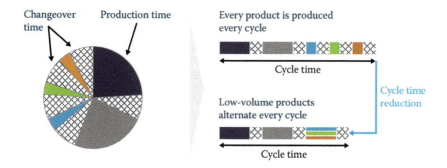

FIGURE 6.16

A shorter cycle time is achieved by not producing every product in every cycle.

BOX 6.4   ASSIGNING OPTIMAL PRODUCTION RHYTHMS IN SHAMPOO BOTTLING OPERATIONS

A mid-size hair care manufacturer decided to schedule its shampoo bottling line operations using the Rhythm Wheel concept. The product portfolio showed a wide range of volumes for every SKU. Some 50 of the SKUs had very small volumes, while the volume of the biggest SKU, for instance, was up to 1000 times higher. This huge difference in product volumes made it impossible to schedule every product in every Rhythm Wheel cycle, because conducting a changeover for low-volume products was not profitable. The company therefore wanted to evaluate the benefits of a High-Mix Rhythm Wheel.

First, the product portfolio was segmented into "A," "B," and "C" products. For the optimal design of the High-Mix Rhythm Wheel, various production rhythms were then tested by means of simulation. Similar to the former planning approach, fixed lot sizes were defined for the low-volume SKUs.

The company found that, in comparison with the solution without production rhythms, it could reduce the cycle time from 32 to 9 days (−72%)! Inventories of high-volume "A" products could be reduced by 64%, since those SKUs were produced much more frequently (see Figure 6.17). Inventories of low-volume "C" products on the other hand increased by 41%, since those SKUs were scheduled less often. But the total inventory value of all SKUs could be reduced dramatically by about 47% with the High-Mix Rhythm Wheel design.

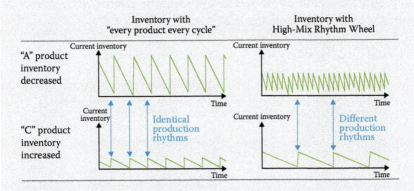

FIGURE 6.17
Inventory is reduced by the introduction of varying production rhythms.

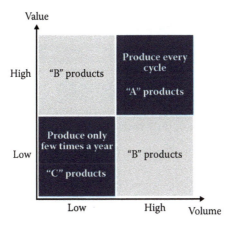

FIGURE 6.18
Volume and value determine how frequently a product should be produced.

As a general guideline, products with high volume and high value ("A" products) should be scheduled more often, as it is important to produce them in smaller lot sizes and keep inventory levels low, which reduces working capital. Products with low volume and low value ("C" products) should be scheduled less often to save changeover time by producing them in bigger batches (see Figure 6.18).

The optimal balance of production rhythms across heterogeneous products with varying volumes and values can be calculated analytically. Appropriate IT tools to support Rhythm Wheel design (see Chapter 10) handle the underlying optimization problem efficiently.

6.1.3.2 Implementation of Production Rhythms

Once optimal production rhythms have been assigned to every product, the question is how these rhythms can be implemented in execution. Here, we discuss three methods that have been developed for implementing production rhythms, each suiting a different set of requirements of the product portfolio and operations strategy: the Blanco Space, Fixed Frequencies, and Minimum Make Quantities (see Figure 6.19).

6.1.3.2.1 Blanco Space

When using the Blanco Space option, some portion of the capacity on the Rhythm Wheel is reserved for scheduling low-volume products and products characterized by erratic demand. Such products are then

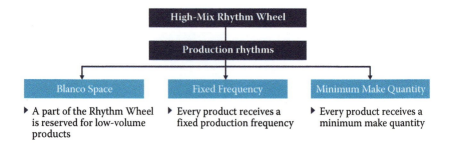

FIGURE 6.19
Production rhythms can be implemented via three methods.

scheduled manually in every cycle as required by the production planner. In this way, the cycle of "A" products is not disturbed, while products with highly unpredictable demand are scheduled flexibly when they are needed.

This method is probably the simplest one, at least from a design perspective. One drawback of using this method is that it can be used only if there are relatively few low-volume products. If the ratio of "C" products to "A" products is too high, the Blanco Space becomes so large that the advantage of the Rhythm Wheel with its predefined sequence is lost.

6.1.3.2.2 Fixed Frequencies

When using the Fixed Frequency option, each product receives a fixed production frequency which is strictly adhered to. Products are scheduled exactly in the predefined order but with flexible production quantities. If a product has an optimal production rhythm of three cycles, for example, it will be produced only every third cycle. This means that the time interval between two production runs is fixed. The production quantity on the other hand can be chosen flexibly, depending on how much of the specific product is required.

The advantages of the Fixed Frequency method are a very transparent schedule and a stable production pattern. Because of the fixed frequency, a planner knows exactly in which cycle the product is to be produced next, and can communicate this to internal or external customers. In addition, required input material can be delivered on a JIT basis. The disadvantage is that products with high production rhythms have to wait many cycles until they are produced again. In situations with highly unpredictable demand, this might lead to relatively high safety stock requirements compared with the Minimum Make Quantities method.

6.1.3.2.3 Minimum Make Quantity

The third option, Minimum Make Quantities, implements a minimum production quantity for low-volume products that lasts for several cycles. The exact cycle in which a product is to be produced remains flexible. If the optimal production rhythm is, for instance, three cycles, then the minimum make quantity is chosen to cover expected demand of three cycles. If a period of high demand occurs, the product may be scheduled again after only two cycles. If a period of low demand occurs, the product takt might be four or more cycles.

The advantage of this method is flexibility regarding the point in time at which production occurs, which is well-suited to variable demand. The drawback is that it is harder to predict in which cycle a product is to be produced the next time. This makes it more difficult to communicate exact finishing times for the next production run to internal or external customers.

6.1.3.3 Balancing Production Rhythms

When production rhythms are defined, it is important to prevent too many low-volume products from being produced in the same cycle. This can happen, for example, if the stock levels of many low-volume products fall below the reorder point within the same Rhythm Wheel cycle. If all products were then produced in the same cycle, this would result in one exceptionally long cycle time (see left-hand side of Figure 6.20).

To prevent large cycle time fluctuations caused by production rhythms, a balancing approach is required. Factoring rules (see above) determine which products may be produced in the current cycle. They distribute products equally over several cycles and balance the cycle times. Factoring ensures that cycle times stay within certain boundaries and that it is possible to achieve predefined takt in production (see right-hand side of Figure 6.20).

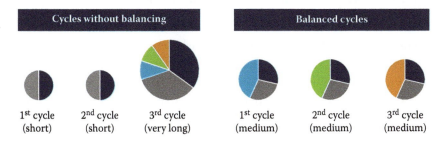

FIGURE 6.20
Unbalanced and balanced Rhythm Wheel cycles.

Summary

In this section, we have shown how your company can configure each of the three Rhythm Wheel types. We have seen that several steps are necessary to achieve the optimal cycle design that best fits the company's product portfolio, its supply chain characteristics, and its operations strategy.

First, we showed how to configure parameters for the Classic Rhythm Wheel, optimizing the product sequence and the fixed cycle time. We saw that the cycle is determined as a trade-off between cycle and inventory costs and needs to be feasible from a capacity perspective.

Second, we explained how to determine the parameters for the Breathing Rhythm Wheel. Minimum and maximum cycle time boundaries must be set in order to keep the cycle time within certain boundaries. Then, we introduced several factoring types, which ensure that the defined cycle boundaries are maintained and production is leveled.

Finally, we described the High-Mix Rhythm Wheel, in which not every product is produced in every cycle. We showed the underlying logic of deriving optimal production rhythms and three distinct implementation methods. We explained how to balance production rhythms to leverage the full potential of the High-Mix Rhythm Wheel.

The next step, after designing optimal Rhythm Wheels on the production side, is designing the replenishment side. In the following section, we see that the Rhythm Wheel has a significant impact on the determination of replenishment parameters.

6.2 SETTING UP THE PARAMETERS FOR LEAN REPLENISHMENT MODES

Inventory optimization by efficient replenishment design remains one of the key challenges of supply chain management. Typically, inventories can be found along the entire supply chain in the form of raw materials, semi-finished products, and finished goods. Considering all stages from purchasing raw materials to regional distribution, a substantial amount of capital is tied up in stocks even in supply chains of moderate size.

However, simply eliminating inventories is not a solution, as they play a very important role in running a supply chain efficiently. Inventory allows

FIGURE 6.21

Various sources of lead time and variability require maintaining stocks in the supply chain.

the leveling of production by buffering variability in stocks rather than in capacity. Furthermore, inventories are required for bridging the manufacturing time and transportation lead times, for preventing supply chain stages from running dry when there are supply disruptions, and for maintaining a high customer service level despite demand variability (see Figure 6.21).

Determining the right stock levels at the various supply chain stages is therefore a key SCM objective. Companies often end up with excess stocks of some products while stock-outs occur for others. It is therefore important to adopt a structured approach to determining appropriate inventory level for every product. In this section, we introduce a structured approach to choosing replenishment parameters:

- We describe the relevant stock components.
- We show how to set the parameters for the LEAN modes IRL and Buffer Management.

6.2.1 How Stocks Are Structured for Variability and Uncertainty

For LEAN replenishment planning, five stock components need to be distinguished (see Figure 6.22). Relevant personnel need to understand the role of each of these components to set the replenishment parameters correctly.

6.2.1.1 Cycle Stock

Cycle stock covers expected demand during replenishment takt. Expected demand is average demand, which occurs over a certain period, on the

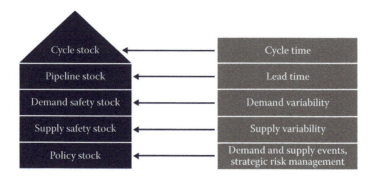

FIGURE 6.22
The inventory level can be structured into five stock components.

part of either internal or external customers. Average daily demand can, for example, be derived from forecasts or from historical sales.

The replenishment takt is the time interval between two orders of the same product. In conjunction with the Rhythm Wheel concept, the replenishment takt equals the product takt. If the cycle time is, for instance, 10 days and the production rhythm of the product is two cycles, then the product takt is 20 days. The cycle stock parameter is now chosen to cover expected demand during the replenishment interval of 20 days.

Although in this case the cycle stock inventory component is chosen to cover 20 days, this does not mean that there are always 20 days of inventory on hand. On average only 10 days of cycle stock is on hand, which is half of the total cycle stock (see Figure 6.23).

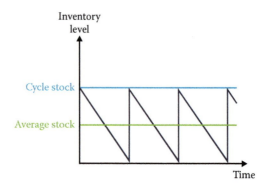

FIGURE 6.23
Half of total cycle stock represents an average that is physically on hand.

6.2.1.2 Pipeline Stock

Pipeline stock covers expected demand during lead time. Lead time is the average required time for order processing, manufacturing, quality control, transportation, inbound handling, and so on. In other words, the lead time is the total time between product order and order arrival.

It is important to understand that pipeline stock is typically not physically on hand at the stockkeeping point, but rather is in transit on the way to that point. Nevertheless, this stock parameter needs to be considered when configuring the IRL and Buffer Management parameters to determine the right replenishment quantity.

6.2.1.3 Demand Safety Stock

Typically, demand is not constant but varies over time. Cycle stock and pipeline stock are therefore not necessarily sufficient to ensure that all demand is fulfilled. Demand safety stock covers demand variability for a given target service level during the replenishment takt and lead time.

A useful unit for measuring demand variability is the standard deviation. The higher the standard deviation, the more safety stock is needed to maintain a target service level. The service level, or more precisely the alpha service level, is one common measure of the number of replenishment cycles in which demand can be satisfied completely. A target service level of 99% means, for example, that in 99 out of 100 replenishment cycles no stock-outs will occur. The higher the chosen target service level, the higher is the safety stock. The relationship between target service level and safety stock is depicted in Figure 6.24. Higher service requirements can increase the required safety stock dramatically.

6.2.1.4 Supply Safety Stock

Supply safety stock covers situations in which orders are not delivered in full or on time. Gaps between the ordered and delivered quantity can occur if, for example, the scrap rate is higher than usual or some portion of a shipment is damaged during transportation. Gaps between expected lead time and actual lead time are summarized in lead time variability. Lead time variability can mean a delay in manufacturing or transportation, for example, if the Rhythm Wheel cycle time is highly variable, a machine breaks down, or the truck transporting the goods is stuck in a

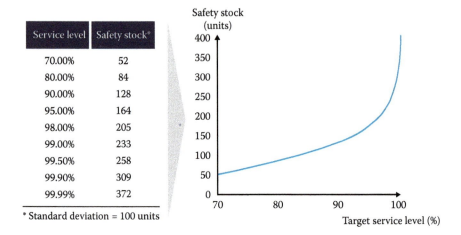

Service level	Safety stock*
70.00%	52
80.00%	84
90.00%	128
95.00%	164
98.00%	205
99.00%	233
99.50%	258
99.90%	309
99.99%	372

* Standard deviation = 100 units

FIGURE 6.24
Safety stock increases above average with the target service level.

traffic jam. The higher the uncertainty about supply quantity and supply lead time, the more stock has to be held to buffer supply variability.

6.2.1.5 Policy Stock

Demand and supply safety stock cover only regular variability. "Regular" in this context means that exceptional demand peaks and exceptional supply disruptions are not included. It is policy stock which covers exceptional demand and supply events—and it can also be used for strategic risk management.

Exceptional demand events that need additional stock might include special marketing campaigns or tender orders. A portion of demand must be served from stock, because production capacity is limited and not all of the required quantity can be produced immediately. To avoid overstressing capacities or running out of stock, the required policy stock for such events should be built up in advance over a certain period. The same holds true for exceptions from the supply side. Policy stock is often kept as a strategic decision for robustness of the supply chain, if, for example, a supplier can no longer deliver its goods, or a production site breaks down. Furthermore, policy stock needs to be built up in advance to cover such events as scheduled asset maintenance or a plant shutdown in which the entire production plant is closed, for example, due to summer or Christmas vacation.

Except for the decision on how much policy stock to build up, a company needs to define how far in advance the process of building up policy stock should begin. If policy stock is built up shortly before the event occurs, additional inventory costs are very low. However, capacity utilization might be overstressed and additional shifts might be necessary to produce the required amount in a short time, further increasing the labor cost. If policy stock is built up too early, inventory costs are high, because precious working capital is tied up in dead stock. Refer to Section 6.4 for further information and a practical example of the adaptation strategy.

We have thus far divided stocks into five components, namely cycle stock, pipeline stock, demand safety stock, supply safety stock, and policy stock. In the next section, we show how replenishment parameters are calculated based on those five stock components.

6.2.2 Right-Size the Parameters to Enable Consumption-Based LEAN Replenishment

The replenishment mode determines when and in which quantity orders are placed. LEAN pull replenishment modes include IRL, Buffer Management, and MTO (see Figure 6.25).

Figure 6.26 provides an overview of the required replenishment parameters and stock components for each replenishment mode. In the following, we describe the IRL and Buffer Management replenishment modes and exclude the MTO mode since it does not require stock (Figure 6.26).

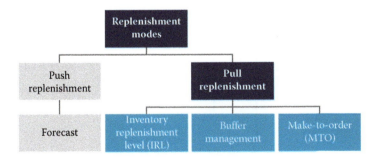

FIGURE 6.25
LEAN pull replenishment modes.

Replenishment mode	Replenishment parameters				Stock components				
	IRL	Min. level	Max. level	Reorder point	Cycle stock	Pipeline stock	Demand safety stock	Supply safety stock	Policy stock (optional)
Inventory replenishment level	✔				✔	✔	✔	✔	✔
Buffer Management		✔	✔	✔	✔	✔	✔	✔	✔
Make-to-order									

FIGURE 6.26
Each replenishment mode requires a particular set of parameters.

6.2.2.1 Inventory Replenishment Level

The IRL mode takes its name from the inventory replenishment level, which is a predefined target stock. It is built up from all five stock components. During every review of stock, the current inventory level is checked, compared with the IRL, and an order to cover a missing quantity is placed to refill the stock (see Figure 6.27). The IRL policy fits the Rhythm Wheel production mode naturally. During every Rhythm Wheel cycle, immediately before a product is produced, the current inventory level of that product is checked. The production quantity is then determined simply as the difference between the IRL and the current inventory level.

Cycle stock covers expected demand within one Rhythm Wheel cycle or some multiple of it if production rhythms are defined for a product. Safety stocks buffer demand and supply variability during the cycle time. We present a detailed example of IRL parameterization in Box 6.5.

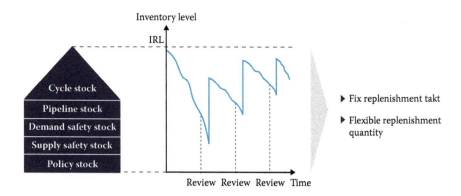

FIGURE 6.27
Using IRL mode, a flexible replenishment quantity is ordered after a fixed replenishment interval.

BOX 6.5   CALCULATING THE IRL PARAMETER IN COMBINATION WITH THE HIGH-MIX RHYTHM WHEEL

The IRL consists of the cycle stock, the pipeline stock, the demand safety stock, the supply safety stock, and, optionally, the policy stock. Here each of the components is calculated for the example of an analgesic drug with normally distributed demand, which is produced on an asset with the High-Mix Rhythm Wheel with fixed frequencies.

Cycle stock: The drug has an expected demand of 1000 packs per day and is produced on a Rhythm Wheel-scheduled resource. The Rhythm Wheel cycle time is 10 days and the product is scheduled every second cycle, which means that the replenishment takt is 20 days. The required cycle stock to cover expected demand during the replenishment takt is then 1000 packs/day × 20 days = 20,000 packs.

Pipeline stock: The drug has a manufacturing lead time of 2 days, a quality control time of 4 days, a transportation time of 4 days, and an inbound handling time of 2 days. The total lead time adds up to 12 days. The required pipeline stock is then 1000 packs/day × 12 days = 12,000 packs.

Demand safety stock: With a replenishment takt of 20 days and a lead time of 12 days, demand safety stocks need to cover uncertainty over a total period of 32 days. The required service level is 99%, which translates into a safety factor of 2.33. The standard deviation of demand is 100 packs/day. The demand safety stock is now calculated as

$$\text{Safety factor} \times \text{standard deviation} \times \sqrt{uncertainty\ period}$$
$$= 2.33 \times 100 \times \sqrt{32} = 1318 \text{ packs.}$$

Supply safety stock: The standard deviation of the delivery quantity is 500 packs (meaning that, in 95% of deliveries, the quantity does not deviate more than ±1000 packs). The lead time variability has a standard deviation of one day (meaning that, in 95% of cases, lead time does not deviate more than ±2 days). The required supply safety stocks for an service level of 99% are then calculated as

Safety factor ×

$$\sqrt{(standard\ deviation_{quantity})^2 + (standard\ deviation_{lead\ time} \times demand\ rate)^2}$$
$$= 2.33 \times \sqrt{(500)^2 + (1 \times 1000)^2} = 2605\ packs.$$

Policy stock: The pharmaceutical packaging line is scheduled to be closed for one entire week due to scheduled maintenance work. This means that the policy stock is built up to cover the expected demand and its variability during 1 week. The policy stock is then calculated as 7 days × 1000 packs/day + 2.33 × 100 × $\sqrt{7}$ =7000 + 616 = 7616 packs.

The IRL parameter is now the sum of the five stock components: IRL = 43,539 packs.

Note however that average stocks on hand are far lower than the IRL parameter: average stock = 0.5 × cycle stock + demand safety stock + supply safety stock + policy stock = 21,539 packs. The difference in quantity between the IRL and the average stock on hand is either in transport or has already been consumed by customers.

The IRL mode can be used in conjunction with the kanban concept as well. In a kanban production system, the question is how many kanbans (cards) to hold for each product. The number of kanbans can easily be calculated as the IRL divided by the lot size of a single kanban. This ensures that there are sufficient kanbans to cover production lead times and to buffer demand and supply variability.

6.2.2.2 Buffer Management

In Buffer Management, the inventory level is held within minimum and maximum boundaries. To stay above the minimum boundary, a reorder point (ROP) is defined. When the current inventory level falls below the ROP, a replenishment signal is triggered. The main difference between this and IRL mode is that with Buffer Management the replenishment interval is flexible but the replenishment quantity is fixed. Figure 6.28 shows an inventory graph for a Buffer Management replenishment policy.

In Buffer Management, stocks are divided into three zones: a green zone, a yellow zone, and a red zone. The green zone is typically defined as the area between the maximum inventory level and the ROP. As long as

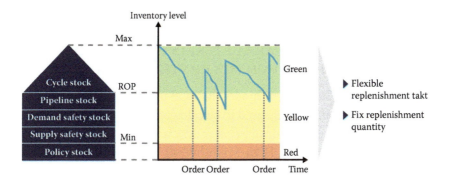

FIGURE 6.28

With Buffer Management, a fixed replenishment quantity is ordered as soon as the ROP is reached.

the inventory level is within this zone, no replenishment orders are placed. The size of this zone is calculated to equal cycle stock. The yellow zone is the area between the ROP and the minimum stock level. As soon as the yellow zone is reached, a replenishment order of a fixed quantity is triggered. The size of the yellow zone equals the pipeline and safety stocks. In this way, the yellow zone covers expected demand during the lead time, as well as the regular demand and supply variability. Note that safety stock is actively used to buffer variability. The red zone is the area between the zero line and the minimum stock level, which equals the policy stock. Only in exceptional demand and supply events, which are not covered by safety stock, should the inventory be allowed to drop into this area.

Besides providing a useful visualization device, the colored areas help to evaluate the setting of inventory parameters. If the actual inventory level stays within the green zone most of the time, it is an indicator that inventory parameters are set too high and inventory could be reduced. If, on the other hand, inventory reaches the red zone quite often, a little more inventory should be held. In both cases, the evaluation can be used for the next tactical renewal process in order to continuously improve the setting of replenishment parameters.

Like IRL, Buffer Management can be used in conjunction with the Rhythm Wheel as a production mode. Whenever inventory drops below the ROP, a replenishment order is sent to the Rhythm Wheel. The yellow zone covers expected demand during the lead time until the product arrives. If a High-Mix Rhythm Wheel is chosen, the fixed replenishment quantity covers demand for several cycles. Box 6.6 shows an example of a calculation for Buffer Management.

BOX 6.6 CALCULATING THE BUFFER MANAGEMENT PARAMETERS IN COMBINATION WITH THE HIGH-MIX RHYTHM WHEEL

In the Buffer Management mode, the ROP and the minimum and maximum inventory levels need to be calculated. Consider again the example of the analgesic drug (see Box 6.5), which in this case is produced on an asset with a High-Mix Rhythm Wheel with minimum make quantities.

Green zone: The green zone is calculated to equal cycle stock. In this example, the green zone covers demand for two Rhythm Wheel cycles. With a cycle time of 10 days and a production rhythm of two cycles, the green zone is calculated to last for 20 days, which equals 1000 packs/day × 20 days = 20,000 packs.

Yellow zone: The yellow zone needs to buffer demand during the lead time as well as demand and supply variability. As an order has a total lead time of 12 days and additionally may have to wait a maximum of 10 days to be produced in the next cycle, the pipeline stock is calculated as 1000 packs/day × 22 days = 22,000 packs.

Demand safety stock needs to cover demand variability during the period of uncertainty. This period is equal to one Rhythm Wheel cycle of 10 days (since the drug may be produced in every cycle) plus the total lead time of 12 days, which equals 22 days. With a safety factor of 2.33 for a 99% service level, and the standard deviation of demand of 100 packs/day, the demand safety stock is calculated as

$$\text{Safety factor} \times \text{standard deviation} \times \sqrt{uncertainty\ period}$$

$$= 2.33 \times 100 \times \sqrt{22} = 1093 \text{ packs.}$$

Supply safety stock is calculated in the same way as in the IRL mode and equals 2605 packs. The yellow zone in total is therefore 22,000 + 1093 + 2605 = 25,698 packs.

Red zone: This zone consists of the policy stock. It is calculated the same way as for the IRL and equals 7616 packs.

The Buffer Management parameters are now calculated: The minimum level equals the red zone, which is 7616 packs. The ROP is the sum of the yellow and red zones and equals 25,698 + 7,616 = 33,314 packs. The maximum inventory level is calculated as the sum of all three zones and equals 33,314 + 20,000 = 53,314 packs.

Note, however, that the average stock on hand is far lower than the maximum inventory level parameter: The average stock equals $0.5 \times$ cycle stock + demand and supply safety stocks + policy stock = 21,314 packs. The difference in quantity between the max inventory level and the average stock on hand is either ordered but not yet produced, in transport, or has already been consumed by customers.

Summary

In this section, we have addressed several replenishment design aspects. First, we identified the main drivers for setting the replenishment parameters within the supply chain. Cycle times, lead times, and demand and supply uncertainties force a supply chain to hold inventories. We then introduced five stock components. This structured approach makes it possible to precisely calculate the parameters for each replenishment mode.

At this stage of the process, the IRL and Buffer Management modes, which determine how replenishment signals are generated, have been parameterized based on the five stock components. In IRL mode actual inventory is compared with a defined target stock, which is the sum of all five stock components. In Buffer Management, the stock components are divided into three zones, resulting in a reorder point and minimum and maximum inventory levels.

It is important to understand that, so far, we have introduced only single-stage concepts. Only one Rhythm Wheel and only one stock-keeping point have been considered at a time. We have not considered the impact of a certain Rhythm Wheel design or replenishment mode parameterization on upstream and downstream supply chain stages, performing only isolated optimizations. This delivers good results, but bringing effective end-to-end supply chain management to your company requires a broader perspective. The next section addresses the multistage synchronization of production and replenishment parameters along the supply chain.

6.3 SYNCHRONIZE PARAMETERS TO ACHIEVE AN END-TO-END LEAN SUPPLY CHAIN

Synchronization is the time-wise and quantitative coordination of events for operating all stages of a system in unison. If the system is a supply chain or supply network, synchronization means coordinating supply chain operations such that the entire system can work in the same rhythm or takt. In LEAN SCM, synchronization is the process of aligning parameters along supply chain stages, such that the overall performance of the system is increased to achieve a high service level, smooth material flow, and low inventory levels, which enables your company to deliver the right product at the right time at the right place in the right quantity.

To help your company appreciate the benefits of LEAN SCM, we distinguish three levels of supply chain maturity, which differ in degree of global coordination and synchronization at various stages along the supply chain (see Figure 6.29). We call the first level the "uncoordinated supply chain" which consists of a series of independently operating planning stages. Each supply chain stage plans and optimizes its production parameters locally in isolation, based on individual, stage-specific forecast input, ignoring any effects of such planning on upstream and downstream operations. With successive processing and propagation of information at each stage, variability is amplified from one stage to another—creating the well-known bullwhip effect. This maturity level is typically managed and coordinated through site-specific MRP-generated planning runs in today's ERP systems.

We call the next level of supply chain maturity the "aligned push supply chain." In this case, local production plans are derived from consolidated forecasts and propagated simultaneously to all supply chain stages in parallel. In global organizations, this is achieved through the implementation of advanced planning systems (APS), integrating the local MRP-generated views into a global supply chain visibility layer. This capability of propagating demand through a single global planning platform prevents demand variability from being amplified, but even in the best-case variability remains at the same level along the entire supply chain. Although this maturity level represents progress in terms of coordination and variability management, tactical parameters are still determined locally and replenishment is based on push principles.

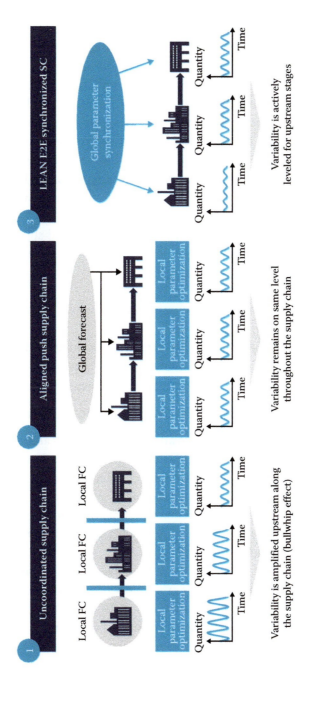

FIGURE 6.29

Three maturity levels in the evolution of supply chain coordination.

The highest level of supply chain maturity—the ideal state—is the "LEAN end-to-end synchronized supply chain." At this level of maturity, LEAN supply chain parameters such as cycle times and inventory target levels are synchronized globally and then propagated to each supply chain stage. Such global LEAN SCM coordination of tactical parameters leads to synchronized material flows along the supply chain, which reduces inventories and total end-to-end lead time. Moreover, through Rhythm Wheel planning in manufacturing operations, variability propagation is actively reduced. In this case, the designed cycle-time boundaries of Breathing Rhythm Wheels function as variability "funnels" and only leveled demand signals are propagated to the upstream stages. The leveling of production quantities and capacity utilization increases output and lowers the cost of production.

In this section, we show how to synchronize the tactical supply chain parameters. In LEAN SCM, two types of parameters need to be aligned—production and replenishment parameters:

- We investigate how production and replenishment parameters are synchronized using a global takt.
- We describe opportunities to further reduce working capital by synchronizing stock components along the supply chain.

6.3.1 Synchronize Supply Chain Cycle Times to a Global Takt

The goal of takt synchronization is to allow material to flow along the supply chain as smoothly as possible. This means that products flow along the supply chain without piling up between supply chain stages or waiting in inventories. Waiting caused by unsynchronized production is waste and should be eliminated. A steady and even material flow reduces end-to-end lead time, which reduces stock levels and makes it possible to react more quickly to customer demand.

Let us consider a boat-racing team as a metaphor that can easily explain the effect of takt synchronization. If every individual member of the boat team rows in isolation, all members row at a slightly different tempo. This lack of synchronization leads to a poor result in a boat race. By watching and feeling the rhythm of other team members, athletes automatically adjust their speed to coordinate with the rest of the team. This synchronizes their efforts and produces far better results. If the crew includes a coxswain,

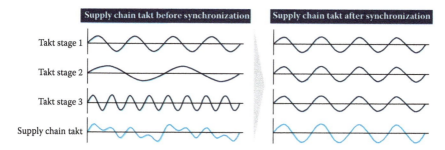

FIGURE 6.30
A global takt synchronizes supply chain stages.

all the rowers adhere to his or her takt, which further synchronizes the racing team and guides it to success.

This metaphor can be easily transferred to a supply chain. If all stages along the supply chain work independently in isolation, the result is an unsynchronized supply chain takt (see left side of Figure 6.30). To synchronize the supply chain, a global takt to which all stages adhere is necessary (see right-hand side of Figure 6.30). Such synchronization can be implemented, for example, by a global supply chain organization that controls the global takt.

In this section, we introduce an effective approach to synchronizing production and replenishment parameters to takt (see Figure 6.31). First, we describe the key factors that influence the determination of a global takt and the trade-off between cycle and inventory costs. Then, we explain the need for aligning the global takt with the bottleneck stage in the supply chain. Finally, we explain the synchronization of the supply chain parameters to the global takt.

FIGURE 6.31
Approach to synchronizing production and replenishment parameters along the supply chain.

6.3.1.1 Determination of Global Takt

A global takt is the pace at which supply chain stages control their material and information flows. A takt is always understood as the time interval between two identical operations. This could be the time between two production runs of a specific product or the time between two transports from or to the same location. The optimization of a global takt involves a trade-off between cycle and inventory costs in light of technical constraints. Figure 6.32 shows the main factors that influence cycle and inventory costs.

Cycle costs arise for actions such as transports and changeovers and are negatively affected by a fast takt. The faster the takt of transports and changeovers, the higher the cycle costs, since more transports and changeovers occur. On the other hand, inventory costs are positively affected by faster takt times. The more often a stockkeeping point is replenished, the fewer stocks have to be held because the time interval between two replenishments is shorter so the stocks have to cover lower demand between replenishments.

However, before setting a global takt, the possible constraints caused by bottlenecks in the supply chain must be considered, since they determine the fastest possible takt.

6.3.1.2 Bottlenecks as Pacemakers of a Supply Chain Takt

When creating flow along the supply chain, all supply chain stages have to produce in takt. But before creating the takt for the supply chain, one very

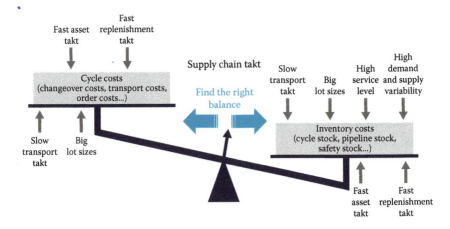

FIGURE 6.32
The optimal global takt is derived as a trade-off between cycle and inventory costs.

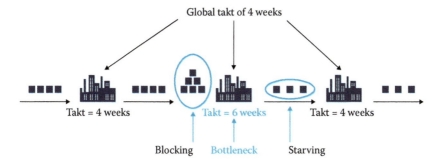

FIGURE 6.33
Supply chain bottlenecks must be considered to achieve smooth material flow.

important constraint has to be considered, namely bottlenecks in the supply chain. Supply chain stages with insufficient capacity might not be able to produce in takt. The slowest of all supply chain stages is the bottleneck which, if we return to the metaphor of the boat racing team, is like the slowest rower in the boat. The boat can move through the water only as fast as the weakest athlete can row. Similarly, a supply chain takt can only be as fast as the worst bottleneck along the chain. In other words, the chain is only as strong as its weakest link. If upstream supply chain stages produce output more quickly than the bottleneck, inventory piles up in front of the bottleneck stage. In shop-floor terminology, this is called blocking. For downstream stages, producing output more quickly than the bottleneck over a long period of time is not possible, because material from the upstream stages is missing, which is called starving. Figure 6.33 illustrates what happens when the global takt is too fast for the bottleneck stage.

To prevent structural bottlenecks in a supply chain, capacity leveling in network design, as described in Chapter 4, is essential. After adjusting the global supply chain takt to the bottleneck constraint, the takt parameters can be synchronized.

6.3.1.3 Synchronization of Takt Parameters to a Global Takt

Multiple parameters must be synchronized to a global takt along the entire supply chain. Those parameters are linked to production assets, transport schedules, and replenishment signals. The goal is to match the asset takt, the transport takt, and the replenishment takt (see Figure 6.34). By deriving lot sizes from such a synchronized takt, it is possible to achieve smooth material flow with low inventory levels in the supply chain. We

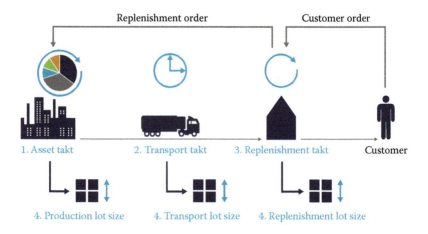

FIGURE 6.34
Takt parameters and lot sizes are synchronized to a global takt.

next explain how the production and replenishment parameters are synchronized to a global takt.

6.3.1.3.1 Synchronization of the Asset Takt

When using the Rhythm Wheel as a production mode, the cycle time determines the takt at the production asset. If the cycle time is, for example, 2 weeks, then the asset takt is also 2 weeks, since a new production run starts every 2 weeks. Insofar as the goal is to achieve the same takt along the entire supply chain, we want to achieve the same cycle time at all production assets. By implementing the same Rhythm Wheel cycle time at all supply chain stages, the production assets along the entire supply chain are synchronized.

In some cases, however, it is not possible to run all supply chain stages on the same cycle time, because Rhythm Wheel cycles at several stages can differ considerably. This might be the case, for example, if changeover times or production rates differ widely. However, synchronization can be applied if there is a cycle time multiple or ratio such that the cyclic signals share a repeating rhythm over consecutive cycles. This means that the Rhythm Wheel cycles can be chosen as multiples of each other if it is not possible to adjust them. In that case, although not all supply chain stages are producing in the same takt, the rhythms of all stages fit the global takt. Figure 6.35 shows three scenarios: in the first, the asset takt is unsynchronized, in the second it is coordinated with multiples of the global takt, and in the third it is perfectly synchronized with the same cycle times. As the figure shows, it is only in the last two cases that a repetitive rhythm is

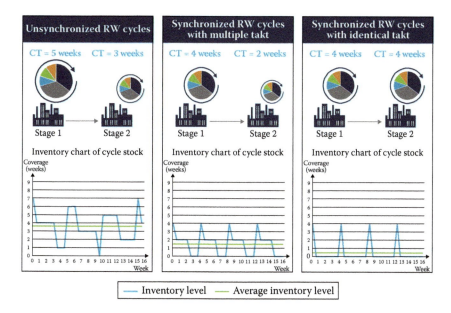

FIGURE 6.35
Identical takt reduce inventory levels along the supply chain.

achieved in the supply chain, which allows a company to reduce average inventory levels.

Despite the perfectly planned synchronization of cycle times, some safety stocks are needed to buffer delays during a cycle and to keep the synchronization upright (see Figure 6.36). Therefore, cycle time attainment

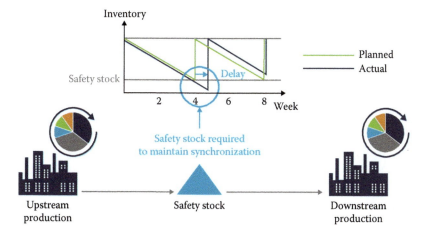

FIGURE 6.36
Safety stock is required to buffer against cycle time fluctuations.

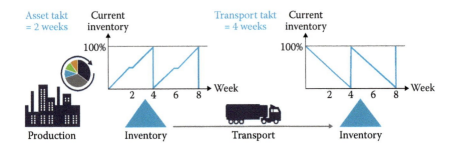

FIGURE 6.37
Unsynchronized transport and asset takt lead to high inventory levels.

and cycle time variation are two very important KPIs to monitor when measuring the synchronization, as they track how far the actual Rhythm Wheel cycle time has deviated from the planned cycle time (see Chapter 7 for more details on those KPIs).

6.3.1.3.2 Synchronization of the Asset Takt with the Transport Takt

Alignment of the asset takt with the transport takt is essential for synchronized material flow. Figure 6.37 illustrates the impact of unsynchronized asset and transport takt on inventory levels. If the takt rates differ, the benefits of a short asset takt of 2 weeks at the upstream supply stage are partly lost, because inventory is built up over 3 weeks until the material is transported to the next stage.

The full benefits of parameter synchronization can be captured in this example if both the asset and the transport component have a takt of 2 weeks (see Figure 6.38). Inventory levels at both supply chain stages can then be reduced significantly. Note that it is important not to mistake the

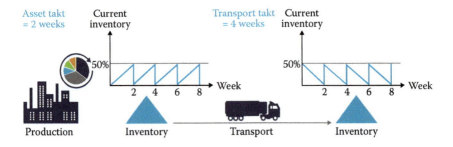

FIGURE 6.38
Synchronized transport and asset takt make significant inventory reductions possible.

transport takt with transport duration: the takt determines how often a transport takes place, while the duration states how long a transport takes.

6.3.1.3.3 Synchronization of the Asset Takt with the Replenishment Takt

The replenishment takt is the time between two orders. Those orders can come from either internal or external customers. In both cases order behavior should be synchronized with the asset takt, which means that orders for products or raw materials should be triggered in the same takt as that of the production asset.

Let us consider an example of unsynchronized asset and replenishment takt: the asset takt of an upstream supply stage is 1 week, but orders from the receiving stage occur only every month. In the first 3 weeks of the month, the supply stage receives no orders and does not require production. In the last week of the month, however, a huge order arrives, which covers demand for an entire month. The result is that no capacity is required in the first 3 weeks but it is heavily stressed in the last week of the month (see Figure 6.39).

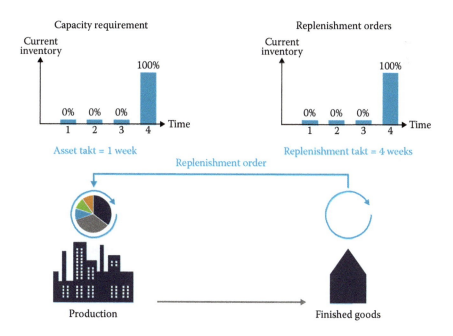

FIGURE 6.39

Unsynchronized asset and replenishment takt lead to wide fluctuations in capacity requirements.

Here, we see that unsynchronized asset and replenishment takt can lead to widely fluctuating capacity requirements. To achieve stable capacity requirements, those two parameters need to be synchronized.

6.3.1.3.4 Synchronization of Lot Sizes with Takt Parameters

In a synchronized supply chain, lot sizes need to be synchronized with the global takt as well. This is a new way of thinking for most companies, since traditionally lot sizes are determined first, and then the intervals of replenishment and production are adjusted: if lot sizes are large, order and production intervals are long; if lot sizes are small, orders and production occur more frequently. In both cases, there is no constant takt interval across supply chain stages. This leads to optimal lot sizes from a local perspective, but not from a global perspective (see Figure 6.40).

To achieve the best performance for the entire supply chain, lot sizes need to be calculated based on the supply chain takt. This ensures that the supply chain can work in takt and that lot sizes are harmonized along the entire supply chain. Even though lot sizes might be suboptimal from a local perspective, end-to-end supply chain costs are minimized from a global perspective (see Figure 6.41).

So far, we have seen that establishing and maintaining a global takt is the key to supply chain synchronization. Synchronization of production and replenishment parameters with a global takt reduces the end-to-end production lead time along the supply chain and reduces inventory levels. However, the takt is not the only component of the supply chain that should be synchronized. The same holds true for the stock components. They should be allocated more efficiently along the supply chain than is possible based on single-stage calculation. Considering interactions between

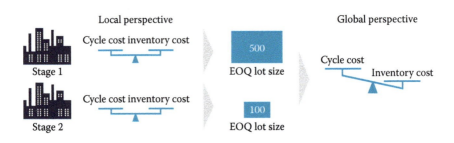

FIGURE 6.40

Local economic order quantity (EOQ) optimization leads to misaligned lot sizes in the supply chain.

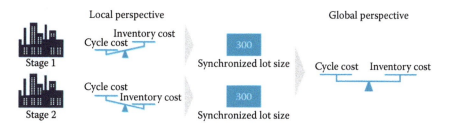

FIGURE 6.41

Lot sizes are synchronized by aligning them with a global takt.

supply chain stages shifts certain stock components either upstream or downstream.

6.3.2 Build on Dynamic Inventory Target Setting to Smooth Cycle Time Oscillation

Inventory target level planning is a key priority for LEAN SCM. This means that a dynamic inventory target setting process needs to be included in for actively managing variability. Remember, one part of the market variability is managed on capacities through the Breathing Rhythm Wheels, but the other part, the volatility outliers outside the designed cycle time boundaries, has to be absorbed through the "planned (safety) stocks" in the supply chain. Dynamic adoption of the inventory target levels throughout the planning process is a buffer mechanism to smooth the Rhythm Wheel cycle time oscillation on the manufacturing assets and to maintain takt. Therefore, inventories—including the demand and supply safety stocks— have to be actively used in supply chain planning to mitigate the effect of demand and supply variability and to enable synchronization of end-to-end product flows. Compared to the conceptual approaches of ERP and APS technology in today's planning practice, LEAN SCM aims at a paradigm change in supply chain planning. Therefore, to realize the vision of LEAN SCM, the systematic synchronization of stock components must be actively addressed from a global end-to-end perspective. This is achieved best by multistage inventory optimization, including stock allocation and the right target level setting, as we will explain and demonstrate in this chapter.

Traditional single-stage inventory planning optimizes stocks locally at each stage or location along the supply chain. As a consequence, each stage of the supply chain holds more stocks than are required. Significant

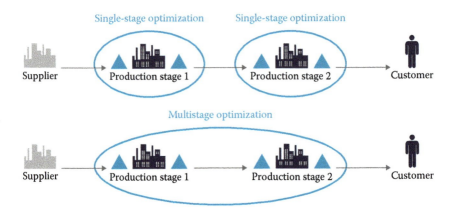

FIGURE 6.42

Multistage inventory optimization considers all inventories along the supply chain simultaneously.

benefits can be realized by jointly optimizing the inventories along the whole supply chain, which takes into account dependencies between stages and leads to optimal inventory allocation. Multistage inventory planning approaches can typically generate a 15–25% inventory reduction while simultaneously preserving or even increasing service levels. The key difference between single- and multistage inventory optimization is illustrated in Figure 6.42.

The multistage inventory optimization approach addresses three major issues, which we discuss in this section:

1. Where to hold stocks in the supply chain
2. How to right-size inventory positions
3. How to balance the four effects impacting efficient safety stock allocation

6.3.2.1 Where to Hold Stock in the Supply Chain

The first question to be answered through multistage inventory optimization is where stock should be kept in the supply chain. Do you need stock at every supply chain stage or can you consolidate stock for several stages? And if stock is required at a certain stage, do you need inbound and outbound stocks, or only one of these two? The answer to these questions depends on the operating replenishment mode, as we have shown in Chapter 5. If, for example, an MTO replenishment mode is chosen, customer orders are produced only when requested and are not delivered from outbound stock. Therefore, outbound stock is not required in MTO mode. If a JIT concept is used, suppliers have to deliver the required pre-materials to the production site precisely when they are needed. In this case, no inbound stock is required at the site. Following the concept of vendor-managed inventory (VMI), the supplier is in charge of refilling the inbound inventory at a customer site. With VMI the supplier does not need to keep outbound stock, because production can be coordinated with replenishment at the customer site. As these examples show, whether or not holding stock is required depends heavily on the underlying replenishment mode.

6.3.2.2 How to Right-Size Inventory Positions

When the stockkeeping points along the supply chain have been defined, the next step is to right-size the various stock components from an end-to-end perspective. We have explained single-stage sizing of stock components in Section 6.2. From a multistage perspective, which takes interactions between supply chain stages into account, those calculations differ slightly.

6.3.2.2.1 Cycle Stock

In a synchronized supply chain, cycle stock is determined by the global takt rather than by a single-stage trade-off between cycle and inventory costs (see takt synchronization in Section 6.3.1). In case of a slow takt rate, cycle stock needs to be high, since it has to cover demand over a long period. If the takt is faster, replenishment and production occur more frequently and cycle stock can be reduced. Adjusting cycle stock to the global supply chain takt ensures that there is sufficient stock at each supply chain stage to cover the time interval between two replenishments.

6.3.2.2.2 Policy Stock

Policy stock, which is held in preparation for exceptional demand and supply events, does not need to be held at every supply chain stage. If, for example, only a few critical suppliers of raw material exist and the probability of supply shortages is relatively high, policy stock should be kept upstream in the supply chain. In case one of the key suppliers cannot deliver the required raw material, the rest of the supply chain can still conduct business normally. If raw material supply is not an issue, but production and delivery processes along the supply chain face possible interruptions due to events such as political unrest or regular natural disasters, policy stock should be kept downstream as well. If one link in the middle of the supply chain is cut off, the downstream production processes can still produce output until the broken link is restored.

6.3.2.2.3 Safety Stock

One of the key challenges in inventory management is the global determination of safety stock along the supply chain. A distribution center, for instance, needs safety stock to cover demand and supply uncertainty. Supply uncertainty in turn depends on the level of safety stock at the upstream production site. If the production site increases its safety stock, the distribution center can reduce its own safety stock, and vice versa. Furthermore, safety stock is the key to buffer variability and to achieve stable cycle times at the production assets—safety stock is required to smooth the Rhythm Wheel cycle time oscillation and maintain the asset takt. Therefore, allocating safety stock optimally allows a company to significantly reduce the amount and costs of inventory for attaining the target customer service level. To understand why multistage optimization leads to better results, it is useful to recognize four effects on optimal safety stock allocation.

6.3.2.3 How to Balance the Four Effects Impacting Efficient Safety Stock Allocation

Typically, safety stock represents a great opportunity for inventory optimization, because for certain products they account for more than 50% of all stock. Despite this large impact on total inventory costs, safety stock has received little attention in the past. However, we observe many companies beginning to redirect their optimization efforts to safety stock.

The key question is typically whether the required inventories should be held upstream or further downstream in the supply chain. For cost-efficient placement of safety stock, four effects must be considered: The pooling effect, the value effect, the lead time effect, and the service level effect. Figure 6.43 illustrates the trade-off between these effects. Advanced multistage optimization tools (see Chapter 10) are capable of identifying the optimal configuration by taking into account all four of these effects

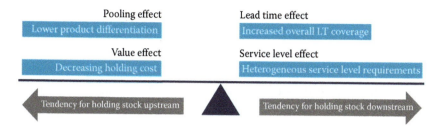

FIGURE 6.43

Optimal multistage safety stock allocation as a trade-off between four effects.

as well as additional constraints such as storage capacities and customers' required lead times.

6.3.2.3.1 Pooling Effect

Process industry supply chains typically exhibit a large increase in the number of SKUs toward downstream stages due to global distribution structures and strong product differentiation. For supply chains with such divergent material flows, the overall inventory level can be reduced by relocating safety stock to upstream stages, as depicted in Figure 6.44. As short-term demand fluctuations from the various downstream stages are typically balanced out at least in part, lower safety stock is sufficient for the joint buffering of variability from downstream SKUs. This effect is known as demand risk pooling. The impact of the demand pooling effect depends on the number of downstream SKUs that can be jointly buffered,

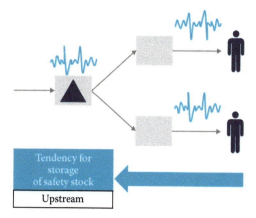

FIGURE 6.44

The pooling effect favors holding safety stock upstream.

the level of demand variability at the downstream stages, and the correlation between these demands.

6.3.2.3.2 Value Effect

Because the total costs associated with stocks along the supply chain are linked directly to the average holding cost per unit, it makes sense to consolidate safety stock at stages at which holding costs are low. Holding costs are determined mainly on the basis of the value of held stock, which generally increases along the stages of the value-added process. As shown in Figure 6.45, the value effect therefore suggests keeping safety stock at upstream stages, where capital costs are lower. In fact, multistage inventory optimization typically leads to partial consolidation of safety stock at upstream stages that precede stages that feature significant increases in product values.

6.3.2.3.3 Lead Time Effect

The lead time effect describes what is also known as "risk pooling over time." The purpose of holding safety stock is to cover fluctuations in supply as well as demand variability for the expected replenishment takt and lead time. Hence, the volume of safety stock required is directly linked to the time that passes between the trigger of a replenishment order and the availability of the respective material. However, because short-term demand and supply fluctuations are usually balanced out over time, the required volume of safety stock does not increase linearly with the replenishment time. When, for example, the time it takes to replenish an inventory doubles, significantly less than twice the volume of safety stock is required to provide the same level of customer service. Consequently, comparatively lower safety stock is required if stock buffers are consolidated at downstream locations in a multistage supply chain, as illustrated in Figure 6.46.

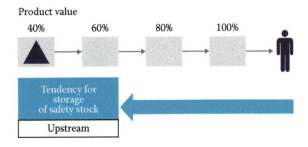

FIGURE 6.45
The value effect favors holding safety stock upstream.

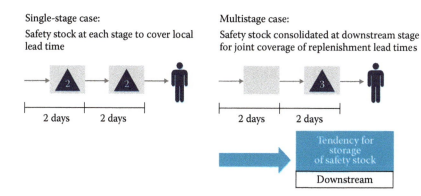

Single-stage case:

Safety stock at each stage to cover local lead time

Multistage case:

Safety stock consolidated at downstream stage for joint coverage of replenishment lead times

FIGURE 6.46

The lead time effect favors holding safety stock downstream.

6.3.2.3.4 Service Level Effect

When adopting a systematic inventory planning approach, the size of the required safety stock buffers is determined based on observed variability and the target service level. Clearly, end-customer service levels may not be the same for all downstream stages. If heterogeneous service levels have to be taken into account, centralizing safety stock can be unfavorable, as the use of a centralized inventory leads to rather complex inventory control policies or unnecessarily high inventory levels. Therefore, it is generally beneficial to buffer high service requirements locally to manage heterogeneous service levels (see Figure 6.47).

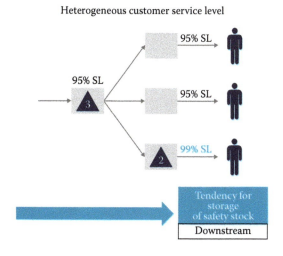

FIGURE 6.47

The service levels effect favors holding safety stock downstream.

BOX 6.7 CASE STUDY: MULTISTAGE INVENTORY OPTIMIZATION IN A PHARMA BLOCKBUSTER SUPPLY CHAIN

The supply chain of a top-selling branded pharmaceutical company, with multi-billion-dollar annual revenue figures, was included in a comprehensive inventory optimization project. One objective was to identify the optimal allocation of safety stock along the supply chain. As service levels were already high, the primary focus of the project was on reducing working capital.

As is typical in the pharmaceutical industry, the company's supply chain operations were highly globalized, with significant product fan-out at downstream stages. Owing to production scheduling, time-consuming clearance processes, and the geographical separation of the production sites, lead times were rather long—more than 10 months in total. Overall, 95 unique SKUs at eight inventory locations were taken into account. Although an advanced inventory planning process was already in place, safety stock was spread along the entire supply chain due to applying the single-stage approach.

As a result of multistage optimization, safety stock was consolidated at a small number of stages along the supply chain. Stock buffer of finished products at the packaging sites were required to ensure off-shelf availability. In addition, safety stock was consolidated at the inbound sites of the API production and formulation. The total working capital tied up in safety stock was reduced by 19%.

We illustrate how a blockbuster pharmaceutical supply chain balanced the four effects of optimal safety stock allocation in Box 6.7. The multistage optimization approach led to a total decrease in inventories of 19% by optimally allocating safety stock in the supply chain.

Summary

Synchronization is the time-wise and quantitative coordination of events that makes it possible to operate a system in unison. The purpose of synchronization is to enable production at lower cost and reduce working capital by harmonizing the tactical parameters across all supply chain stages. Stable material flow levels capacity utilization and reduces the total lead time for a product along the end-to-end supply chain. For true synchronization, adopting an end-to-end perspective is crucial, since optimization of single-stage problems does not lead to global multistage optimization of the supply chain.

In LEAN SCM, two types of parameters need to be synchronized: production parameters and replenishment parameters. These parameters are synchronized to a global supply chain takt. The determination of the global takt reflects a trade-off between cycle and inventory costs, which is similar to the trade-off that must be negotiated at a single stage. Moreover, the bottleneck stage of a supply chain has to be considered as a constraint, since it determines the fastest possible global takt.

Following synchronization with the global takt, the stock components of the replenishment parameters must be adapted. Stock can be allocated efficiently along the supply chain to smooth the Rhythm Wheel cycle time oscillation and maintain takt if a global perspective is taken. Not all supply chain stages need to carry equal stocks, but stocks can be distributed intelligently to cover supply and demand risks.

Since demand and supply conditions along the supply chain may change, synchronized production and replenishment parameters need to be reviewed periodically. Synchronization achieves the desired benefits only if the tactical parameters are up to date. The next section describes the tactical renewal process by which all parameters are refreshed to fit current supply chain conditions.

6.4 THE TACTICAL RENEWAL PROCESS TO PARAMETERIZE LEAN SUPPLY CHAINS

Conditions along the supply chain change frequently. Changes in demand patterns, product allocations across production sites, resource availability, and so on all have substantial impacts on supply chain performance. In LEAN SCM, the tactical supply chain parameters preconfigure the supply chain and provide the framework for smooth execution. Therefore, companies must ensure that the production and replenishment parameters are always up to date. Regular renewal of parameters ensures the best match with current supply chain conditions and enables fluent execution.

The tactical renewal process is the process of reviewing and renewing the tactical supply chain parameters, comprising a predefined set of tasks and meetings in which key stakeholders from various business units in a company agree on the optimized parameters in light of internal and external factors. Tactical renewal covers all relevant areas of production and replenishment planning, such as setting and synchronizing production and replenishment parameters. The tactical renewal process is structured into three main phases: the input phase, the renewal phase, and the output phase (see Figure 6.48).

In the input phase of tactical renewal, all information that is relevant to determining the tactical supply chain parameters is gathered from the strategic, tactical, and operational levels. In the renewal phase, the production and replenishment parameters are computed based on the input information that has been gathered and synchronized across the supply chain. In the output phase, the tactical parameters are communicated to the relevant organizational stakeholders and then released in the company's planning-related IT systems.

It is important to regularly review and adjust the tactical parameters to current conditions along the supply chain. How frequently it is necessary to run the tactical renewal process depends on the dynamics of a given supply chain. In a very dynamic environment, product allocations to production sites, supply conditions, and demand patterns might change rapidly. In such cases, the tactical renewal process should be conducted at short intervals, as frequently as monthly. If the supply chain environment

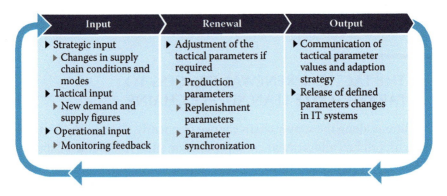

FIGURE 6.48
The tactical renewal process consists of three main phases.

is relatively stable, the tactical renewal process can be conducted less often. A quarterly review might be sufficient. Generally speaking, the more volatile the supply chain conditions are, the more frequently the renewal of tactical parameters should be conducted. However, if a sales and operations planning process is established, it is recommended to review the parameters in the same frequency.

In addition to renewing tactical parameters on a regular basis, certain events may also trigger the tactical renewal process, since some events require immediate changes in the production and replenishment parameters. Typically, such events are difficult or impossible to predict. An example of an event that requires tactical renewal is a natural disaster, such as the eruption of the volcano with the unpronounceable name Eyjafjallajökull (or E15 for simplicity) in Iceland in spring 2010. Volcanic ash disrupted air traffic across Europe, which led to a significant change in supply conditions.

In the following sections, we describe:

- What information base you need during the input phase.
- How to renew parameters in the renewal phase.
- How to align the parameters in the output phase.
- Who in the organization is involved in maintaining the parameters for a LEAN supply chain?

6.4.1 What Information Base You Need

The first phase of the tactical renewal process addresses the collection, consolidation, and validation of data and information that is relevant to adjusting parameters. Input for the tactical renewal process is gathered from all relevant business perspectives. In this way, the tactical renewal process functions as an essential link between strategic, tactical, and operational LEAN SCM (see Figure 6.49). The input phase ends with an evaluation of current performance to assess the need for parameter adjustments.

Input from the strategic level is essential for successfully configuring supply chain parameters. Strategic input can include, for example, changes in product flows along the supply chain or changes in the allocation of products to production assets. New transportation modes, for example a change from air to sea shipping, require adjusting lead times, which in turn affects the production and replenishment parameters. Also, decisions made in the strategic renewal process—for example, a change from kanban to the Rhythm Wheel—require a completely new set of tactical supply chain parameters.

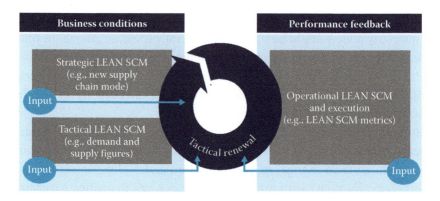

FIGURE 6.49
The tactical renewal process links various business perspectives.

Once the strategic input is gathered, tactical information that is relevant needs to be consolidated. The forward view provides demand and capacity figures for the coming planning period. Demand figures are derived from aggregated forecasts that are enriched with market intelligence in light of demand trends and seasonality. Available capacities along the supply chain need to reflect special supply events such as Christmas or summer vacation periods. This input may for example come from an already established sales and operations planning (S&OP) or similar process.

Operational LEAN SCM provides input mainly in the form of a monitoring feedback loop that enables a company to analyze and evaluate past performance of established supply chain parameters. The gathered insights influence the determination of the parameters for the next planning period. If, for example, the Rhythm Wheel cycle time constantly hits the maximum cycle time boundary, then the cycle time parameters should be chosen differently. If unused inventory is piling up at stockkeeping points, the replenishment parameters can be lowered. For more on the relevant production and replenishment KPIs for analysis, see Chapters 7 and 9.

After completing the data- and information-gathering activities, it is crucial to validate the obtained input and ensure consistency by checking for data errors.

6.4.2 Establish Regular Renewal of Planning Parameters

In the renewal phase, the tactical supply chain parameters are revised based on the strategic, tactical, and operational input described above. The renewal phase begins with a decision as to which production and replenishment parameters need to be adjusted and which parameters can remain at their previously defined values. This decision is based on consolidated information from the input phase and the subsequent evaluation of supply chain performance during the previous planning period. Then, production and replenishment parameters are adjusted and synchronized (see Figure 6.50). For special demand and supply events, production and replenishment parameters additionally need to be adapted over time. In this case, a dynamic adaptation strategy is defined for the tactical parameters.

Determining production parameters depends on the underlying production mode, as described in Section 6.1. If the Rhythm Wheel is chosen as the production mode, the specific Rhythm Wheel parameters, like sequence and cycle time, are maintained. The task of configuring Rhythm Wheels is greatly supported by a Rhythm Wheel design tool, which is integrated in the company IT system landscape.

Replenishment parameters are determined in accordance with the chosen replenishment mode: IRL or Buffer Management. Stock components are calculated as described in Section 6.2. A stock parameter configurator tool should calculate the stock parameters with all data integrated (see Chapter 10).

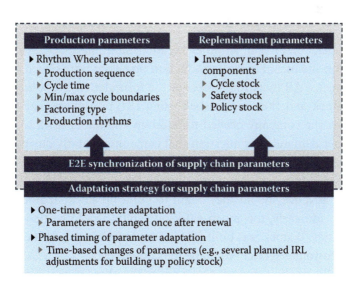

FIGURE 6.50

Production and replenishment parameters are adjusted, synchronized, and adapted over time.

To achieve true end-to-end synchronization of production and replenishment parameters along the supply chain, the parameters need to be determined globally as described in Section 6.3. Material flow along the supply chain is synchronized by aligning the Rhythm Wheel cycle times with the global takt. Safety stock is optimized by intelligently allocating it along the supply chain using a multistage inventory optimization approach.

In addition to determining production and replenishment parameters, a strategy for adapting parameters to new conditions must be specified. An adaptation strategy determines how far in advance and in how many steps changes in production and replenishment parameters are implemented. This can either be a one-time adjustment of the new tactical parameters or a phased timing of changes. A phased timing is required, for example, if special demand events dramatically increase demand in the near future, or if extraordinary supply events hinder production and require a buildup of stock. Box 6.8 describes how various adaptation strategies were developed by a pharmaceutical manufacturer. In this example, the production site had to build up stock to cover demand during a 2-week summer vacation, in which production was closed. On the one hand, the company wanted to

---

BOX 6.8   PREPARING FOR SUMMER VACATION AT A PHARMACEUTICAL FORMULATION SITE

As happens every year, summer vacations at a pharmaceutical formulation site meant that all production processes were closed down for 2 weeks. All incoming orders at the formulation site were filled from stock and then replenished using IRL pull mode. For those 2 weeks, the company needed to build up additional stock to meet customer demand. The question that came up regularly during this period was how to prepare for the plant shutdown. How much stock was needed to cover demand during the shutdown? And how much time in advance should they begin to build up stock? If stock was built up too early, too much working capital would be tied up in inventories. If stock was built up too late, expensive overtime and a loss of customer service would be the consequence.

To build up the required stocks optimally, three parameter adaptation strategies were simulated. The first strategy began the stock build-up very early, more than 6 months before the shutdown. The

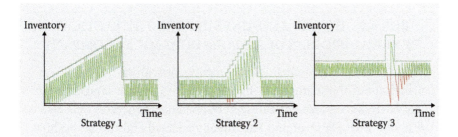

FIGURE 6.51
Evaluation of strategies to prepare for a plant shutdown.

IRL was raised over many small steps, such that production quantities were increased only marginally. The second strategy began the build-up of stock 3 months in advance. The IRL then needed to be adapted in only a few steps in order to reach the required stock level to cover demand during the shutdown. The third strategy began the build-up only 1 month before the 2 weeks of vacation. The IRL was adapted in only one large step (see Figure 6.51).

Strategy 1 provided the benefits of a smooth buildup. Upstream supply chain stages were nearly unaffected and service levels were easily maintained. Yet because stock was built up very early, huge amounts of working capital were tied up and caused high inventory costs. Strategy 3 on the other hand required almost no additional inventory cost, but resulted in a loss of service, because capacity was not sufficient to cover such an increase in production quantities for all products. Strategy 2 was judged to be most likely to balance the inventory cost and service levels efficiently. Additional inventory costs involved in the shutdown were 59% lower than they would have been under strategy 1, and the achieved service level was 99.5% which is much higher than it would have been under strategy 3.

start building up stock very late, since inventory levels should be kept low; on the other, customer service was not to be jeopardized, which favored starting comparatively early.

We cannot stress often enough that IT systems improve the renewal phase and reduce effort considerably. In Box 6.9, you can read how a global chemical company aims to improve its planning processes, and especially the tactical renewal process, by the means of IT add-ons.

BOX 6.9   HOW IT ADD-ONS SUPPORT THE TACTICAL RENEWAL PROCESS OF A GLOBAL CHEMICAL COMPANY

The global specialty chemicals company providing this case was proud about the maturity of its supply chain system landscape. It had harmonized and standardized the business processes on a central SAP ERP platform and had already enabled end-to-end visibility through the SAP SCM global planning solution. However, they did not stop with this achievement and aimed to further improve the supply chain performance. A major "Oscar" SCM program was kicked-off to improve agility towards the increasing market volatility and to increase reliability to their customer commitments. The program was very much focused on production and supply chain planning practices, with the intention of better leveling the material flow and capacity utilization, and to change the attitude towards inventory usage and risk-buffering in the planning process. LEAN SCM Planning principles have been applied and first performance improvements could be reached quickly. However, the tactical renewal process still lacked efficiency. The key challenge referred to the pre-configuration and renewal of Rhythms Wheels and dynamic adjustments of IRL across the supply chain. In most cases, it was time consuming, since it was handled either manually or with the help of locally developed Excel solutions. Soon, the organizational cry for further IT support pushed the company to invest into additional SAP SCM Add-ons. (see Figure 6.52).

On local plant level, first the *Rhythm Wheel Designer* has been added to the SAP SCM PP/DS application. It supports the planner in preconfiguring the *product sequence* (grades) and manufacturing quantities for the *designed cycle times* of the Rhythm Wheel. This data is later on joined with the actual pull replenishment signals during the local *LEAN Planning Heuristic* run, to generate only consumption-based production orders. To ensure that market volatility, passed through the replenishment signals into operations, is not propagated unfiltered to the manufacturing lines and even further upstream in the supply chain, the concept of cycle time boundaries is used. Therefore, the *LEAN Factoring Heuristic* has been added to the standard SAP SCM PP/DS functionality to ensure that cycle time boundaries are adhered

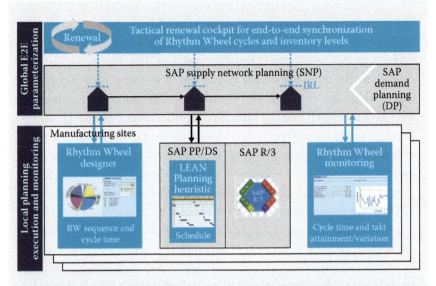

FIGURE 6.52
LEAN SCM Planning IT architecture of a global chemical company.

to. With the operational short-term factoring application, all Rhythm Wheel-included production demands are trimmed, when necessary, to fit into the designed boundaries. This reduced the variability in capacity utilization significantly and aligned the takt along the supply chain. The factoring functionality finally enables the flexibility to react faster to short-term supply disruptions and market needs.

But there are obviously permanent deviations between the forecast-based Rhythm Wheel design parameters and the later executed, con-sumption-based Rhythm Wheel production schedule. Monitoring of acceptable, minor deviations as well as unacceptable, major devia-tions, above the defined parameter boundaries, had to be automated to ensure that the LEAN SCM Planning concept is accepted in the company. The added *Rhythm Wheel Monitor* provides the function-ality for the real-time parameter oscillation control such as *cycle times*, run to target, capacity utilization and *inventory replenishment level (IRL)* development. This feature allows the planner to evalu-ate continuously to which extent he really adhered to the optimized preconfigured Rhythm Wheel set-up of the manufacturing line. The same local parameter monitoring data—cycle time oscillation and

IRL trend development—are passed to the global supply chain level to allow for end-to-end supply chain synchronization.

On a global level, the company considers the *Tactical Renewal Cockpit* as the most important IT add-on to manage the end-to-end supply chain synchronization. On the one side, the dynamic inventory target setting calculations for *the IRL-levels* need to be supported by the capability of a multi-echelon optimization. On the other side, the inventory calculations have to be aligned with the *multistage cycle time configurations.* The Tactical Renewal Cockpit provides decision support with regard to this end-to-end parameter configuration (parameterization) of the global supply chain by defining and synchronizing IRLs and cycle times across the supply chain network. It also indicates how to adjust the parameter configuration in case of supply chain events such as planned plant shutdowns or large tenders. Furthermore, simulation features have been embedded to provide the opportunity to evaluate the impact of different scenarios on the future supply chain performance. The integration of the cockpit to SAP DP and SAP SNP was important to the company because it kept manual intervention at a minimum level.

By completing its existing SAP system landscape, the company has significantly improved its tactical renewal process in terms of time and quality. Especially, the improved capability for an end-to-end supply chain parameterization is considered as outstanding achievement. Based on this remarkable step forward, the chemical manufacturer is confident to seize even more benefits during its ongoing *LEAN SCM journey.*

The renewal phase ends with the organizational confirmation of both adjusted parameters and parameters that have not been changed during renewal. It is important to assign clear lines of responsibility for each set of parameters. The responsible person (or role) should sign off on the agreed-upon values for the parameters for which he or she is accountable. In most cases, only a small subset of parameters needs to be modified during the renewal phase. Therefore, it is important to confirm the values of parameters that have not been altered across the organization. This avoids problems caused when several planning environments exist and no one really knows when planning parameters have been finally reviewed or changed.

6.4.3 Alignment of Planning Parameters for the LEAN Supply Chain

Following the renewal phase, all production and replenishment parameters have been determined, synchronized, and given a defined adaptation strategy. In the output phase of the tactical renewal process, two actions need to be taken:

- Communication of results to the relevant stakeholders
- Release of the renewed parameters via the company's IT systems

6.4.3.1 Communication of Renewed Parameters

Effective communication of renewed parameters is as important as determining them in the first place. To win acceptance of such changes within a company, it is essential for employees to understand why the tactical parameters have been changed. Furthermore, explaining the parameter adaptation strategy can create awareness of exceptional events that will occur in the near future, and help relevant stakeholders understand how the company plans to prepare for them. During this process of communicating a change in tactical parameters and explaining the company adaptation strategy, valuable stakeholder feedback should be collected and used as input for the next tactical renewal process. This helps to continuously improve supply chain performance.

6.4.3.2 Release of Renewed Parameters in the IT Systems

In parallel, the confirmed tactical parameters are released via the company's IT systems. The affected data fields are updated to guarantee frictionless and automated execution. By releasing the tactical parameters via the IT systems, up-to-date information is made globally available and aligned across the entire organization. Supply chain units can then trust the confirmed parameters and schedule their operations accordingly. Detailed information about the implementation of LEAN tools in the existing IT landscape can be found in Chapter 10.

Figure 6.53 summarizes the steps taken during the tactical renewal process. The figure shows the key activities and decisions involved in each of the three phases as well as links to other planning processes in the supply chain such as the strategic renewal process and S&OP processes. The three phases—input, renewal, and output—are depicted on the left, while the

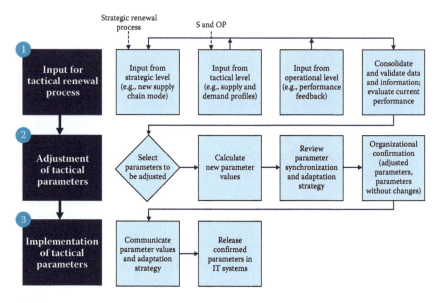

FIGURE 6.53
Overview of the process flow involved in the tactical renewal process.

various tasks involved in the process are shown on the right. Each of the tasks is carried out by at least one organizational role. The roles that participate in the tactical renewal process are described in the next section.

6.4.4 Who Is Involved in Keeping the Supply Chain LEAN through Synchronized Parameters?

In the tactical renewal process, three distinct roles deliver input to the process, make decisions regarding the renewal of parameters, and implement and communicate the relevant decisions to the relevant stakeholders. These roles are those of the supply chain planner, the local planner, and the market planner (see Figure 6.54). While the supply chain planner's view focuses on the end-to-end supply chain and value stream, ensuring global optimality of lean supply chain planning, the local planner's tasks center on a single site and a number of (manufacturing) assets. To guarantee that the supply side matches the market and demand requirements, the market planner acts as an interface between operations and the supply chain on the one hand and the marketing organization within the planning processes on the other. All three roles contribute unique perspectives and deliver information from separate business units. Although all three

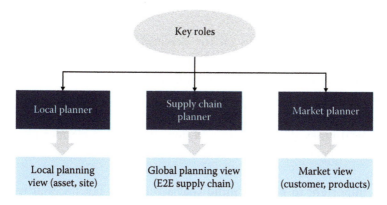

FIGURE 6.54
Overview of the roles participating in the tactical renewal process.

pursue specific goals, together they contribute to the overall success of the tactical renewal process.

6.4.4.1 Supply Chain Planner

The Supply Chain Planner represents the supply chain in the tactical renewal process. He or she assumes global responsibility over stocks and service levels and contributes the end-to-end perspective of the supply chain, typically for at least one product family. The supply chain planner leads the tactical renewal process and coordinates alignment with S&OP or similar processes and activities. In the tactical renewal process, the supply chain planner supports the renewal of tactical parameters and their adaptation strategies and is responsible for synchronizing parameters across manufacturing assets as well as defining inventory targets, especially policy stock.

6.4.4.2 Local Planner

The local planner focuses on the production site, providing information about current capacity and future constraints at production assets, and supports the renewal of tactical parameters from a local perspective. It is the responsibility of the local planner to implement the agreed-upon production and replenishment parameters at the site. The performance of the local planner is measured by local service levels, fulfillment of agreed-upon inventory targets, and LEAN metrics such as cycle time attainment (LEAN metrics are described in Chapter 7).

6.4.4.3 Market Planner

The market planner focuses on the demand side, managing markets, countries, products, and customers, serving as the interface between operations and the sales and marketing organization. The market planner provides consolidated forecasts, demand variation profiles, and service levels, which have been agreed upon with the commercial units, to the tactical renewal process. In addition, the market planner provides market intelligence and information about future exceptional demand events.

It is important to keep in mind that a role does not necessarily correspond to a job position for one individual or function. A role is rather a descriptor of an associated set of tasks that are relevant to LEAN SCM. Further details on the abovementioned roles and guidelines for their organizational alignment are presented in Chapter 8.

Summary

The tactical renewal process is the framework for updating the tactical supply chain parameters. Over the course of this process, production and replenishment parameters are adjusted to current demand and supply conditions. To keep the supply chain parameters up to date, the tactical renewal process is conducted on a regular, typically a monthly, basis. Rhythm Wheel design tools and stock parameter configuration tools in the IT systems support the tactical renewal process and reduce the renewal effort significantly.

In the tactical renewal process, the first step is to gather input from the strategic, tactical, and operational levels of a company. The production parameters, such as the Rhythm Wheel cycle time, and replenishment parameters, such as the IRL, are then renewed. These are then synchronized with each other and, if necessary, an adaptation strategy to prepare for special demand and supply events is defined. Finally, the parameters and the adaptation strategy are communicated to the relevant stakeholders and released via the company's planning-related IT systems.

The various tasks are carried out through distinct roles involved in the tactical renewal process. The supply chain planner leads the tactical renewal process and is supported by the market planner and the asset planner. Since input and feedback are gathered from a range of business perspectives, the tactical renewal process has an integrating function within supply chain management and creates acceptance among the relevant supply chain stakeholders.

CHAPTER SUMMARY

In this chapter, we determined the tactical parameters of the LEAN SCM concepts. The supply chain parameters preconfigure the supply chain, allowing execution to take place automatically. This simplifies planning and scheduling, and makes it more efficient and more transparent.

In LEAN SCM, two types of parameters need to be determined: production parameters and replenishment parameters. First, we showed how to configure the production parameters for the Classic, Breathing, and High-Mix Rhythm Wheels.

Second, we explained how to set replenishment parameters, by first introducing general stock components and their functions to the supply chain. We then described how those components set the parameters for the IRL and Buffer Management replenishment modes.

To achieve the benefits of a lean end-to-end supply chain in your company, the production and replenishment parameters should be synchronized along the entire supply chain. The concept of a global takt was introduced to achieve a steady material flow. We then explained how to allocate stocks efficiently along the supply chain to reduce total inventory.

To keep the synchronized supply chain parameters up to date, we introduced the tactical renewal process. In this process, production and replenishment parameters are adjusted to the current demand and supply conditions.

After having preconfigured the supply chain parameters, in the next chapter, we show how they determine operational scheduling. We explain how the production and replenishment parameters guide execution in day-to-day business operations, and how the monitoring of LEAN SCM KPIs ensures that the processes are running as designed.

7

Operational LEAN Supply Chain Planning Execution

In this book, so far, we have explained the selection of LEAN production and replenishment modes as well as tactical parameterization and synchronization from a global supply chain perspective. The parameter-driven design of LEAN Supply Chain Planning is now complete and ready for operational planning, which is the central topic of this chapter.

In what follows, we explain how to apply and execute LEAN planning methods in day-to-day business operations. We leave the design perspective behind and enter the dynamic—and unfortunately very often unpredictable—operational horizon. This is the world in which your company operates. A host of questions arise here and are answered in this chapter:

- How are production and replenishment modes integrated and what does planning and scheduling look like in practice when applying LEAN SCM?
- How can the LEAN principle of leveled production be followed successfully in daily business operations and what tools does LEAN SCM provide to cope with and mitigate variability within the supply network?
- How can a company ensure that operational planning and scheduling follow the defined tactical design and parameterization, and, equally important, how can it identify if adjustments in strategic and tactical design are required?

Section 7.1 centers on planning and scheduling with Rhythm Wheels. It explains how a replenishment trigger report links production and replenishment modes and how the Rhythm Wheel translates demand signals into a production schedule.

Section 7.2 focuses on production leveling and explains how planning and scheduling can help to increase stability and predictability and mitigate nervousness along the supply chain. Here we show how the concept of cycle time boundaries can prevent Rhythm Wheel cycles that are either too long or too short.

Section 7.3 describes a monitoring framework for LEAN SCM that considers two key elements of LEAN Planning: design feasibility regarding parameterization of production and replenishment modes, and adherence to this design in a company's daily business. We introduce a set of performance metrics to provide your company with a basis for revalidating and adjusting the strategic or tactical configuration of its supply network. Potential root causes of deviations as well as corrective actions are explained for each of the performance metrics.

7.1 HOW TO EXECUTE PLANNING AND SEQUENCING WITH RHYTHM WHEELS

No matter which Rhythm Wheel design or replenishment mode is in place at your company, the basic concept of LEAN planning and scheduling with Rhythm Wheels is always the same. Products on an asset are scheduled in the optimized Rhythm Wheel sequence with leveled production volumes in every production cycle.

We now consider the source of information that informs the Rhythm Wheel as it drives production decisions and determines required production. This source is the replenishment trigger report, a direct link between replenishment and production. The replenishment trigger report aggregates demand signals coming from the market or downstream supply chain stages and connects the production and replenishment modes, as illustrated in Figure 7.1.

The demand information that is provided by the replenishment trigger report is then used by the Rhythm Wheel to generate an optimal changeover production schedule for both the short- and long-term planning horizons.

The following sections explain in detail how replenishment and production modes interact in operational planning and how demand signals are processed by Rhythm Wheels.

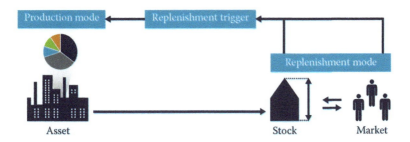

FIGURE 7.1
Interaction of the production and replenishment modes.

7.1.1 The Replenishment Trigger Report as a Link between Production and Replenishment

The replenishment trigger report is the main source of information for production and the basis for the generation of the production schedule. It displays the production decision and replenishment quantity for each of the products on the Rhythm Wheel.

A production decision determines whether or not a product has to be produced in the current production cycle. If a product is not scheduled for production in a Rhythm Wheel cycle but it should be according to the Rhythm Wheel design, we say the product is skipped. Skips might happen when there is no need for replenishment due to reduced or no consumption, such as in periods with unexpectedly low demand. If demand has occurred, the result is typically a positive production decision that is called a make decision. In this case, the product is scheduled according to the Rhythm Wheel sequence.

The replenishment quantity—in other words the volume of goods that needs to be produced based on a positive production decision—is the basis for generating the required production orders. Depending on the replenishment mode, however, there are obviously several ways to determine the replenishment quantities (see Chapter 6). Figure 7.2 illustrates how IRL, Buffer Management, MTO, and make-to forecast (MTF) can all be used to determine replenishment quantities and generate a replenishment trigger report. However, the replenishment trigger report is used with all replenishment modes to trigger production and link production modes with the report, providing the required information regarding replenishment quantities.

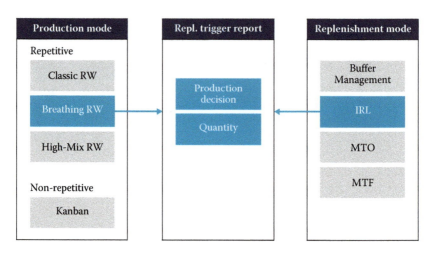

FIGURE 7.2
Linking replenishment and production modes with the replenishment trigger report.

As we have mentioned, we do not consider all of these combinations in this chapter but rather explain the operational planning process and its underlying logic only for the combination of the Breathing Rhythm Wheel and the IRL replenishment mode, as indicated in Figure 7.2. If your company uses another RW design or another replenishment mode for a given product line, the explanation provided here will guide the application of LEAN planning with such choices.

Using inventory replenishment levels, incoming customer orders are typically satisfied from supermarket stock. Supermarket stock levels are sized to cover the expected demand during an average production cycle. Additionally, safety stock is kept to buffer both supply and demand variability, minimizing the risk of stock-outs.

Production decisions are made based purely on pull. That is, they are consumption-driven decisions, based on actual customer demand. Inventory levels of supermarket stock are reviewed at certain points in time. If they fall below the defined IRL, replenishment is required and production will be triggered to refill the inventory to the target level. Figure 7.3 illustrates inventory levels and replenishment quantities for IRL replenishment over time.

While make quantities are always calculated by comparing inventory levels with IRLs, variations in planning horizons require unique solutions to determine the appropriate inventory level.

With a short-term planning horizon, production orders should cover real consumption only. Production is triggered solely by demand; in other

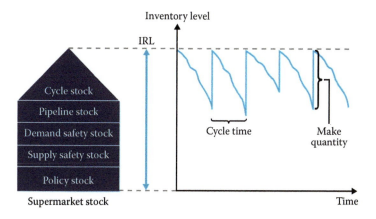

FIGURE 7.3
Behavior of inventory levels using IRL replenishment.

words, there is no production in a cycle without demand. If the inventory on hand falls below the defined IRL, production has to replenish the difference between IRL and current inventory. Otherwise, production is skipped, that is, the SKU will not be produced in the upcoming production cycle.

In contrast to what happens in short-term planning, the planning objective with a medium-to-long-term planning horizon is mainly to enable accurate procurement while balancing asset capacities. It is clear that order generation for such a horizon cannot be based on actual inventory levels. Instead, projected stock levels are used, reflecting the expected future demand and supply profiles. The projected inventories are based on a given day's current inventory levels and future expected requirements and receipts. However, the logic of order generation remains the same here as with a short-term horizon: if the projected inventory falls below the IRL, a new planned order will be generated.

It is, however, important to note that such a forecast should be used only with a medium-to-long-term horizon and never with a short-term horizon, under which production is purely pull driven. This is achieved by eliminating all forecasted elements from a production order immediately before releasing it to execution (see Figure 7.4).

Figure 7.5 shows how a replenishment trigger report looks and the information that should be included. The report displays the IRL and current inventory level for each of the five products on the Rhythm Wheel (A–E). Production decisions are indicated in the fourth column, followed by the corresponding replenishment quantity. In this example, product D was

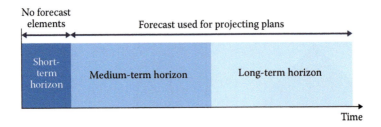

FIGURE 7.4
Forecasts are used only to project the production plan, not for execution.

Replenishment trigger report

Product	IRL	CI	Decision	Quantity
A	1000	500	Make	500
B	800	450	Make	350
C	700	400	Make	300
D	700	700	Skip	0
E	800	400	Make	400

FIGURE 7.5
Example of a replenishment trigger report with production decisions and quantities.

not consumed during the previous production cycle. As a consequence, production of product D is skipped in the upcoming cycle. However, as it is entirely possible that some consumption of D will occur before production of A, B, and C (which are prior to D in the Rhythm Wheel sequence) is finished, the replenishment trigger report needs to be updated frequently, at least immediately before a product's turn on the wheel.

Once a complete replenishment trigger report has been issued, with production decisions and quantities for every product for the next production cycle, the next step is to schedule production orders with the Rhythm Wheel.

7.1.2 Handling of Demand Signals with Rhythm Wheels

Planning and scheduling with Rhythm Wheels are generally based on the replenishment signals provided by the replenishment trigger report. Following the report's make-or-skip decisions and replenishment quantities, planned orders for the upcoming Rhythm Wheel cycles are scheduled according to the predefined Rhythm Wheel sequence.

Figure 7.6 shows the production schedule for the next Rhythm Wheel cycle for the replenishment trigger report example shown above in

Replenishment trigger report

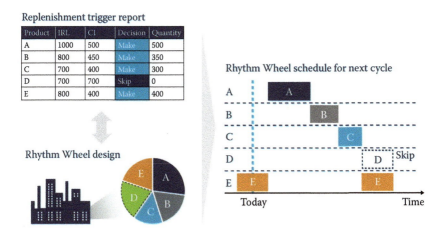

Product	IRL	CI	Decision	Quantity
A	1000	500	Make	500
B	800	450	Make	350
C	700	400	Make	300
D	700	700	Skip	0
E	800	400	Make	400

Rhythm Wheel schedule for next cycle

Rhythm Wheel design

FIGURE 7.6
Translating the replenishment trigger report into a production schedule.

Section 7.1. All products on the Rhythm Wheel except for D are to be produced in the following production cycle. According to the negative production decision in the replenishment trigger report, D will not be produced but rather will be skipped in the next cycle. The production quantities equal the replenishment quantities given in the replenishment trigger report. This means, for example, that a production order of 500 units of product A will be scheduled. The production schedule for the short-term horizon directly follows actual consumption and with IRL replenishment only quantities that have been taken out of supermarket stocks are produced.

As already mentioned, to make it possible to procure necessary components and raw materials as well as to view future-oriented capacity, planned production orders need to be created not only for the short-term horizon or for the next Rhythm Wheel cycle but also for the mid-to-long-term horizon.

The example above underlines the point that, when using a Breathing Rhythm Wheel, production quantities are typically not fixed but are dynamically adjusted depending on the replenishment trigger report. In other words, the Breathing Rhythm Wheel is strongly oriented toward actual consumption and adheres to fluctuating replenishment quantities. Therefore, if the replenishment trigger report provides a make quantity of 500 units for product A, a production order for 500 units will be scheduled according to the Rhythm Wheel sequence; if the replenishment quantity is 700 units, then a production order for 700 units will be generated.

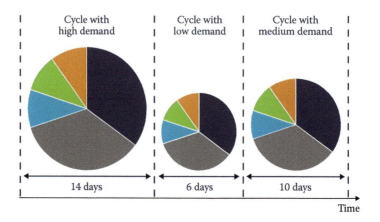

FIGURE 7.7
Breathing Rhythm Wheel with dynamic cycle length.

As a consequence, the Rhythm Wheel cycle time depends entirely on the overall cycle demand. In periods of low demand, production lots will be smaller, resulting in a shorter overall cycle time. In periods of high demand, production lots will be larger, leading to a longer overall cycle time.

When speaking about the Breathing Rhythm Wheel with its dynamic character, the term "breathing" here metaphorically refers to the swelling and shrinking of the cycle time depending on the required production quantities, as illustrated by Figure 7.7.

However, to achieve leveled production with constant capacity utilization, the Rhythm Wheel cycle times should be kept as stable as possible. Although production quantities should principally follow actual consumption, the breathing of the wheel should be limited to a certain extent to mitigate nervousness both in manufacturing and along the end-to-end supply network.

Summary

In the first section of this chapter, we focused on the interaction of production and replenishment modes on an operational level. We introduced the combination of IRL replenishment and the Breathing Rhythm Wheel with an illustration to describe how the replenishment trigger report connects the production and replenishment modes and makes it possible to translate demand data into a production schedule.

Using IRL as a replenishment mode, consumption-driven replenishment signals are generated based on target inventory levels. These signals are then transferred to production where corresponding production orders are generated and scheduled by the Rhythm Wheel. Because these production orders are based on replenishment quantities from the replenishment trigger report, the Breathing Rhythm Wheel behaves dynamically regarding production quantities and cycle times.

7.2 HOW TO LEVEL PRODUCTION WITH FACTORING

Avoiding the transfer of nervousness from the market to production is an integral element of LEAN SCM. Thus, a blind translation of variable replenishment quantities from a replenishment trigger report into production orders might contradict the principle of leveled production.

In other words, if the replenishment quantities are calculated in a dynamic way and fluctuate too widely, production needs to be partly decoupled from the replenishment trigger report. In this way, the variability is prevented from entering production and creating nervousness along the supply chain. The following section explains how this can be achieved using cycle time boundaries.

7.2.1 Use Cycle Time Boundaries to Stabilize the Asset Takt

Cycle time boundaries, which we introduced in Chapter 6, help to mitigate nervousness and achieve stable production. Production stability means that production is leveled within a given range. In other words, the Rhythm Wheel cycle time is kept within the predefined upper and lower cycle time limits, as shown in Figure 7.8. If the actual cycle time tends to violate either of these boundaries, the production cycle should be adjusted accordingly to keep the overall cycle time within the allowed range.

By limiting production flexibility, demand outliers are adjusted, which moves the buffering of variability from production to inventories, as illustrated in Figure 7.9, which in turn smoothes capacity utilization. Safety stock is used actively to cover major fluctuations in both demand and supply. However, variability is still buffered in capacity but only within the defined range. As a consequence, rush orders or short-term adjustments of the production plan are allowed only to a limited extent, namely until the

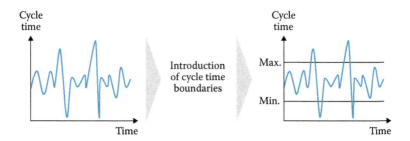

FIGURE 7.8
Introducing cycle time boundaries to mitigate nervousness.

cycle time boundaries are reached. Demand peaks outside these limits are satisfied by safety stock, which is refilled in upcoming production cycles with lower overall demand. Thus, cycle times and capacity utilization are smoothed and stabilized and the potential of repetitive production patterns can be leveraged more effectively.

To keep the cycle length within the defined range, upper and lower factorings come into play, as shown in Figure 7.10. Upper factoring involves shortening the Rhythm Wheel cycle in response to violations of the maximum cycle time boundary. Lower factoring is applied if the actual cycle time falls below the minimum cycle time boundary. In this case, idle time is added to lengthen the production cycle as required. We explain both approaches in the following sections of this chapter.

Although a suboptimal setting can lead to backlogs and poor service levels, cycle time boundaries are a highly powerful means of controlling and stabilizing production. If used appropriately, cycle time boundaries help

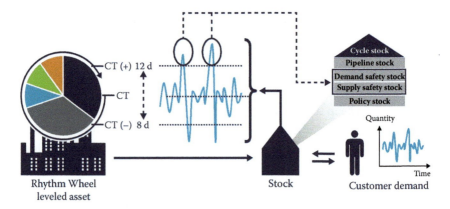

FIGURE 7.9
Cutting of demand outliers and active use of safety stock.

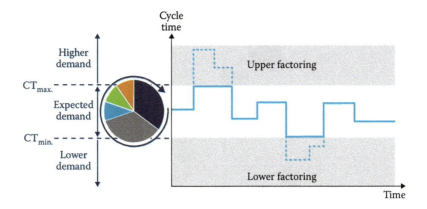

FIGURE 7.10
Upper and lower factoring.

to reduce the variability of capacity and stock requirements. Furthermore, they increase the predictability of production and as a consequence prepare the ground for end-to-end synchronization of the supply chain.

7.2.2 Use Upper Factoring When the Cycle Becomes Too Long

The general idea of upper factoring is to reduce the production quantities in cycles with abnormally or unexpectedly high demand in order to keep the length of the Rhythm Wheel cycle below the upper cycle time boundary. Production quantities that are demanded by the replenishment trigger report but cannot be produced in one cycle because they would violate the upper cycle time boundary are moved to upcoming production cycles. Missing quantities are filled by safety stock, which is held for exactly this purpose. As a consequence of factoring, production is partly decoupled from incoming demand and leveled over time.

In the example shown in Figure 7.11, the upper cycle time limit is exceeded. The last production order of the current cycle as shown in the example—product E—lies outside the allowed maximum cycle time time. As a consequence, the production quantities of A, B, C, D, and E should be adjusted to shorten the cycle time.

By reducing the production quantity of each product within the Rhythm Wheel cycle, overall cycle time is reduced and kept within the upper boundary. Since the behavior of production cycles is actively controlled and cycle time peaks are completely eliminated by this method, extraordinarily long production cycles and high cycle time variability can be avoided.

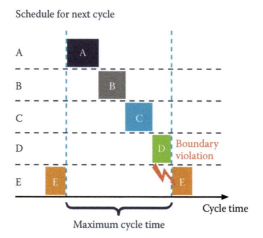

FIGURE 7.11
Violation of the upper cycle time boundary.

Other upper factoring methods have already been described briefly in Chapter 6. In the following, IRL factoring is explained to show the impact on the production schedule, which is essentially the same for all factoring methods mentioned in this book.

IRL factoring uses the so-called straightening factor to adjust the production cycle and keep it within the allowed range. This straightening factor is calculated based on the extent to which production exceeds the upper cycle time boundary. If the actual cycle time is, for example, 20% above the upper limit, the straightening factor for each product is set such that the overall cycle time is reduced by 20%. To shorten the production cycle, the target inventory levels of all products on the Rhythm Wheel are adjusted according to the straightening factor, as illustrated by Figure 7.12.

A new replenishment trigger report is then created to consider the adjusted IRLs and recalculate the make quantities for the shortened

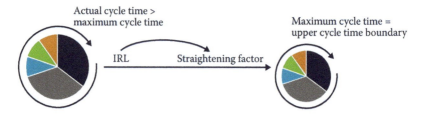

FIGURE 7.12
IRL factoring and the straightening factor.

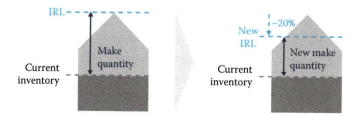

FIGURE 7.13
Recalculating production quantities following IRL adjustment.

Rhythm Wheel cycle. Since the target inventory levels have been reduced, the production quantities are reduced as well, as illustrated by Figure 7.13.

Smaller production quantities result in the required reduction of the cycle time. Since the straightening factor depends directly on the extent to which the upper boundary has been exceeded, the adjusted cycle time after factoring will exactly meet the upper cycle time boundary, as illustrated in Figure 7.14.

It is important to note here that IRL factoring does not result in general changes in IRLs. The described adjustments based on the straightening factor are short term and are valid only for the current production cycle, which in our example was originally too long. For upcoming Rhythm Wheel cycles, the original IRLs are used again—at least as long as the cycle times remain within the allowed range.

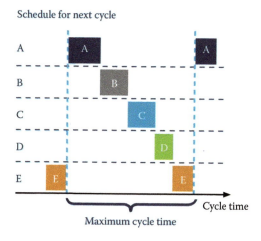

FIGURE 7.14
Shortened Rhythm Wheel cycle following IRL factoring.

7.2.3 Use Lower Factoring When the Cycle Becomes Too Short

If the minimum cycle boundary is violated, idle time is used to maintain the balance between production and the demand side. Generally speaking, production lots are sized to cover average demand within the planned cycle time. If the actual cycle time is considerably shorter because little or no demand occurred for some products, the next production cycle should probably not be started immediately. To prevent unnecessary accumulation of inventories or production in uneconomically small batches and support synchronization with suppliers, the next production cycle should be started only after the minimum cycle time has elapsed. Idle machine or labor capacity at the end of the factored cycle should be used for maintenance, training, continuous improvement, or other frequently required activities.

In the example shown in Figure 7.15, the lower cycle time limit has been violated; the actual length of the production cycle is shorter than the defined minimum cycle time. As a consequence, lower factoring is applied, which means that idle time is added at the end of the cycle to artificially lengthen the production cycle and adjust the cycle time to the minimum, as indicated in Figure 7.16. The following Rhythm Wheel cycle begins with product A only after the minimum cycle time of the current cycle has elapsed.

By enforcing the minimum cycle length, adequate time is provided for consumption by either customers or downstream stages of the supply chain. In other words, enforcement of the minimum cycle length ensures that working stock of all products, which are based on the planned or average cycle time, can be consumed to a certain extent before new production

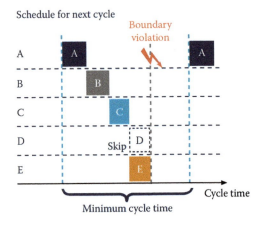

FIGURE 7.15
Violation of lower cycle time boundary.

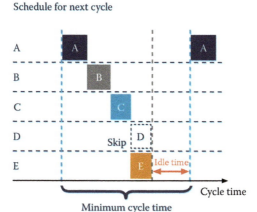

FIGURE 7.16
Lengthened Rhythm Wheel cycle with idle time after lower factoring.

runs are triggered. This enables a company to avoid uneconomically small production batches and disproportionately high changeover times while stability and predictability remain high.

Summary

The second section of this chapter has focused on demand variability. We have explained how LEAN Supply Chain Planning concepts level production and mitigate nervousness on an operational level. Although production should generally follow demand, it needs to be at least partly decoupled from demand if incoming variability is too high.

To control the behavior of the Rhythm Wheel, cycle time boundaries are implemented, leveling the fluctuation of production quantities and cycle times within a certain range. Upper and lower factoring come into play to implement the concept of cycle time boundaries.

The extent to which variability should be allowed in production depends on both business objectives and context. However, a suboptimal setting of cycle time boundaries can lead to backlogs and poor service levels. Thus, it is absolutely necessary to monitor planning and scheduling activities constantly in order to ensure that the benefits of leveled production with repetitive patterns are achieved as intended without negatively influencing time, service, or quality. The following section provides an overview of suitable performance metrics for parameter-driven SCM.

7.3 EFFECTIVE MONITORING OF PLANNING EXECUTION IN LEAN SCM

To provide a backward look at the performance of LEAN planning, a company must choose the right performance metrics. Such KPIs should not only focus on the overall performance of the supply chain, but also provide a basis for evaluating the quality of the various supply chain control parameter settings. Both the tactical and strategic renewal processes need a comprehensive picture of the health of the value chain areas to revalidate and realign the supply chain configuration and tactical parameterization.

In other words, global and local planning have to be able to determine whether adjustments of the production and replenishment parameters are required to match the behavior of planning and scheduling in light of market conditions and the characteristics of the supply network.

A monitoring framework for operational KPIs should answer the following questions for your company:

- Did we meet our short-term targets in terms of production times and quantities?
- Did we meet the planned Rhythm Wheel cycle times?
- How nervous was planning?
- Did we need to intervene manually?
- Did we hold inventories in the right quantities to reach our service targets?

In the next section, we describe a holistic monitoring framework for operational LEAN SCM, following in its structure both the requirements of parameter-driven planning as well as the differentiation between the production and replenishment modes.

7.3.1 What Should Be Monitored?

Operational performance metrics for LEAN SCM are oriented toward the tactical parameters that are used in LEAN planning to design and control the behavior of production and replenishment. The two classes of metrics we introduce here follow the distinction between the replenishment and production modes, as illustrated in Figure 7.17. The first set of KPIs focuses on production, namely evaluating the design and performance of

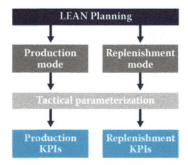

FIGURE 7.17
Classification of operational performance metrics for LEAN SCM.

the chosen Rhythm Wheel mode. The second group of metrics addresses supply and replenishment along the supply chain.

Generally, two key elements of LEAN planning should be considered when setting up an operational KPI framework: the feasibility of the design and parameterization of the production and replenishment modes, and adherence to this design in the daily business.

Logically, a company needs to know if design and parameter settings were chosen such that it is able to produce the right quantity of goods to satisfy demand and reach the inventory targets. Sometimes, however, poor input data regarding both the supply and demand sides, such as unreliable demand forecasts and production rates, can result in an infeasible Rhythm Wheel design and suboptimal tactical parameter setting. In such cases, adjustments must be made at once or in the next renewal cycle on both the tactical and strategic levels.

Adherence to design—at least as far as circumstances and external factors allow it—is crucial for avoiding nervousness and enabling synchronization along the supply network, which are key principles of LEAN SCM. Since the design and configuration of the production and replenishment modes at one stage of the supply chain are the fundamental bases for design and tactical parameterization at other stages, every operational deviation at one stage will directly impact both upstream and downstream stages, reducing stability and predictability within the end-to-end supply network.

In the following section, we explain the most important operational metrics for LEAN SCM. As mentioned above, they are classified into two groups, focusing on the production and replenishment side of LEAN planning.

7.3.2 Operational LEAN Production KPIs to Monitor Asset Performance

The first set of operational performance metrics focuses on the production side of LEAN SCM. The most important KPIs for production are cycle time attainment (CTA), run to target (RTT), and cycle time variation (CTV). The combination of these three performance indicators provides a good overview of production performance, providing answers to key questions such as:

- Did we adhere to the planned cycle time (CTA)?
- How widely did the cycle times fluctuate (CTV)?
- Did we produce the required quantities (RTT)?

CTA, CTV, and RTT can be supplemented with additional production-oriented KPIs to enable a more detailed analysis of performance, such as:

- Number of skips per cycle or product.
- Factored orders and quantities per cycle and product.
- Distance to cycle time boundaries per cycle.
- Manual production adjustments, for example, changes in the designed Rhythm Wheel sequence.
- Production interval variation per product.
- In the following section, we explain CTA, CTV, and RTT in detail.

7.3.2.1 Cycle Time Attainment

The CTA metric tracks the cycle length over time and evaluates the difference between planned and actual cycle time. It shows how consistently an asset runs according to the designed Rhythm Wheel time and is thus a crucial metric for maintaining synchronization within the end-to-end supply network. Sustainable synchronization is possible only if all supply chain stages or assets adhere to the designed takt.

The CTA is computed as the ratio of average cycle time to planned cycle time. The time needed to complete each Rhythm Wheel cycle is measured from the start of the run of the first product in the Rhythm Wheel sequence to the start of the run of the same product in the next cycle. If there is no demand for this specific product, the time is measured from or to the run of the next product in the Rhythm Wheel sequence.

The cycle times for consecutive Rhythm Wheel cycles are typically visualized on a process behavior chart, as shown in Figure 7.18.

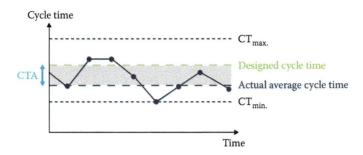

FIGURE 7.18
The cycle time attainment metric.

It is important to understand that cycles that are too short are just as bad as cycles that are too long. Every deviation from the planned production cycle reduces stability and predictability and increases nervousness along the supply chain. Therefore, it is crucial to monitor rhythm times constantly to be able to identify changes on the demand or supply side as early as possible.

If the cycle times are, for example, longer than designed, a variety of potential causes on both the demand and the supply sides could be responsible. The overall demand at the asset might have been higher than expected, perhaps due to demand peaks for some products, or process lead times might have been excessive, perhaps due to unexpectedly long changeover times or low productivity.

If deeper analyses show that the long cycle times have not resulted from short-term events but are rather the product of future trends, an adjustment of the planned cycle time and the cycle time boundaries is required and should be considered in the next iteration of the tactical renewal process.

7.3.2.2 Run to Target

The RTT tracks compare the required replenishment and the actual production quantities in every Rhythm Wheel cycle. In other words, RRT is a quantity-oriented metric, answering the key question: "Did we produce the right amount of goods as requested by the replenishment trigger report?"

RTT shows how consistently the quantities are produced according to the replenishment trigger report (Figure 7.19). This KPI evaluates how efficiently planned make quantities and cycle time boundaries have been

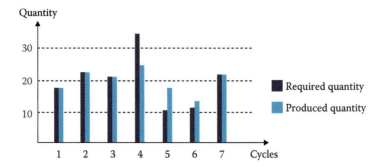

FIGURE 7.19
The run to target metric.

chosen. The RTT metric will indicate, for example, whether products are constantly under-produced such that the required replenishment quantities are not delivered by production. This can result in compromised service levels, since the demanded quantities are not replenished in the required way. A potential source of such a scenario could be that estimated demand per cycle does not match reality. Potential reasons could be either changes in the demand pattern or poor forecasting accuracy.

7.3.2.3 Cycle Time Variation

CTV evaluates how widely the cycle time fluctuates. While CTA is intended to indicate the degree of adherence to the plan by measuring the distance to the planned cycle time in every production cycle, CTV focuses solely on the variability of the cycle time, as illustrated by Figure 7.20.

Therefore, CTV is defined as the coefficient of variation of the cycle time, which is the ratio of the standard deviation of the cycle time to its mean.

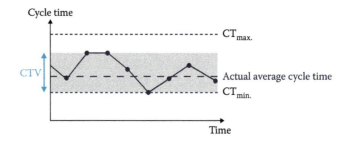

FIGURE 7.20
The cycle time variation metric.

Transparency regarding the behavior of cycle times is essential for determining the optimal amount of safety stock that has to be held to cover variability of both demand and supply. As a general principle, we recommend orienting the safety stock toward the longest cycle time measured in order to mitigate the risk of stock-outs due to extraordinarily long production cycles.

Moreover, the cycle time boundaries should be set in accordance with fluctuations of the cycle time. The greater the cycle time variation, the less strictly should the cycle time boundaries be set. Thus, existing variability is at least partly buffered at the asset rather than being buffered entirely in inventories.

Experience shows that the CTV depends heavily on the heterogeneity of the product portfolio at the asset. The more heterogeneous is the mix, the higher the cycle time variation typically will be. The allowed ranges need to be defined accordingly on a case-by-case basis. Generally, too much variation can be the consequence of higher-than-expected demand variability, unbalanced production rhythms, or cycle time boundaries that are too lax.

7.3.3 Operational LEAN Replenishment KPIs to Evaluate Inventory Parameterization

The second set of operational performance metrics focuses on the replenishment side of LEAN SCM. The most important KPIs for replenishment are service level (SL), target inventory attainment (TIA), and dead stock ratio (DSR). The combination of these three metrics provides an accurate picture of replenishment performance, answering key questions such as:

- Did we reach our service targets (SL)?
- Did we reach our inventory targets (TIA)?
- Did we set adequate inventory targets (DSR)?

Additional replenishment-oriented KPIs can be tracked to enable a more detailed analysis:

- Forecast accuracy
- Safety stock ratio
- Shelf-life ratio
- Manual replenishment adjustments, such as adjustment of replenishment quantities
- MTO ratio

In the following section, we explain the key metrics for replenishment—SL, TIA, and DSR—in detail.

7.3.3.1 Service Level

SL indicates the extent to which customer orders have been satisfied. Typically, two SLs can be found in practice, the alpha and the beta SLs.

The alpha SL is an event-oriented performance criterion. It measures how many of the customer orders arriving within a given time interval are delivered in full without shortages. The alpha SL is defined as the ratio of the number of fully delivered orders to the total number of orders.

The beta SL is a quantity-oriented performance measure describing the proportion of total demand within a reference period that is delivered without delay. It is defined as the ratio of the delivered quantity to the overall ordered quantity per period.

Generally, both SLs can be calculated on an aggregated level per asset or per SKU for a more detailed analysis.

If the SL falls out of the permitted range, it may be time to reevaluate the setting of the replenishment parameters. Potential reasons for poor SLs include a suboptimal setting of inventory target levels, wrongly sized safety stock components, or a mismatch between replenishment cycle times and various stock components.

7.3.3.2 Target Inventory Attainment

Just as the CTA metric measures adherence to design on the production side—targeting the difference between planned and actual cycle times—so the TIA metric focuses on adherence to the design of the inventory parameters on the replenishment side. More precisely, TIA measures the extent to which inventory targets have been realized, by comparing planned inventory levels with actual inventory levels, as illustrated in Figure 7.21. Again by analogy to CTA, it is defined as the ratio of the average as-is inventory to the planned average inventory.

A poor TIA generally indicates a mismatch between supply and demand. To identify the root causes of such a mismatch, a more detailed analysis is typically required. Since it is not possible to draw conclusions regarding the feasibility of the chosen target inventory levels based on TIA alone, we include another metric in the framework, the DSR.

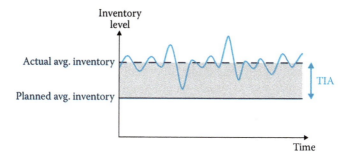

FIGURE 7.21
The target inventory attainment metric.

7.3.3.3 Dead Stock Ratio

Dead stock is the part of the inventory that has not been moved, which makes it the lowest inventory level to be monitored in the period under consideration, as illustrated by Figure 7.22.

Although the absolute dead stock level indicates the amounts of superfluous inventory and excess working capital that contribute to inventory holding costs, there is additional utility in comparing dead stock to average inventory to complete the picture. Thus, we suggest incorporating the DSR into the operational KPI framework for LEAN planning. The DSR is defined as the ratio of dead stock to the target average inventory level.

A high DSR generally indicates that a company has set inadequate inventory targets at the strategic level or has set the tactical replenishment parameters—such as target inventory levels or safety stock—suboptimally.

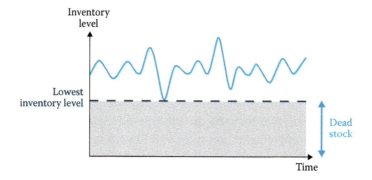

FIGURE 7.22
The dead stock ratio metric.

Typically, poor forecasting accuracy makes it even more challenging to find the optimal settings of the inventory parameters.

Quite often, high levels of dead stock suit the traditional safety-oriented mindset that is still deeply embedded in organizations that have just begun the journey to LEAN SCM. Replenishment quantities are often increased manually as a consequence of planners' personal need for security, resulting in longer replenishment cycles with higher inventory levels.

Summary

In this section, we have introduced and explained several performance metrics that are well-suited to parameter-driven SCM. We described an operational KPI framework, focusing on two key elements of parameter-driven planning: first, the feasibility of the design and parameterization of production and replenishment modes and, second, adherence to this design in daily operations. These KPIs will help your company evaluate the settings of the various supply chain control parameters and realign the configuration of the supply chain. We provide a higher-level set of KPIs for end-to-end supply chain performance management in Chapter 9.

Following the differentiation of production and replenishment modes, we distinguished two classes of operational performance metrics. CTA, CTV, and RTT are the key metrics on the production side, tracking and evaluating cycle times and production quantities. On the replenishment side, the most important KPIs are SL, TIA, and DSR, which monitor service and inventory levels over time.

CHAPTER SUMMARY

In Chapter 7, we have focused on the operational implementation of LEAN SCM, the implementation stage that follows tactical parameterization of replenishment and production modes (Chapter 6), and described how LEAN planning methods can be applied in practice.

Section 7.1 explained how production and replenishment are integrated into operational planning, and how demand signals are generated, processed, and finally translated into a production schedule. As long as demand variability remains low, there is no need to intervene in the functioning of the Breathing Rhythm Wheel. However, since leveled production is a key principle of LEAN SCM, mitigation measures are needed occasionally to actively prevent variability from entering production in cases marked by widely fluctuating demand.

In Section 7.2, we addressed this need directly, outlining how to achieve leveled production. To this end, we explained cycle time boundaries as well as upper and lower factoring, which can enable your company to cope with and mitigate variability within the supply network. While upper factoring reduces production quantities to keep the cycle length below the upper boundary, lower factoring includes idle time to artificially lengthen the cycle to the lower boundary.

In Section 7.3, we provided a monitoring framework for operational LEAN SCM to ensure that planning and scheduling follow the defined tactical design and parameterization, and of equal importance, that the strategic and tactical design remain feasible. The operational performance metrics should be oriented toward the tactical parameters that are used by LEAN planning to design and control production and replenishment. Hence, a set of KPIs was provided for both the production and replenishment sides of operational LEAN SCM.

Implementing the right operational planning process is an essential step on your company's journey to a synchronized and properly configured supply network. However, it is important to keep in mind that these planning processes must be embedded in an adequate organizational framework to achieve sustainable results. Accordingly, we address the design of organizational structures that create a favorable environment for LEAN Planning and Scheduling in Chapter 8.

Part III

What to Implement and Transform for LEAN SCM

8

Build an Organization for LEAN SCM

Over the course of this book, we have introduced the concepts and processes that are vital to LEAN SCM. Ultimately, however, the success of LEAN SCM depends on people and the organizational set-up in which they work. For a company's personnel, LEAN SCM represents a significant change in the way in which the supply chain works. To ensure that LEAN principles are implemented effectively and yield sustainable benefits over the long run, it is vital to develop understanding, acceptance, and commitment to LEAN SCM throughout your company's supply chain organization. Experience shows that even the most carefully conceived concepts will not deliver benefits and will not be sustained if the organization does not support the change and accept the new way of working.

So, what needs to change in your company to make LEAN SCM a reality, and how can you support the change? A growing tree as shown in Figure 8.1 provides a helpful metaphor. The trunk and branches—representing "hard" factors such as organizational structures, processes, and formal role descriptions—comprise only the visible part of the tree. But strong roots are needed to support the tree: experience shows that the "soft" factors that lie below the ground—especially the buy-in of the entire organization and the behaviors of staff and senior leaders—are essential underpinnings of LEAN SCM. The tree and its roots need to be nurtured with the right supporting tools and techniques to support the transformation to LEAN SCM.

This chapter provides guidelines for establishing an effective LEAN supply chain organization. While there is no one-size-fits-all approach to creating one, there are several characteristics that increase the odds of success.

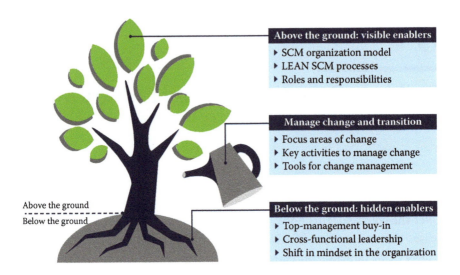

Above the ground: visible enablers
- SCM organization model
- LEAN SCM processes
- Roles and responsibilities

Manage change and transition
- Focus areas of change
- Key activities to manage change
- Tools for change management

Below the ground: hidden enablers
- Top-management buy-in
- Cross-functional leadership
- Shift in mindset in the organization

Above the ground
Below the ground

FIGURE 8.1
The LEAN Tree Model for transforming the supply chain organization.

This chapter explains:

- Why leadership, commitment, and shifts in mindset and behaviors are the important hidden enablers below the ground
- Which processes, roles and responsibilities, and organizational structures need to be in place as visible enablers above the ground
- How to successfully manage the change and transition to LEAN SCM

Following the LEAN Tree Model, Section 8.1 starts below the ground, demonstrating how the hidden enablers of the LEAN paradigm can be developed in order to ensure that LEAN SCM is deeply rooted in all levels of your company. It addresses the importance of top management buy-in, cross-functional leadership, and a shift in the mindset of the entire supply chain organization to introduce LEAN SCM successfully.

Section 8.2 examines the LEAN Tree Model above the ground, and discusses the visible enablers: the organizational structures, processes, and roles and responsibilities that need to be aligned to support LEAN SCM.

Section 8.3 highlights the importance of change management in order to make a successful transition to LEAN SCM. A brief overview of key areas, useful approaches, and best practices for change management is provided.

8.1 BELOW THE GROUND: THE PREREQUISITES FOR LEAN SCM

What changes in a company when LEAN SCM is implemented? For most companies, the simple answer is "a lot." To estimate the impact of LEAN SCM on your company, keep in mind the following:

- Implementing LEAN SCM is a major transformational program for building a company's future supply chain—it therefore requires significant buy-in and mobilization of top management.
- LEAN SCM involves an integrated end-to-end process approach which impacts all involved functional units—it therefore requires strong leadership and commitment from functional management.
- LEAN SCM is a paradigm change in the way production and replenishment planning is conducted along the supply chain—it therefore requires a shift in accountabilities, mindsets, and behaviors in the SCM community.

Depending on how your company conducts SCM today, moving to LEAN SCM can mean big changes in the way your supply chain organization works. Consequently the entire organization must actively participate in the change, recognizing that it is often the less visible enablers below the ground that determine success or failure (see Figure 8.2).

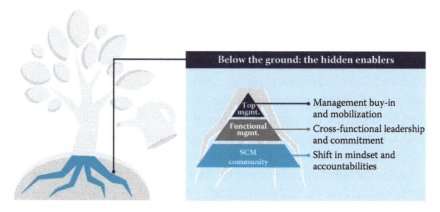

FIGURE 8.2
The LEAN Tree Model—hidden prerequisites.

8.1.1 Management Buy-In and Mobilization for LEAN SCM

Adopting LEAN SCM requires targeting nothing less than a quantum leap in supply chain performance. Reducing inventory by more than 30% and improving asset utilization by more than 20% while improving customer service are not uncommon, as the industry case studies in Chapter 12 show. Achieving such results, however, requires rebuilding the supply chain and its organization. To achieve such a mammoth task, there are two prerequisites:

- Unconditional top-management buy-in
- Mobilization of the entire organization

8.1.1.1 Top-Management Buy-In

The first step to obtain top-management buy-in is to focus on the benefits to a company of LEAN SCM. A realistic and well-founded benefit case is needed. An effective, proven means of developing such a case is using a simulation-based approach that demonstrates and evaluates the impact of LEAN SCM. With a computerized simulation model, even the most complex supply chains can be analyzed. Box 8.1 provides an example showing how the benefits of LEAN SCM were demonstrated for a top 10 pharmaceutical company.

Once the benefits have been clearly presented, the principles of LEAN SCM need to be explained. It is important that top management has a clear, basic understanding of the principles of LEAN SCM and its planning concepts. This will enable it to provide authentic, credible support to facilitate the changes required.

Good understanding will help top management recognize that the move to LEAN SCM represents a paradigm shift. This will help managers appreciate the magnitude and scope of the change impact of LEAN SCM. It must be clear that the implementation of LEAN SCM is not a "one-off" project, but instead requires a holistic program that typically takes several years to fully embed in the SCM organization.

However, it normally takes some time for an organization to adapt completely and achieve the full potential of LEAN SCM. In addition, management should recognize the need to sustain LEAN SCM processes once they are established.

BOX 8.1 CREATING A LEAN SCM
BENEFIT CASE WITH SIMULATION

A leading pharmaceutical company was concerned about applying LEAN SCM, in particular the Rhythm Wheel production planning approach. Some members of the management team initiated the discussion and were therefore major supporters. However, there were still some skeptical voices who questioned the benefits of the new supply chain planning approach. To dispel their doubts, the management team agreed to set up a simulation to evaluate the performance of the Rhythm Wheel concept.

Therefore, a holistic simulation scenario was modeled which considered three supply chain stages and a value stream of 250 finished products within a time horizon of one year. Based on the results of the simulation, the performance of the Rhythm Wheel concept was evaluated. As shown in Figure 8.3, major improvements were demonstrated compared with the currently implemented planning approach. Despite enormous inventory reduction potential, customer service levels as well as OEE increased.

In total, the results of the simulation left no doubt that major improvements could be achieved with LEAN SCM. A key argument

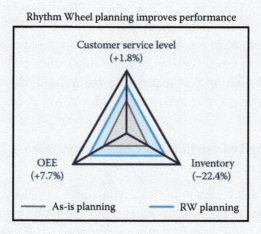

Rhythm Wheel planning improves performance

Customer service level
(+1.8%)

OEE
(+7.7%)

Inventory
(−22.4%)

—— As-is planning   —— RW planning

FIGURE 8.3
Results of the supply chain simulation confirmed the benefits of LEAN SCM.

for conducting simulation analysis is that it evaluates competing approaches under realistic conditions. Accordingly, LEAN SCM advocates were easily able to persuade the doubters. Management therefore decided unanimously to develop a detailed LEAN SCM implementation concept.

8.1.1.2 Mobilization of the Organization

Once top management has bought in to implementing LEAN SCM, the next step is to reach out to stakeholders and mobilize the wider organization. Many functions, such as manufacturing, sales, marketing, and finance, are involved in or impacted by a company's supply chain. LEAN SCM relies on close cross-functional cooperation. Hence, all functions need to be engaged and fully committed for successful implementation of LEAN SCM. Comprehensive change management with structured stakeholder management and proper organizational development is required to achieve this.

It is important that the LEAN SCM program does not get lost amidst other initiatives or desires to follow the latest SCM fads. Top management must display coherent leadership and consistency of direction in order to succeed with LEAN SCM implementation. Tenacity and a strong sense of purpose are needed. Aligning initiatives with LEAN SCM discourages factions within an SCM organization from rushing from one supply chain philosophy to the next, thereby avoiding dissonance that often results. Thus, top-level leadership is required to ensure both the adoption and ongoing development of as well as support for LEAN SCM.

8.1.2 Ensuring Leadership and Commitment across Functional Borders

In the LEAN SCM paradigm, leadership must be able to work effectively across functional boundaries and unite differing perspectives. For example, in the tactical renewal process described in Chapter 6, company entities representing market demand, the supply chain, and manufacturing need jointly to determine the right inventory levels for the supply chain. The supply chain must move away from operating and thinking in functional silos based on local optimization only. Close collaboration is required to achieve end-to-end excellence. To overcome organizational barriers, two features are vital:

- Cross-functional leadership
- Commitment of the organization

8.1.2.1 Cross-Functional Leadership

Moving to LEAN SCM represents a radical change for many employees. For them, this will not be just a technical change in how they work: there will also be an emotional response to the change, typically associated with feelings of insecurity, uncertainty, and anxiety since jobs and roles will be affected. Often these feelings create resistance and can lead to rejection of new ways of planning and working. Because LEAN SCM relies on collaborative action across functional boundaries, it is important to acknowledge and communicate with employees who are reluctant to go along with the change. This means active leadership and support for the change are required. Without this, it is often observed that some staff will fall back on old ways of working and habitual behavior patterns. Time and effort must be invested to explain the change, train the staff in the new ways of working, and support and monitor the effectiveness of these efforts.

However, leadership is required not only at the top level of an organization: leaders are important at all levels. For example, it is important to bring experienced supply chain planners on board who can lead the change and help to roll out LEAN SCM in their areas. Leaders and potential leaders need to be developed at middle and junior levels representing key functions so that they can contribute to the successful implementation of LEAN SCM.

8.1.2.2 Commitment of the Organization

To build up commitment in the relevant supply chain functions, the question "what's in it for me?" needs to be addressed. One important aspect in this context is stakeholder management. Only if the benefits of LEAN SCM that will accrue to specific functions are clearly communicated is commitment likely to evolve.

Although some functions may see disadvantages in LEAN SCM and some groups or individuals may regard the change to LEAN SCM as diminishing their roles, it is better to be clear about what the change means. It is important to communicate effectively and provide support to help people cope with the change. Realism and honest expectations from management are needed. Staff will normally accept the "bigger picture" that the end-to-end approach of LEAN SCM will deliver better supply chain performance with greater benefits to both the customer and the company than working in functional silos. Thus, senior cross-functional backing is a prerequisite for strong buy-in to optimizing overall supply chain performance rather than focusing on individual unit performance.

8.1.3 Shift in Mindsets and Accountabilities in the SCM Community

Beyond top management and functional leaders, by far the most important facilitator for implementing LEAN SCM is convincing the SCM planning organization to embrace the change. The people routinely planning and operating supply chain functions are the backbone of any planning activities: their support and commitment are essential for LEAN SCM to succeed. With the introduction of LEAN SCM, how they work will change dramatically. To implement LEAN SCM successfully, your company's SCM organization will need:

- A shift in mindsets
- A shift in accountabilities

8.1.3.1 Shifting the Mindsets of the SCM Community

A shift in mindsets is required to assimilate LEAN principles and adopt a new way of planning. For example, planners must understand how variability is to be managed under the new paradigm: they must learn that demand variability is buffered proactively by consumption of appropriate inventories rather than through increasing production quantities at the expense of manufacturing asset utilization. The traditional response when a market planner notices that demand for a product is higher than expected is to react nervously and place emergency orders to production sites even when there is sufficient safety stock to buffer such demand fluctuations. In contrast, market planners following LEAN SCM principles

	Production	Replenishment
Tactical planning	Configuration of Rhythm Wheel parameters (sequence, cycle time)	Configuration of inventory parameters (order-up-to level, reorder point)
Operational planning	Scheduling of production orders in pre-defined sequence and cycle time	Creation of replenishment orders in pre-defined intervals or quantities

FIGURE 8.4
Separation between tactical and operational planning in LEAN SCM.

would avoid such costly and inefficient action, remain calm, and continue planning according to the defined parameters. Only if higher demand persists outside the normal variability range would the planner trigger the renewal of the replenishment parameters at the tactical level.

Within LEAN SCM, tactical and operational planning are separated (see Figure 8.4). In tactical planning, supply chain parameters such as a Rhythm Wheel sequence are defined. In operational scheduling, execution is planned within these pre-configured parameters: the asset planner simply follows the previously designed sequence. If short-term demand now arrives from the market, the planner still sticks to this sequence and schedules the orders into the next available slots. In this way, all products are produced in a timely way and production is optimized. The schedule is no longer driven by the market planners who shout loudest. The pre-configured planning in LEAN SCM provides strict separation of tasks and clearly defined priorities for the SCM organization.

The role of demand forecasting also changes in LEAN SCM. Our experience shows that the manufacturing and supply chain community often undergoes a major shift in mindset from "sell what we are producing" to "produce what we are selling." While forecasts were traditionally used as triggers for push-driven production (a "classic MRPII" approach), in the new planning paradigm the forecasts are used on the tactical level. As described in Chapter 6, forecasts are used to determine planning parameters for pre-configuration. At the operational level, production is planned based on tactical parameters and triggered only by real consumption. This connects the company and its supply chain directly with the voice of the customer.

8.1.3.2 *Shift in Accountabilities*

Planners need to apply new planning concepts such as the Rhythm Wheel and pull replenishment, and follow new processes such as the tactical renewal process. These task changes mean a shift in accountabilities:

- Traditionally local tasks are conducted at a global level—for example, alignment of replenishment parameters to production takt.
- Previously decentralized tasks are centralized—for example, the governance of LEAN SCM concepts and planning processes.
- Local planning tasks are moved to the supply chain organization—for example, the alignment of production takt across sites.
- Commercial sales organization tasks are moved to the supply chain organization—for example, setting finished goods target inventory levels.

Such a shift in accountabilities may feel daunting to many employees. This is why it is so important for staff to gain a thorough understanding of the concepts behind LEAN SCM and learn that these approaches have proven successful. Extensive training in applying new concepts, new processes, and even new IT tools is essential for successful implementation of LEAN SCM.

Summary

Without engagement and buy-in on the part of the people who work in a company's supply chain organization, LEAN SCM will not work properly. Employees at all levels of the supply chain organization must be engaged to ensure that LEAN SCM processes, tools, and concepts are successfully established and applied. Top management needs to embrace the principles and concepts of LEAN SCM, and then give support and sponsorship as the concepts are communicated throughout the company and the supply chain organization. Functional management must provide strong leadership to overcome functional barriers to LEAN SCM within the organization. Cross-functional commitment is vital both to making LEAN SCM work and to optimizing the benefits to the company of end-to-end planning. Last but not least, the most important part of the organization to embrace LEAN SCM is the entire supply chain planning community. A shift in mindset is required to adopt the LEAN way of planning and accept the new accountabilities.

These are the hidden "roots" required to provide a secure foundation for LEAN SCM. We now move "above the ground" to introduce the more visible enablers.

8.2 ABOVE THE GROUND: THE VISIBLE ENABLERS FOR LEAN SCM

In this section, we outline the formal structure of the supply chain organization with its processes, roles and responsibilities, and present the organizational design principles that are needed to support LEAN SCM. In our LEAN Tree Model, these elements represent the visible, "above-the-ground" part of an organization.

Best practice for LEAN SCM involves establishing a supply chain organization that focuses on a company's value streams, practices end-to-end supply chain process management, and is driven by customer demand. To shift from a traditional model organized around functions such as production and logistics to a process-centric supply chain organization that supports LEAN Supply Chain Planning effectively, a company must introduce planning processes that are tailored to LEAN SCM. Both the organization and its processes need to be aligned with this new approach. Therefore, in this section, we focus on organizational models, the integration of LEAN SCM processes with the existing business process framework, and the roles required to conduct the processes in the new planning paradigm (see Figure 8.5).

8.2.1 What Is the Right SCM Organization Model for LEAN SCM?

Creating and sustaining an effective SCM organization is a key enabler for LEAN SCM. It is crucial to organize SCM in a more integrative and collaborative fashion from an end-to-end perspective. Therefore, it is important to understand:

- Which organizational supply chain model is best suited to LEAN SCM.
- The best way to transition to this organizational model.

Above the ground: the visible enablers

▶ Transformation of the supply chain organization model

▶ Integration of LEAN SCM with other processes

▶ Mapping of roles and responsibilities to LEAN processes

FIGURE 8.5
The LEAN Tree Model: elements above the ground.

8.2.1.1 Typical Supply Chain Organizational Models in Process Industries

In process industries, organizational models can vary considerably across companies even within sectors such as chemicals, life sciences, and fast-moving consumer goods (FMCG). However, there are three basic and distinct types of organizational models that process industry companies tend to follow: the decentralized local model, the centralized hub & spoke model, and the coordinated network model (see Figure 8.6).

In the decentralized local model, the SCM organization is typically highly decentralized and local units act very independently. This can allow them to align well with an entrepreneurial sales organization and be very responsive to local market needs. However, these same attributes also raise challenges. High customer focus at the local level often results in

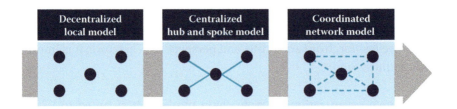

FIGURE 8.6
Supply chain organization models and key influence factors.

high inventories, a less competitive cost structure, and a weak end-to-end supply chain. In short, the focus on local optimization comes at a cost to the overall supply chain, especially in globally operating companies with a global supply chain set-up.

The centralized hub & spoke model, with its directive focus on efficiency and standardization, brings both advantages and limitations. Usually introduced to provide strong central oversight and coordination, it has intense cost-optimization focus, a harmonized supply chain that is strongly coordinated and integrated, and is capable of driving network rationalization and end-to-end performance. But these positive attributes also present challenges. Implementing the centralized model requires strong change management if the previous culture has been one of decentralized independence. Considerable effort may be required to obtain alignment and compliance with the new operating model. When established, concerns can arise that the centralized model is too remote from customers and lacks operational agility and entrepreneurialism. When trying to achieve rapid growth in new markets, which is an imperative today given the growing importance of emerging markets, this model can be too constraining. Despite these issues, it is a model that should be broadly adopted by most companies as a first step away from the decentralized local independence model—and it can also provide a strong platform for moving toward the coordinated network model.

The model best suited to LEAN SCM is the coordinated network model. In this model, the central supply chain organization provides broad coordination while allowing greater freedom to execute at the local or regional level. Furthermore, it matches the requirements shown in the previous section regarding collaborative behaviors and alignment across planning process levels.

The coordinated network model balances cost and service and fosters the entrepreneurialism required to support dynamic growth in developing markets. For the model to be effective, it requires good governance and shared processes, objectives, and behaviors. The local structure in this model can be helpful for developing effective relationships with marketing and sales affiliates and improving customer focus. In the LEAN SCM context, this model allows for the local creation of a concrete Rhythm Wheel sequence for a production site, but within the centrally standardized and agreed-upon Rhythm Wheel concept for the entire company. Without this standardization, end-to-end synchronization of the supply chain, as described in Chapter 6, would not be possible.

8.2.1.2 Transition of the Supply Chain Organization Model Requires Harmonization

In the past, the desire to lower inventory and reduce costs was the key driver for making changes to the SCM organization. Today, the need for greater customer responsiveness and more agile product supply throughout the entire network dictates the change. Although process industry companies have been steadily moving away from the decentralized model to the centralized hub & spoke model, so far very few have evolved to create a networked SCM organization that is based on collaborative interdependence.

Experience shows that it is extremely hard for companies to transition successfully from the local independent model to a coordinated SCM network without first having introduced key elements of the centralized hub & spoke model. The reason is clear: the necessary standardized and harmonized processes and roles are lacking. For example, the Rhythm Wheel concept needs to be standardized centrally, so that all sites follow the same concept, rather than having several localized interpretations across the company. Critical success factors for an interdependent model include a sound understanding of SCM, process standardization, strong awareness of customer requirements, aligned rewards, and harmonized roles.

Good process governance, visibility and understanding of global and local requirements, and integrating the entire supply chain organization are key success factors for LEAN SCM. Efficient coordination and governance of supply chain planning lead to excellent performance due to planning processes that consider the current business reality and challenges in a volatile and uncertain world. In the next section, we describe the LEAN SCM processes that a process-centric organization should follow.

8.2.2 Integration of LEAN SCM Processes with the Existing Planning Processes Framework

LEAN SCM employs regular renewal of LEAN parameters for supply chain planning as well as systematic review of production and replenishment modes. Renewal is performed in response to changes in internal or external business requirements. When supply chain planning works in this way, daily planning and scheduling run smoothly, reducing complexity and costs.

The strategic and tactical renewal process lies at the heart of the LEAN Supply Chain Planning landscape (see Figure 8.7) ensuring alignment and

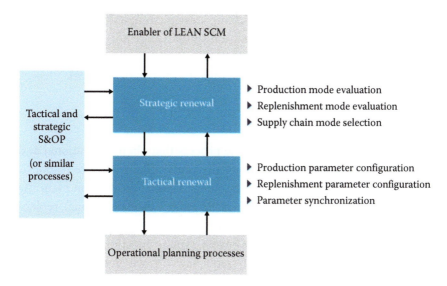

FIGURE 8.7
Renewal processes form the core of the LEAN SCM Planning process landscape.

coordination at all planning levels in an organization. The most important link to other processes is the interface with the tactical and strategic sales and operations planning processes (S&OP).

8.2.2.1 Strategic Renewal Process

The strategic renewal process is designed to provide regular reviews of the production and replenishment modes used in the supply chain. We recommend establishing a regular review cycle that fits your company's supply chain modes and product segments to assess the need for improvements or changes due to altered strategic conditions. The frequency of this process should reflect the dynamic of the company's business environment. Typically it should take place at least yearly to ensure that it fits into the planning process for the annual budget cycle.

As described in Chapter 5, selection of the right production and replenishment modes is a basis for LEAN SCM Planning. The strategic renewal processes address the types of production and replenishment modes that need to be followed. The chosen production modes determine what, how many and when products should be manufactured, while the replenishment mode defines the logic dictating when the products should be moved between stages and when production should be triggered. Because these decisions affect the entire supply chain, strategic renewal should be integrated with other processes that have a similar scope for impacting such operations as major product lifecycle events, technological innovations, and geographical or M&A-driven changes to the business model.

8.2.2.2 Tactical Renewal Process

Tactical renewal is a regular, routine process that aligns operational and tactical supply chain planning along the supply chain. Its core responsibility is to confirm and, if necessary, adjust tactical production planning parameters (e.g., production sequence, Rhythm Wheel cycle time), replenishment parameters (inventory replenishment components such as cycle, safety, and policy stocks), and to ensure synchronization of parameters along the entire supply chain. The frequency with which tactical renewal should be undertaken depends on supply chain characteristics such as market volatility, stage of product life cycles, and market growth. Tactical renewal will typically be set in motion every 1–3 months. A clearly defined interface to monthly or quarterly S&OP (or similar processes that conduct the respective tasks) ensures that the latest information about demand and supply profiles, which is aligned between the commercial units and operations, is used for planning purposes.

8.2.2.3 Ensuring Process Governance

Effective process governance is an important prerequisite to sustaining the planning processes for LEAN SCM. It ensures the required cross-functional interaction and collaboration for establishing end-to-end supply chain planning. Process governance is thus an essential aspect of the successful implementation of LEAN SCM, as it:

- Ensures adherence to designed end-to-end planning process standards.
- Facilitates clear process understanding on the part of employees
- Continuously improves LEAN SCM processes.

Process governance not only ensures that the designed processes are adhered to, but it can also identify current weak points in the implemented LEAN processes and continuously improve these. This can be achieved through regular assessments within the renewal processes and feedback from participating employees. Where agreed to, the LEAN processes may be adjusted to fit the requirements of an organizational unit, but the process governance body should ensure that a company-wide standard is maintained. This guarantees that each organizational unit adheres to LEAN SCM principles, concepts, and processes, although some degrees of freedom are granted.

In the next section, we introduce the key roles regarding LEAN SCM and describe their recommended responsibilities associated with the renewal processes. We emphasize that a given role does not need to be translated directly into a complete job description, even though both may be identical. A role is a descriptor of an associated set of tasks and may be performed by one or more employees, and one employee can play several roles.

8.2.2.4 Roles within the Tactical Renewal Process

The roles considered in the tactical renewal process should combine three important dimensions of supply chain planning: the local view of assets, the global view along the supply chain as a whole, and the view from the market. All three views must be taken into account in the tactical renewal process to properly align operations with products and customer needs. The three key dimensions can be addressed through the roles of the local planner, the supply chain planner, and the market planner as shown in Figure 8.8.

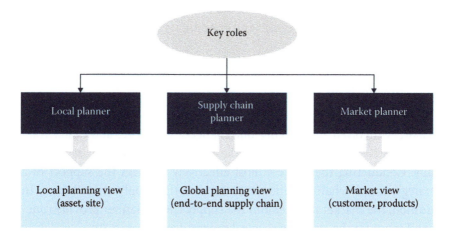

FIGURE 8.8
Key roles for the tactical renewal process.

Local planner	Supply chain planner	Market planner
▸ Local view of tactical parameter renewal and adjustment strategies ▸ Calculation and implementation of tactical parameters at asset(s) (Rhythm Wheel parameters) in accordance to agreed-upon targets ▸ Provides (local) information/ view of capacity and capabilities of assets	▸ Leads tactical renewal process and coordinates alignment to S&OP ▸ Represents supply chain in strategic renewal process ▸ Global view of tactical planning and synchronization of lean parameters across assets ▸ Definition of inventory targets (in particular policy stocks) ▸ Facilitator function for LEAN Supply Chain Planning along supply chain	▸ Interface of operations to sales and marketing organization ▸ Provides or consolidates forecasts, demand variation profiles, and service levels to asset planners and supply chain planner ▸ Provides market information and view for tactical and strategic renewal processes

FIGURE 8.9

Role descriptions for the tactical renewal process.

The supply chain planner's view encompasses the whole supply chain and value stream, and focuses on optimizing the end-to-end chain. The local planner's tasks center on a single site and a number of (manufacturing) assets. The local planner ensures that pre-determined planning parameters meet local requirements and is responsible for their implementation. To balance supply with market and demand requirements, the market planner acts as an interface between operations and sales & marketing. (Note that both the supply chain planner and the market planner also support the strategic renewal process.) Figure 8.9 details the recommended scope of the planning roles, and describes their respective tasks within the tactical renewal process.

Of course, additional planning tasks may be assigned to the three roles outlined above. In particular, they could be assigned responsibilities within a supply chain's S&OP processes and continuous improvement programs.

8.2.2.5 Roles within the Strategic Renewal Process

As explained in Chapter 5, the scope of the strategic renewal process is greater than the scope of its tactical counterpart in terms of length and resources involved in carrying out changes. Therefore, additional roles and resources are required to run the strategic renewal process successfully. Conducting the process most effectively requires the support of a strategic supply chain excellence center and the supply chain board. In addition, the market planner and the supply chain planner provide functional input for the strategic renewal process (see Figure 8.10).

In contrast to the roles of the market planner and the supply chain planner, which are assigned to individuals, both the strategic supply chain

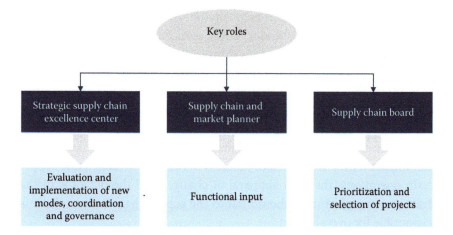

FIGURE 8.10
Key roles for the strategic renewal process.

excellence center and the supply chain board will have a broader scope of responsibilities and therefore those responsibilities should be delegated to a team or organizational unit. Figure 8.11 summarizes the key tasks within the strategic renewal process.

In general, improvement projects at the strategic level typically last longer (from initial analysis to final implementation) and require in-depth analysis as well as greater effort and expertise to manage the implementation project. For example, implementation of a new replenishment mode at a supply chain stage (e.g., the introduction of pull) will require a formal project and resources to achieve a successful outcome.

A supply chain excellence center can be made responsible for the in-depth analysis and review of a supply chain, the generation of strategic

Strategic supply chain excellence center	Supply chain board
▶ Leads the information and data-gathering process within the strategic renewal process	▶ Ultimate decisions on production and replenishment modes used in the supply chain
▶ Final analysis, preparation, and definition of strategic alternatives for the selection of modes	▶ Providing the strategic direction for the supply chain (e.g., based on corporate and operations strategy)
▶ Key responsibility for the implementation of change projects concerning modes	▶ Accountability for the definition and implementation of change projects regarding modes

FIGURE 8.11
Role descriptions for the strategic renewal process.

options, and the successful implementation of intended changes of production or replenishment modes. Those working in such a supply chain excellence center need to have a good business understanding and a broad supply chain perspective (e.g., covering a range of supply chain functions). In addition, they need to be credible and respected by other groups, good at working cross-functionally, and possess the necessary managerial and leadership skills to carry out those activities.

Since the entire strategic renewal process addresses questions of significant impact for the supply chain in terms of resources, budget, and supply chain operations, the final approval of changes and the outcome of the strategic renewal process should fall to the supply chain board, which is responsible for overall SCM.

8.2.3 Mapping Roles and Responsibilities to Renewal Processes

Defining the roles for LEAN Supply Chain Planning is the first step in designing renewal processes. To finalize that design, it is necessary to map the roles to the required tasks. In the following sections, we illustrate how to use the RACI (*responsible, accountable, consulted,* and *informed*) methodology to map the roles to the particular tasks within the tactical and strategic renewal processes. The RACIs we present are based on the process flow diagrams and tasks presented in Chapters 5 and 6. They help to define how the introduced roles for LEAN SCM should interact in the renewal processes and describe their participation in activities.

8.2.3.1 Mapping of Responsibilities within the Tactical Renewal Process

Within the tactical renewal process, three roles—the local planner, the supply chain planner, and the market planner—should be assigned to the various process steps for continuous review of tactical parameters. Figure 8.12 illustrates a RACI for the tactical renewal process, showing the tasks on the vertical axis, ordered according to the process flow, with the three roles on the horizontal axis.

8.2.3.2 Mapping Responsibilities within the Strategic Renewal Process

The strategic renewal process builds on four roles—the strategic supply chain excellence center, the supply chain board, the market planner, and

Phase	Task	Supply chain planner	Local (asset) planner	Market planner
Input for tactical renewal process	Input from strategic level	A, R	I	(R),C
	Supply and demand profiles	A, C	R	R
	Performance management	A, R	R	C
	Consolidate and validate data, evaluate performance	A	C	C
Adjustment of tactical parameters	Select parameters to be adjusted	A, R	R	C
	Calculate parameters and determine adjustment strategy	A, (R)	R	
	Review synchronization and modify parameters if necessary	A, R	C	
	Organizational confirmation	A	I	I
Implementation of parameters	Communicate adjustment strategy and new parameters	A, R	R	I
	Implement confirmed parameters	A, R	R	I

R = Responsible, A = Accountable, C = Consulted, I = Informed

FIGURE 8.12
RACI for the tactical renewal process.

the supply chain planner. As we have noted, individual employees fill the roles of market planner and supply chain planner, while both the strategic supply chain center and the supply chain board refer to groups of individuals responsible for certain tasks within the process. Figure 8.13 defines the recommended RACI for the strategic renewal process.

Phase	Task	Supply chain planner	Market planner	Supply chain board	Supply chain excellence center
Input for strategic renewal process	Operations and corporate strategy (alignment)	C		C	A, R
	Supply chain structure and business constraints	R	R	C	A
	Performance measurement	R	C	I	A, R
	Consolidate and validate data, evaluate performance	C		I	A, R
Strategic renewal	Assess need for renewal of modes	C	C	A	R
	Trigger further analysis (pre-assessment)	I		A	R
	Finalize and prioritize improvement areas	C		A	R
	Start proposal and evaluation projects for new modes	C		I	A, R
	Evaluate alternatives and select new modes	C		A	R
Results	Roadmap for implementation	R	C	A	R
	Implementation and adjustment	C	I	A	R

R = Responsible, A = Accountable, C = Consulted, I = Informed

FIGURE 8.13
RACI for the strategic renewal process.

8.2.3.3 Consideration of Capabilities and Resources

In addition to instituting effective process governance through the appropriate definition of roles and responsibilities, a company should also evaluate the necessary technical and managerial capabilities as well as the availability of sufficient resources to enable employees to execute their tasks in the renewal processes.

A clear description of each role is necessary to communicate the respective responsibilities and skill requirements. Role descriptors will also be required by line managers and the human resources department to identify appropriate candidates for each role. Besides the tasks and necessary technical skills and experience, the descriptors should also consider the "softer" behavioral skills that will be required to perform each role effectively (e.g., inter-personal skills, cross-cultural working skills).

It is important to recognize that the RACI analysis shows only where tasks should reside; it does not measure the resource or time requirements for each task. These assessments, however, need to be conducted to evaluate the effort required for each process step and to allocate the resources needed to accomplish each task and activity. It is important to avoid overloads, especially at initial implementation, since in that case some tasks will not be performed effectively and may lead to a reversion to old ways of working.

In this context, we emphasize that appropriate IT support (planning and decision support systems) reduces the workload involved in renewal processes considerably. We introduce the appropriate IT landscape and tools for LEAN Supply Chain Planning in Chapter 10.

Summary

When introducing LEAN SCM, a company should ensure that the appropriate organizational model is in place. The coordinated network model is the best suited for LEAN SCM processes in comparison with the centralized and decentralized model but requires a sound foundation of standardized processes and aligned behaviors.

To run LEAN SCM effectively, a company needs strategic and tactical renewal processes to ensure that supply chain planning meets internal and external requirements as well as seamless integration with other relevant planning processes such as S&OP.

Conducting the strategic and tactical renewal process requires the engagement of a range of roles and functions. For the tactical renewal process, the local planner, the supply chain planner, and the market planner are required; for the strategic renewal process, the supply chain planner, the market planner, the supply chain excellence center, and the supply chain board are required.

Finally, roles and responsibilities need to be defined for each process step needed to carry out the strategic and tactical renewal process. RACI methodology was used to define the degree of involvement of each role in a key process steps.

We have now described the visible enablers of LEAN SCM above and below the ground. In the following section, we explain how to use change management to support the LEAN transformation successfully.

8.3 MANAGING CHANGE AND TRANSITION FOR LEAN SCM

The transformation to truly LEAN SCM can have a significant impact on an SCM organization and the larger company in which it operates. Managing such change successfully is a prerequisite for establishing the necessary structures and processes involved in LEAN SCM. To return to the metaphor of planting and growing a tree as a symbol of successful introduction of LEAN SCM, no one who plants a tree must forget to water it. In this sense, change management can be compared to the water: without change management nothing grows—either above or below the ground. Change management is imperative for completing the transition in the supply chain organization from the current state to LEAN SCM. Figure 8.14 shows the major points we focus on in this section.

8.3.1 Focus Areas of Change Management

Successful implementation of LEAN SCM must address people, processes, systems, and the supply chain organization (see Figure 8.15). Therefore, the objective of change management is to manage change and transition in each of these areas, ensuring that the required capabilities for LEAN SCM are in place.

The transition to LEAN SCM has been carried out successfully if . . .

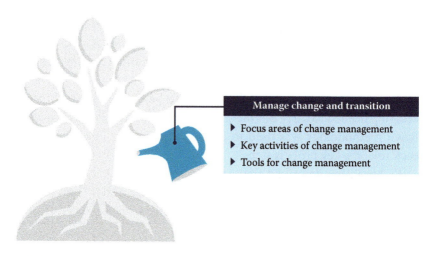

FIGURE 8.14

The LEAN Tree Model: elements of change and transition management.

- All employees in the supply chain are equipped with the right skills, knowledge, and behaviors regarding LEAN SCM concepts (e.g., Rhythm Wheel-based planning).
- The right supply chain organization in terms of structure and process-centricity has been established to support the LEAN SCM processes.
- The entire operations organization follows the customer-driven planning approach and embodies the voice of customers in every activity.

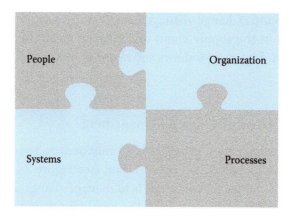

FIGURE 8.15

Change management should develop and align four elements.

- LEAN SCM processes have been established that allow for an end-to-end renewal and synchronization of tactical parameters and strategic supply chain modes.
- All planning processes are supported by adequate planning IT systems and LEAN SCM add-ons (see Chapter 10 for more information on recommended IT add-ons).

During all change activities, it is vital to remember the importance of the human element; focusing solely on the operational and technical aspects does not guarantee successful transformation of the supply chain. Those who are responsible for the implementation of new supply chain structures and principles need to understand the impact it will have on the individuals or groups who need to alter their knowledge, skills, attitudes, and behavior to accommodate the change.

Human beings experience diverse emotions when they are faced with a situation that they perceive is beyond their current capabilities, such as conducting planning by Rhythm Wheels. Emotional responses accompany organizational change and affect acceptance of that change. When transforming an organization, it is helpful to understand the typical emotional cycle of change and the psychological phases and emotions people experience. Understanding an employee's emotional state allows a dedicated change management team to interpret and manage the behavior of those affected by changes like LEAN SCM. The team will also track change measures, such as information and mobilization, to align them with the situational needs of the employees.

When these phases are ignored or not taken seriously, dysfunctional behaviors (e.g., increased anxiety, confusion, miscommunication, low morale, defensiveness, and territoriality) and resistance to change are likely to increase. In particular, the emotional state of an employee is significantly determined by his or her level of information and knowledge, again highlighting the importance of communication and personal capability development.

8.3.2 Key Activities of Change Management

Five activity fields need to be addressed during a LEAN SCM transformation. In practice, the required scope and necessary extent of each activity will vary with the type of supply chain and its business requirements, the organization around the supply chain, and the experience of those in the organization

FIGURE 8.16
Areas to address when introducing LEAN SCM.

with LEAN SCM principles. Nevertheless, we recommend always keeping these five steps in mind when transforming a supply chain (see Figure 8.16).

8.3.2.1 Change Planning and Benefits Tracking

The foundation for change is built by first assessing the readiness of an organization for adopting LEAN SCM. A clear and thoroughly documented picture of the as-is situation enables the development of change approaches tailored to the specific needs of each organizational function. Using the initial analysis, the next step is to develop a change plan that defines the nature and sequence of specific activities as well as the respective resource requirements for facilitating the change.

Tracking progress and benefits throughout the transition phase is important for project control and risk management. In addition, such measurement provides motivation by highlighting benefits even during the period of change. When early reports start to come in describing site or business unit successes (e.g., after applying Rhythm Wheel planning), others are normally prompted to increase their own efforts and demonstrate their capability.

8.3.2.2 Communication and Stakeholder Management

Consistent and active communication from the beginning is crucial for avoiding rumors and misleading information, reducing resistance, and increasing commitment and involvement among employees. A communication strategy should address every part and member of the supply chain organization, while focusing on the specific needs of each group.

Stakeholders can act as multipliers of information or opinion about change within an organization. These key figures may also own the necessary financial, intellectual, or physical resources to bring the LEAN transformation forward. It is natural to associate the influence and impact of stakeholders with seniority and management positions, but key stakeholders can also include opinion leaders, experts, and union representatives. Successful stakeholder management addresses all relevant groups of stakeholders in an individual way concerning the objectives, the way forward, and the projected benefits of LEAN SCM.

8.3.2.3 Leadership and Change Network Management

Strong leadership and mobilization of leaders across the supply chain organization is one of the key success factors for transforming a supply chain in accordance with the LEAN SCM paradigm. During the transformation phase, leaders should be developed at all levels of the organization. Although strong commitment from top management is vital to overall success, leaders should emerge at all hierarchy levels and functional areas relevant to LEAN SCM to support and encourage new behaviors.

A change network consists of leaders and change agents who possess the skills needed to drive change and are able to accept accountability during the transformation phase. A change agent is an individual or group that is responsible for designing, supporting, and implementing the change to LEAN SCM. Such a network can be used as a highly efficient and effective channel for distributing information about the progress of the LEAN transformation journey.

8.3.2.4 Organizational Alignment

The planning processes and organizational design principles presented in Section 8.2 should be considered to develop the supply chain organization as effectively as possible. To establish LEAN SCM processes in the organization, the following key steps are necessary:

- *Process-to-organization mapping:* Develop as-is and to-be process steps, including organizational responsibilities (RACI analysis). Special emphasis should be placed on the design of appropriate renewal processes.

- *Role and job profile definition:* Develop role descriptions for required roles, including tasks, responsibilities, reporting structures, interfaces with other roles, performance metrics, and skill sets.
- *Workforce capacity planning:* Conduct a workload analysis for each role. Agree to workload per role and identify the required number of jobs.
- *Capability assessment:* Conduct structured capability assessments to identify job candidates and training requirements. Such an assessment is a prerequisite for successful capability development.

8.3.2.5 Capability Development

Training is an integral part of capability improvement. Capability assessments should be used to identify any skill gaps and to ensure that training programs are focused on areas where improvements are needed. Typical training topics include such LEAN SCM concepts as Rhythm Wheels and pull replenishment, the strategic and tactical renewal processes, and using IT planning systems with LEAN add-ons. It is often helpful to use a modular training curriculum, consisting of *basic training* to lay the foundations and *specialized training* for (individual) capability improvement to ensure that new processes, roles, and responsibilities can be managed properly.

To increase motivation and involvement in training, a company might choose to reward employees for successful participation, perhaps offering financial incentives or opportunities for promotion when staffing new roles in the organization. Finally, it is important to recognize that capability development should comprise more than just training. To ensure a full and continuous transfer of skills and know-how, a company should consider establishing ongoing review of capability levels. This is best conducted by the supply chain excellence center which can act as steward of the LEAN SCM concepts and manage knowledge pertaining to the processes.

8.3.3 Valuable Tools for Change Management in LEAN SCM

Successful change management requires profound tools. In this section, we highlight several change management tools that go beyond the standard instruments (which are explained in specialized literatures).

The first instrument is supply chain simulation, which already has been mentioned. Simulation is an important step between qualitative concept discussion and its implementation in an organization. With simulation, the impacts of LEAN SCM can be evaluated with a computerized model of the supply chain. One of the key advantages of simulation is that it makes it possible to visualize the full supply chain. Many people can grasp the basic steady state set-up of a supply chain, but few are able to understand its dynamic response and how it behaves in practice. Such visualization includes not only the physical set-up of a supply chain, but also the physical flow of materials within it, planning processes to coordinate the flows as well as real-time performance measurements to evaluate the changes. In this way, it is easy to comprehend the new processes and concepts of LEAN SCM and understand how they affect the supply chain. People are usually more open to change when they can follow the flow of goods through the supply network and understand how LEAN SCM concepts work in practice. Simulation helps people explore the details of the LEAN SCM concept and grasp the subtleties of how it works. The opportunity to question and challenge gets people involved and is an excellent way of building commitment as they realize the effectiveness of the LEAN SCM approach.

The second instrument is interactive training tools to facilitate understanding of how the LEAN SCM concepts work in practice. A simplified supply chain is set up in the interactive training tool with focus on the new concepts. People can then test many alternate configurations and learn how the various parameters interact and impact KPIs. An example of an interactive training tool is the Rhythm Wheel Trainer (see Figure 8.17). The Rhythm Wheel Trainer provides a small-scale model of a Rhythm Wheel-controlled supply chain. It demonstrates the basic interactions between inventory levels and cycle time and provides a first impression of how these parameters interact. A big advantage is that the tool allows people to "play" with the settings by moving buttons and sliders and promptly seeing the effects of their actions. After a while, people understand how service level is influenced by various inventories and cycle times.

Probably the best way to gain hands-on experience with not only LEAN SCM concepts but also the associated processes is to construct a prototype in the IT planning system. A prototype is a fully functioning IT planning system informed by the actual locations, resources, and products of a company. The entire company is not mapped within a prototype; only a certain part of the supply chain with limited scope is modeled. This is usually sufficient to show how LEAN processes work in the company's IT

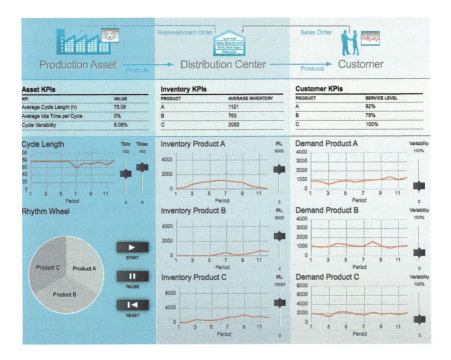

FIGURE 8.17

The Rhythm Wheel Trainer as interactive training tool (www.rhythm-wheel.com).

systems. Often the prototype is set up in one of a company's training systems. The supply chain planning community can then see what the LEAN processes mean for them in everyday business terms. They can test the concepts and processes and learn to understand the changes in tactical and operational planning they should expect in "real life." When people see their own products and resources, they are more willing to accept and identify with LEAN SCM processes. This also helps build familiarity with how the future planning IT system will look and work.

Summary

Managing change and transition is vital when introducing LEAN SCM. This section has highlighted three key facets.

First, we discussed the four focus areas of change management: people, organization, processes, and IT systems. Only when all areas are considered can change management enable a smooth transition to the LEAN SCM paradigm. Focusing solely on operational and technical process aspects does not ensure a successful transformation of the supply chain. Addressing the human element and recognizing the emotional cycle of change are crucial.

We then introduced the five change steps that should be followed during a LEAN SCM transformation—from change planning through communication to capability development.

Finally, we presented supply chain simulation, interactive training tools, and IT prototypes as highly valuable change management tools. These tools help to bring the supply chain to life and make LEAN SCM something real and tangible for staff: they are able to interact with a real supply chain rather than merely discuss concepts and theory. These tools have proved successful in past implementations and have provided a very practical means not only of training but also of convincing the supply chain organization community of the practicality and benefits of LEAN SCM.

CHAPTER SUMMARY

When transitioning to LEAN SCM, organizational transformation needs to be an integral part of your company's lean journey. Therefore, we have focused on the organization for LEAN SCM in this chapter by presenting guidelines and best practices for preparing and managing the transition.

Section 8.1 started "below the ground," suggesting that seemingly "soft factors" are highly important for adapting an organization to LEAN SCM successfully. The section showed the importance of top management buy-in, cross-functional leadership, and a shift in mindset throughout the SCM planning community.

We then moved "above the ground" in Section 8.2, focusing on "hard factors." The section surveyed organizational models that can be found in process industry supply chains and showed how to develop them. Subsequently, we briefly discussed the integration of the renewal process and the S&OP process, followed by the organizational roles required to participate in these processes. Tasks within the renewal processes were assigned to the appropriate roles using RACI methodology.

In Section 8.3, we highlighted the importance of change management when transforming an organization to LEAN SCM. Change management needs to focus on the areas of people, organization, processes, and technology. However, the personal dimension should command special attention by developing the necessary capabilities for LEAN SCM using unique training approaches and sustaining the motivation of the SCM community.

9

Performance Management for LEAN SCM

LEAN SCM promotes a culture of measurement in the supply chain that allows for a parameter-driven planning approach. Therefore, a well-designed and executed performance management system is fundamental to LEAN SCM. The good news first: if your company already has an effective performance management system in place, it does not need to change much to adopt LEAN SCM. It just needs to extend its framework in accordance with tailored LEAN SCM metrics and supporting design principles.

In this chapter, we focus on the key characteristics of successful performance management for LEAN SCM. This includes specific changes that are required to introduce LEAN SCM as well as best practices for effective performance management that provide the right basis for a smooth introduction. This chapter answers the following questions:

- What is performance management for LEAN SCM and how can it be designed?
- What metrics are especially important and suitable for LEAN SCM?
- Which best practices should be considered in performance management?

In Section 9.1, we explain the role of performance management in LEAN SCM and explain which changes in current performance management practices are required. Metrics that are especially tailored to LEAN SCM are presented in Section 9.2. To conclude, we present best practices for the design and implementation of performance management systems in Section 9.3, which greatly facilitate the introduction of LEAN SCM.

9.1 ROLE OF PERFORMANCE MANAGEMENT IN LEAN SCM

Performance management is crucial to linking supply chain planning processes effectively and is thus an integral part of LEAN SCM architecture. Effective performance management has always been an important enabler of successful SCM—the introduction of LEAN SCM further increases its importance. If your company wants to manage volatility effectively and establish a synchronized and consumption-driven supply chain, it should tailor its performance management practices to LEAN SCM principles.

Your company can leverage effective performance management practices by introducing a few very powerful changes that support LEAN SCM. An important focus area for LEAN SCM is the operational level of supply chain planning. Implementing takted and synchronized scheduling of assets with Rhythm Wheels requires the right KPIs for performance management. Most supply chains struggle in this area since it is difficult to translate envisioned operational planning and execution strategies into action.

9.1.1 Key Objectives of Performance Management for LEAN SCM

Well-executed performance measurement and analysis are two major attributes of successful performance management for LEAN SCM. As shown in Figure 9.1, performance management is a prerequisite for

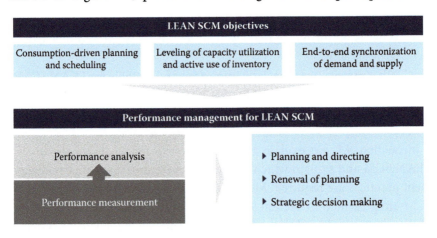

FIGURE 9.1
Key performance management objectives in LEAN SCM.

pursuing central LEAN SCM objectives: consumption-driven planning and scheduling, leveling of capacity, and end-to-end synchronization of the supply chain.

If the appropriate performance indicators are integrated and used in LEAN Planning (especially within the renewal processes), the various planning layers (from execution to strategic planning) can be tightly coupled. Your company's performance management practices will then strongly support strategic planning, enable effective management of people and resources, and allow for the regular renewal of planning based on LEAN SCM principles.

9.1.1.1 *Performance Measurement*

"If you can't measure it, you can't manage it": A simple sentence that makes clear why LEAN SCM encourages the measurement of meaningful metrics to foster supply chain performance. Without measurement of metrics such as CTA, it is simply not possible to evaluate whether the targeted supply chain takt has been achieved. Unlike traditional performance management, which focuses mainly on measuring past performance regarding costs and service, LEAN SCM strives for a high degree of agility and resilience by establishing a forward view of supply chain behavior.

9.1.1.2 *Performance Analysis*

Measuring performance is, however, only the first step to successful performance management. All LEAN SCM performance metrics therefore serve as inputs to renewal processes to ensure regular and systematic performance analysis. A key differentiator of performance analysis within LEAN Supply Chain Planning is its focus on organizational behavior. To ensure adherence to a defined supply chain takt and avoid self-inflicted variability, your company should determine whether its planners are to adhere to predefined production sequences. To this end, LEAN SCM provides appropriate metrics and guidelines for analyzing organizational behavior and improving it when necessary.

9.1.1.3 *Planning and Directing*

Effective set-up and execution of supply chain planning is the heart of LEAN SCM. The parameter-driven concept behind LEAN SCM requires the use of meaningful metrics to effectively plan cycle times, inventory levels, and production sequences. The novel idea here is to split planning into the configuration of parameters and the execution of those configurations in daily operations. To this end, performance management can support pre-configuring supply chain parameters and monitoring how they are executed in daily operations. We show that most changes in a company's system of metrics and performance management practices will take place at the operational level in order to achieve takted and synchronized planning and scheduling of supply chain operations. Across all planning layers, performance management within LEAN SCM is focused on measuring and analyzing end-to-end variability. Consequently, effective performance management strengthens planning capability and supports active management and reduction of variability along the supply chain.

9.1.1.4 Renewal of Planning

Weak adaptability to changing conditions is the Achilles' heel of supply chain planning and execution. Performance management within LEAN SCM supports supply chain planning renewal by systematically monitoring and analyzing current performance and projecting key planning KPIs and parameters into future. In this way, it allows for a direct temperature check of the current supply chain configuration and ensures quick adaptation to changed business conditions. Strategic and tactical renewal processes ensure that performance analysis becomes an integral part of supply chain planning and that required changes in planning and configuration can be triggered immediately. In this vein, the regular renewal of supply chain planning also ensures continuous improvement of information flows throughout the supply chain organization.

9.1.1.5 Strategic Decision Making

Before undertaking LEAN SCM Planning improvements, the physical supply chain set-up should be validated, as it provides the basis for effective strategic, tactical, and executional planning. LEAN SCM provides additional KPIs for this purpose to facilitate the creation of transparency and improve planning conditions. Using those KPIs as part of an end-to-end assessment (see Chapter 4) can guide decisions regarding network design or complexity management. Within strategic renewal processes, the measurement and analysis of performance further improves the supply chain configuration that is in place by providing the means for selecting the most effective production and replenishment modes on a regular basis.

9.1.2 Orchestrating Supply Chain Planning Processes Successfully

Performance management is essential for linking the various planning layers of LEAN SCM. Integrating these planning layers based on LEAN SCM principles can improve both the efficiency and agility of supply chain operations substantially. Without proper linkage between the layers of supply chain management, it is virtually impossible to propagate strategic objectives or enable continuous renewal of planning within the organization. To set up the LEAN Supply Chain Planning process landscape effectively, performance management should:

- Set well-defined top-down targets for planning and directing.
- Receive bottom-up feedback for renewal of planning by tracking and analyzing the performance.

9.1.2.1 Integration of Performance Management into LEAN SCM Planning Processes

Many companies know about the downsides of insufficient alignment and coordination between planning layers. Suppose your company wants to cut cycle times to reduce supply chain inventory and increase market responsiveness. However, although this may be a strategic objective, supply chain planning often struggles to translate cycle time reduction into measureable targets. Without clear and measurable KPIs at the shop floor level, planners continue to reschedule in a haphazard, ad-hoc way, jeopardizing the targeted cycle time reduction.

A key learning from such observations is that effective performance management depends on setting appropriate targets as well as using feedback and a monitoring process to review success when it comes to meeting those targets. For this reason, performance management plays a prominent role in the LEAN SCM architecture and is deeply rooted in all LEAN Supply Chain Planning processes.

In particular, performance management ensures that the strategic and tactical renewal processes can effectively link planning activities by means of top-down targets for planning and directing and bottom-up feedback

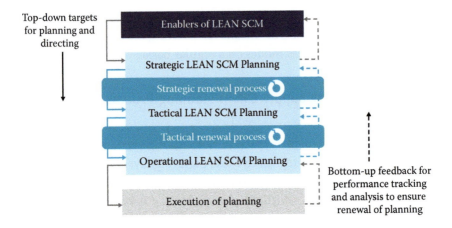

FIGURE 9.2
Performance management links the various planning layers of LEAN SCM.

for performance tracking and analysis (see Figure 9.2). This approach synchronizes supply chain planning across planning hierarchies. Without such synchronization, a company will typically fail to achieve central LEAN Supply Chain Planning objectives: reducing variability in the network and improving asset performance through a takt-leveled and synchronized supply chain.

9.1.2.2 Enabling Effective Configuration and Renewal of LEAN SCM Parameters

When implementing LEAN SCM, the interplay between tactical and operational planning changes considerably. As explained so far, in contrast to monolithic but slow and clumsy planning approaches such as MRP within ERP or APS systems, the LEAN SCM Planning approach is divided into a two-level planning and execution scheme. By determining the parameter configuration at the tactical level and leaving the execution of that configuration to the operational level, the entire planning process can be simplified considerably while providing greater agility and flexibility in the fulfillment of plans.

To achieve such tight coordination between and alignment of tactical and operational planning, performance management plays an important role. To link both layers successfully, an integrated top-down and bottom-up flow of information needs to be established. Traditional planning and scheduling is virtually incapable of synchronizing assets and allowing for

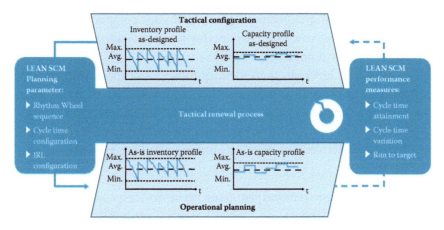

FIGURE 9.3
Performance management for configuration and smooth execution of planning.

demand-driven operations. To resolve these issues and eliminate the need for excessive use of detail-level forecasts, new KPIs and planning parameters should be introduced. Performance management for LEAN SCM has to establish, measure, and analyze those parameters within the tactical renewal process.

Figure 9.3 shows the separation into the tactical configuration and its execution within operational planning. To pre-configure the supply chain at this level, planning parameters such as a (designed) Rhythm Wheel sequence, cycle time, and inventory levels (IRLs) need to be determined and communicated to operational planning. By using LEAN Supply Chain Planning parameters such as cycle time boundaries for a Rhythm Wheel, it is possible to actively measure and manage future supply chain performance. Performance management provides the necessary input for evaluating the stability of a chosen configuration and thus supports identification of required adjustments within the tactical renewal process. LEAN SCM metrics such as CTA and CTV will also support the analysis of organizational behavior and guide the supply chain organization toward truly agile and flexible supply chain planning.

9.1.3 How the LEAN SCM Paradigm Changes Your Performance Management

As we have seen, effective performance management is one of the pillars of a successful introduction of LEAN SCM, forming an integral part of a proper LEAN SCM architecture. However, if your company has already established an effective supply chain performance management, introducing LEAN SCM will not entail a radical change.

The next step, however, is to identify those elements of your current performance management practices that require special attention when transforming the supply chain. We recommend several powerful changes to provide the capability within performance management of supporting LEAN SCM objectives and effectively linking planning processes:

- **Integration into renewal processes:** Performance management should be linked to strategic and tactical renewal processes. This requires measuring and analyzing KPIs that support LEAN SCM objectives. Effective renewal processes will ensure that LEAN SCM KPIs are used to set targets and evaluate performance on a regular basis.
- **Focus on organizational behavior:** LEAN SCM emphasizes sound organizational behavior along the supply chain. To ensure adherence to predesigned Rhythm Wheel sequences and avoid unwanted re-scheduling, for example, the newly implemented behavioral norms should be measured and rewarded by your performance management system.
- **Establish a prospective performance view:** Traditional supply chain performance management has typically taken a strongly retrospective view by focusing on past performance. LEAN SCM performance indicators in contrast provide a prospective view of supply chain behavior. This allows a supply chain organization to focus on preventing fires instead of fighting them.
- **Management of end-to-end variability:** Mastering variability is a key imperative in the VUCA world. Tailoring performance management to this central objective of LEAN SCM requires the use of appropriate metrics as well as their systematic use in planning.
- **Integration of LEAN SCM KPIs:** Traditional KPIs such as customer service level and inventory turnover are still valid and provide insight into your supply chain performance. To fully enable performance management for LEAN SCM and conduct the required changes, however, we recommend the use of several new and very effective metrics such as cycle time for planning.

Summary

Performance management is essential for translating key LEAN SCM objectives—consumption-driven planning and scheduling, leveling of capacity, and end-to-end synchronization—into your company's planning processes. Tailoring performance management practices to LEAN SCM ensures that the various planning layers are tightly coupled and all LEAN Supply Chain Planning processes can be carried out successfully. Especially at the tactical and operational levels, it is important to recognize that new KPIs such as Rhythm Wheel sequence and cycle time should be introduced to establish a parameter-driven planning approach following LEAN SCM principles. As shown in this section, only a few very powerful changes are recommended to tailor your company's performance management system for compliance with LEAN SCM principles.

9.2 HOW TO MEASURE LEAN SCM PERFORMANCE

Choosing the right metrics is the foundation of any successful performance management system. Traditional KPIs such as customer service level and inventory turnover remain valid and provide insight into supply chain performance. However, to implement LEAN SCM properly, we recommend using metrics that are tailored to LEAN SCM objectives. In this section, we provide a condensed overview of the LEAN SCM metrics that should be integrated into your current performance management and planning processes.

Most of the suggested changes regarding metrics are required at the tactical and operational level of supply chain planning. Nevertheless, to fully benefit from demand-driven and synchronized operations, we also provide KPIs that support more effective selection of production and replenishment modes within the strategic renewal process and also facilitate the assessment of the overall LEAN SCM maturity of a supply chain. In this chapter, we highlight those metrics that are, based on our experience, less common or even absent from most supply chain organizations.

9.2.1 Metrics to Link Tactical and Operational LEAN Supply Chain Planning

To link tactical and operational LEAN Supply Chain Planning in the most effective way, the tactical renewal process is a central success factor. Within the LEAN SCM architecture, the tactical renewal process is used to pre-configure the tactical supply chain parameters and provide the framework for smooth execution. To keep pace with changing demand and supply conditions, companies must ensure that their production and replenishment parameters are always up to date and renew them in accordance with business conditions on a regular basis.

As shown in Figure 9.4, the relevant metrics at this planning level can be divided into production and replenishment metrics, along with KPIs for measuring organizational behavior in planning. Chapter 7 describes in detail the application of these KPIs at the planning execution layer. The exclamation mark within the table highlights those KPIs that are, based on our experience in process industries, less frequently adopted in supply chain planning. However, to ensure smooth introduction of LEAN SCM, they should be integrated into supply chain performance management. All of these metrics provide important input for renewal processes and should therefore be measured and reviewed continuously.

On the production side, the chosen production sequence and cycle time of your Rhythm Wheel needs to be evaluated by appropriate metrics. Measuring CTA and CTV as well as RTT provides a comprehensive overview of the stability of production planning and scheduling. The key is here to link the production side with metrics for replenishment planning. To ensure truly demand-driven operations and high customer satisfaction, the attainment of defined service levels and inventory levels should be monitored and evaluated. A metric that is often not explicitly included in traditional supply chain performance management is DSR. By measuring this KPI, it is possible to prevent dead stock from accumulating in the supply chain, causing costly write-offs on inventory.

Finally, bear in mind that a key change in performance management comes from the focus on organizational planning behavior. With the appropriate KPIs for measuring manual adjustments within production and replenishment processes, further insights into planning performance are generated. Monitoring the manual adjustments conducted in execution supports the identification of potential gaps between designed and executed planning and makes it possible to reward behavior that supports the new planning norms.

| Enablers of LEAN SCM |
| Strategic LEAN SCM Planning |
| Tactical LEAN SCM Planning |
| Operational LEAN SCM Planning |
| Execution of planning |

Category	Metric	Description
Tactical production planning	Cycle time attainment (CTA)	Measures cycle length over time and evaluates the difference between planned and actual cycle time
	Run to target (RTT)	Ratio between produced quantity and demanded quantity in every production cycle
	Cycle time variation (CTV)	Ratio between standard deviation and mean of cycle time
Tactical replenishment planning	Service level attainment (SLA)	Ratio between actual service level and target service level
	Target inventory attainment (TIA)	Ratio between actual inventory level and planned inventory level
	Dead stock ratio (DSR)	Share of inventory with no turnover
Planning behavior	Manual production adjustment	Share of manual adjustments, e.g., manual changes of designed production sequence
	Manual replenishment adjustment	Share of manual adjustments, e.g., manual changes of replenishment quantities

FIGURE 9.4

Essential metrics for linking tactical and operational LEAN Supply Chain Planning.

To synchronize operations and cycle times across assets and supply chain locations, it is important to ensure that planning parameters and metrics be measured and analyzed in a standardized fashion; inconsistent use of metrics causes severe problems for supply chain planning.

9.2.2 Metrics for Linking Strategic and Tactical LEAN Supply Chain Planning

Within the LEAN SCM framework, we have emphasized the importance of selecting the right supply chain modes to ensure that a company's requirements are met. In the face of constantly changing business conditions, reviewing production and replenishment modes within the strategic renewal process is essential to ensuring a competitive supply chain. Effective selection of supply chain modes (such as switching from push-to-pull replenishment when customers demand greater flexibility) depends on the facts provided by the performance management system.

Figure 9.5 provides an overview of key metrics that should serve as inputs to the strategic renewal process. KPIs such as operating costs, capacity utilization, service, and inventory have proven their value over many years

Enablers of LEAN SCM
Strategic LEAN SCM Planning
Tactical LEAN SCM Planning
Operational LEAN SCM Planning
Execution of planning

Category	Metric	Description
Strategic production planning	Operating costs	Operating costs at asset (labor, machine), split into fixed and variable components
	Capacity utilization/OEE	Ratio between capacity consumption and available capacity
Strategic replenishment planning	Service level	Event-oriented α-service level and quantity-oriented β-service level (or fill rate)
	Inventory level	Total inventory development at stock-keeping point
End-to-end supply chain planning	Variability amplification ("Bullwhip" effect measurement)	Normalized measure for variability amplification, defined as variability of placed orders divided by variability of received orders
	Global lead time	Time required to convert raw materials into the finished products delivered to customer
	Supply chain takt	Takt time per asset/site defined as the average time period between orders of the same product
End-to-end supply chain agility and resilience	Up-/downside flexibility (of processes & resources)	Measures capability of expanding/decreasing output of capacity within a certain time or cost range
	Time to recover (from unplanned disruptions)	Measures unexpected elapsed time between a risk event and recovery to normal operational conditions

FIGURE 9.5

Essential metrics to link strategic and tactical LEAN Supply Chain Planning.

and are thus valuable input. However, although those metrics have been widely used for many years, they involve trade-offs that merit attention. Typically, it is impossible to achieve high service levels, low inventory, and full capacity utilization at the same time—therefore, clear priorities associated with distinct targets must be set within the strategic renewal process.

Again, we denote with exclamation marks those KPIs that are rarely used in supply chains but are essential to mitigating variability and synchronizing the end-to-end supply chain following LEAN SCM objectives. For true end-to-end planning, variance amplification must be measured along the entire supply chain. Most supply chains struggle with this since their performance management systems have not analyzed the extent to which volatility is self-inflicted due to ineffective configuration of supply chain modes. For instance, push replenishment at the wrong supply chain stage can cause substantial variability that jeopardizes overall supply chain performance.

Analyzing the synchronization of the end-to-end supply chain requires monitoring the extent to which individual assets or stages adhere to the globally defined supply chain takt. Measuring and analyzing global lead time yields a holistic overview of supply chain responsiveness and the time needed to create value for customers. By complementing these KPIs with metrics for the upside and downside flexibility of processes and resources, the agility and resilience of the current configuration of supply chain modes can be evaluated explicitly.

If supply chain performance falls below expectations based on the KPIs listed in Figure 9.5, one potential root cause is that a company has chosen unsuitable supply chain modes. In addition, a company should also project KPIs into the future to assess the robustness of the supply chain configuration with respect to growth expectations on the market side. To guide your subsequent tactical decisions regarding the cycle time of assets, for example, target values for such KPIs should act as input for tactical renewal processes. This ensures tight linkage between strategic and tactical planning and guarantees consistency of decisions across planning layers.

9.2.3 Metrics for Assessing the Maturity of a Supply Chain for LEAN SCM

The appropriate set-up of a supply chain improves planning considerably. As explained in Chapter 4, no one would build a multi-story house on sandy ground, just as your company should not attempt to improve strategic, tactical, or operational planning within an ineffective physical supply chain set-up. To prepare a supply chain for LEAN SCM, the root causes of variability, complexity, and uncertainty in the supply chain set-up should be eliminated as far as it is economically feasible to do so.

The metrics shown in Figure 9.6 make it possible to assess the LEAN SCM maturity of a supply chain systematically. When assessing a supply chain, the respective metrics can be tied to sources of waste such as over-supply, over-processing, waiting, wasted motion, transport time, inefficient inventory control, defects, and inappropriate customer service. By tracking these metrics, the supply chain organization obtains a comprehensive overview of its level of maturity and the progress it is making in terms of waste elimination along the entire supply chain.

Some of these metrics, such as waiting time or value contribution, might be familiar to you. We see, however, that most companies regard them primarily as KPIs for measuring activity on the shop floor level. For example, it is important to recognize that waiting times are an issue not only for shop floor operations but also for total supply chain performance. Thus, the key here is to apply those metrics at the supply chain level and integrate them into supply-chain-wide performance management. In the following section, we again highlight KPIs that are less often adopted by most supply chain organizations.

The effectiveness with which assets are leveled is indicated as much by the *stability* of asset utilization as by the average utilization. Measuring the KPI information distortion makes it possible to better understand whether the availability or processing of planning data is an issue for an organization. Complexity is a key challenge for many companies. However, the product portfolio is not the only source of complexity—complexity can also be rooted in the supply chain itself. By measuring the number of organizational interfaces and system interfaces involved in the supply chain, for instance, it is possible to assess the supply chain's contribution to complexity. Finally, by introducing a clear inventory classification, inventory remains tied to clear objectives, thereby minimizing inefficient use of working capital.

When conducting a top-down assessment of your supply chain, as described in Section 4.2.2, these metrics help you to enrich the gap analysis indicating the difference between the as-is and the ideal state, which then identifies the starting point for improvement of the supply chain in preparation for adoption of the LEAN SCM paradigm. The KPIs we have introduced here can guide initiatives for network design or complexity management to improve the physical supply chain set-up. Thus, KPIs support both the start of a LEAN SCM journey as well as the ongoing evaluation of achievements.

| Enablers of LEAN SCM |
| Strategic LEAN SCM Planning |
| Tactical LEAN SCM Planning |
| Operational LEAN SCM Planning |
| Execution of planning |

Category	Metric	Description
Supply efficiency (avoidance of oversupply)	Stability of utilization (of resources and processes)	Fluctuation of capacity utilization
	Dead capacity ratio	Share of idle capacity
Processing efficiency (avoidance of overprocessing)	Fulfillment efficiency	Over-fulfillment of customer requirements, e.g., actual lead time to customer shorter than requested one
	Planning efficiency	Over-planning above business requirements in terms of, e.g., planning granularity and frequency
	Value contribution (of activities and processes)	Ratio between customer value-added and total costs
Time efficiency (avoidance of waiting)	Information distortion	Time between data availability and data processing, e.g., for order information
	Waiting time (of activities and processes)	Ratio between waiting time and total time, e.g., for order fulfillment process
Motion efficiency (avoidance of inefficient motion)	Number of organizational interfaces	Complexity of interfaces and responsibilities for process governance
	Number of system interfaces	Number of systems and data formats used in supply chain
Transport efficiency (avoidance of inefficient transportation)	Routing complexity	Number of used routings for one origin–destination pair
	Transportation time ratio	Ratio of transportation time to total lead time
	Landed costs	Total landed costs per distance and per weight unit (e.g., per mile and per pound)
Inventory control efficiency (avoidance of high inventories)	Inventory classification	Share of inventory without clearly defined purpose (e.g., not classified as safety, cycle, or policy stocks)
Processing quality (avoidance of defects and inappropriate service)	Yield factor	Ratio between produced/ordered items and fully functional/received items (percent complete and accurate)
	Customer satisfaction	Expressed by service level and on-time delivery measures
Continuous improvement efficiency	Waste elimination effectiveness	Ratio of eliminated waste to acknowledged waste
	Return on improvement	ROI for improvement activities/projects

FIGURE 9.6

Metrics for assessing the LEAN SCM maturity of a Supply Chain.

Summary

Selecting appropriate metrics lays the foundation for successful LEAN SCM performance management. In this section, we have presented KPIs designed specifically to support strategic and tactical LEAN SCM Planning and ensure smooth execution of planning at the operational level. By integrating such metrics into LEAN SCM renewal processes, your company can effectively pre-configure its supply chain to ensure that the most suitable production and replenishment modes are always chosen. To establish a supply chain set-up that perfectly suits the LEAN SCM paradigm, your company can use the metrics presented here for assessing the LEAN SCM maturity of its supply chain, guiding you in systematically identifying and eliminating the root causes of variability, complexity, and uncertainty in its supply chain set-up.

9.3 FIVE POINTS TO CONSIDER FOR SUCCESSFUL PERFORMANCE MANAGEMENT

Effective performance management is a prerequisite for LEAN SCM. To facilitate the successful introduction of LEAN SCM, certain key building blocks of a performance management system should be in place. As shown in Figure 9.7, the elementary building blocks of successful performance management for LEAN SCM can be separated into two main components: design and implementation. In the following, we highlight best practices for each of these dimensions.

9.3.1 Develop a Balanced and Comprehensive System of Metrics

Performance measurement and thus effective design of the underlying system of metrics is the backbone of any performance management system. Typically, a system of metrics will consist of a range of metrics that reflect the various needs of decision makers in the supply chain organization, reporting either financial (monetary) data or operational data (e.g., cycle times, inventory levels, service levels).

Since it is not feasible to measure every supply chain activity, identifying a reasonable number of metrics that support a company's strategic objectives and meet the requirements of the supply chain is crucial.

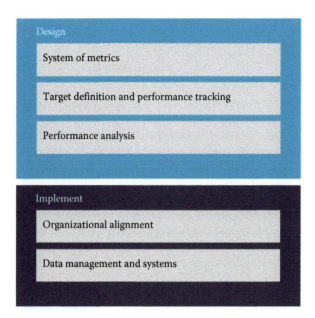

FIGURE 9.7
Building blocks of successful performance management.

Having to track hundreds of metrics in performance models such as SCOR often overwhelms a supply chain organization. Therefore, LEAN Supply Chain Planning by definition encourages the use of a manageable number of parameters to plan and control supply chain activities and performance.

The selected performance and planning metrics should be used and maintained as a holistic system, such that they complement and support each other and provide a balanced picture of overall supply chain performance and allow for end-to-end coordination along the supply chain. The criteria listed in Table 9.1 can serve as a checklist for evaluating the competitiveness of the overall system of metrics; a balanced and comprehensive system of metrics is characterized by strategy focus, end-to-end integration, balancedness, forward orientation, and lean focus.

9.3.2 Effective Target Definition for Performance Tracking

Identifying what needs to be measured is just the first step in developing an effective performance management system; defining appropriate targets is the next step. To be evaluated, metrics have to be linked to target

TABLE 9.1

Criteria for a Balanced and Comprehensive System of Metrics

Criterion	What to check?	
Strategy focus	Does the system of metrics translate the supply chain and business strategy to all decision-makers and is it connected to the reward systems?	✅
End-to-end integration	Does the system of metrics capture all relevant activities and dimensions of performance from both a planning and a monitoring perspective?	✅
Balancedness	Are all performance dimensions and supply chain segments represented by an adequate number of meaningful metrics?	✅
Forward orientation	Does the system of metrics track activities and indicators that allow for assessing future performance and process behavior?	✅
LEAN focus	Does the system of metrics support LEAN SCM and continuous improvement programs?	✅

values, so that the value of the measure can be assessed in terms of meeting expectations. Like the system of metrics itself, a balanced and comprehensive set of performance targets is essential and should be closely linked to the strategic objectives of the supply chain organization.

A stable set of performance targets should avoid conflicts across distinct targets. It is wasteful for the entire supply chain if performance targets in one area jeopardize the objectives set for another area. In particular, it is important to avoid setting planning parameters in the supply chain that conflict with performance targets that are assigned to individuals whose activities depend on those systems.

For instance, a typical conflict might arise between production and customer service. While production aims for high asset utilization with large lot sizes and optimized production sequences, customer service seeks high flexibility in production for serving customer demand. In this example, planning parameters and the procedures used to determine order quantities would have to balance both objectives (see Figure 9.8 for typical trade-offs to be considered).

Defining targets is not a one-time exercise. Targets should be reviewed on a regular basis, and metrics and associated targets should be highly visible and monitored at all levels of the supply chain organization. It should be recalled here that performance tracking not only should address past performance, it should also allow for assessing expected future performance. Finally, effective performance management integrates targets in the reward systems of an organization.

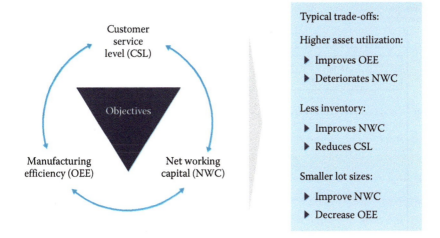

FIGURE 9.8
Trade-offs when balancing supply chain targets.

9.3.3 Systematic and Regular Performance Analysis for Sustainability

If target levels for metrics are not met, it is important to understand the causes of such deviations from targeted performance. Understanding these causes will serve as the starting point for corrective actions and performance improvement through which a company can regain the target performance level for each metric. To define appropriate (counter-) measures, a thorough understanding of what constitutes the observed exceptions as well as a timely response is needed.

The first step in identifying sources of underperformance is to determine whether a target violation is driven by internal issues or caused by external events or changes. This distinction is important to avoid blaming the owner of an unsatisfactory performing metric for an external change that is beyond his or her influence. Second, it is important to understand whether a performance deviation is systematic or caused by erratic or exceptional events. There is a big difference between a one-time event such as the breakdown of a machine and a design flaw in supply chain processes.

Systematic performance tracking and analysis generate substantial value. For example, consider the case of receiving a major tender in Asia for one of your key pharmaceuticals. As shown in Figure 9.9, you need to raise your production output, and thus your inventory to deliver the tender-agreed amount that exceeds normal demand substantially. Performance management makes it possible to raise inventory targets in a controlled

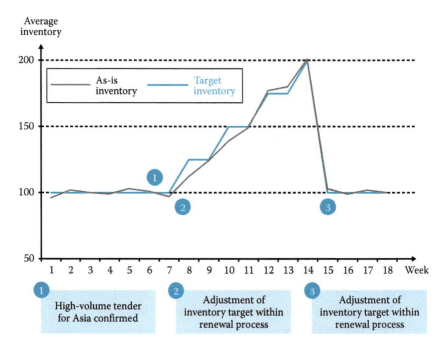

FIGURE 9.9
Supporting supply chain decisions with performance management.

and timely way and communicate the required changes across the entire organization. Once the tender is delivered, effective performance tracking ensures returning to a normal operating mode. This practice is especially powerful if it is integrated into appropriate supply chain processes such as the tactical renewal process for LEAN SCM.

Thus, maintaining systematic and regular performance analysis requires ensuring that related tasks and activities have clearly defined responsibilities and are integrated into supply chain processes. Supply chain processes can be made substantially more effective with the proper use of support systems for analyzing performance and alerting the organization to disruptions. Finally, a company's performance management system should ensure that, after key metrics have been measured and analyzed, the supply chain acts upon the findings.

9.3.4 Create Clear Responsibility for Metrics

Performance management needs to be aligned with supply chain processes to establish a true performance-driven organization. The performance of each key process step in the supply chain organization should be measurable by a set of specific metrics. This is especially important for structured and regularly conducted processes such as the tactical and strategic renewal processes.

Every metric must have clear owners who are personally responsible and accountable. The same holds true for activities related to performance tracking and analysis. To balance potentially conflicting metrics such as service levels and inventory levels in the supply chain organization, we recommend assigning responsibility for important but often conflicting metrics (and thus target fulfillment) to separate managers or organizational units.

To fully anchor performance management in the supply chain organization, the fulfillment of metrics should be incorporated into a company's reward systems. This creates strong incentives for employees to fulfill targeted objectives and therefore requires strong involvement of the human resources department. Finally, performance metrics that are used to evaluate and compensate personnel must not conflict with the planning parameters and targets that have been set for the supply chain.

9.3.5 Use Data Management and IT Systems for Support

To obtain meaningful metrics for performance management, it is crucial to ensure high data quality as well as reliable processes for data processing. To minimize the manual effort involved in both calculating and analyzing metrics, appropriate IT systems should be in place. In particular, a company needs planning systems that support the setting of balanced and strategically linked targets for the supply chain.

In many companies, compiling and analyzing metrics is very time consuming. Information is often manually aggregated from operational data sources and is prone to error and significant delays. If metric tracking is manual, numbers are often calculated incorrectly or inconsistently over time. To avoid such problems, appropriate systems for automated data extraction and processing should be in place.

Furthermore, decision support systems for analyzing metrics and defining meaningful targets are needed by both managers and regular employees. If appropriate metrics fall below their targeted values, such systems should alert users and propose appropriate countermeasures. Integrating advanced business analytics such as simulation makes it possible to analyze even highly complex and challenging supply chain problems and provides clear and well-grounded answers.

In Chapter 10 of this book, we lay out in detail an IT landscape that is suitable for LEAN SCM and illustrate how performance management can be supported by IT tools (e.g., an IT component for Rhythm Wheel monitoring).

Summary

Effective performance management is a prerequisite for the successful introduction of LEAN SCM. In this section, we have introduced important building blocks of performance management that provide that right foundation for LEAN SCM. The section summarizes best practices for designing a balanced and comprehensive system of metrics and ensuring that targets for the supply chain are effectively set and tracked. To ensure that performance management leads to the right actions and guides supply chain improvement, we have highlighted the value of systematic and regular performance analysis. Besides adopting a good design, your company should also pay attention to the right implementation of the performance management system. This requires organizational alignment around performance management and providing effective data management and supporting IT systems.

CHAPTER SUMMARY

In this chapter, we have explained why performance management is an integral part of LEAN SCM, providing the basis for running all LEAN Supply Chain Planning processes effectively. Nevertheless, LEAN SCM does not require starting from the scratch and building an entirely new performance management system for your company's supply chain. If your company already has a well-designed and implemented performance management for its supply chain, only a few changes are required. You need only to extend your company's framework using tailored LEAN SCM metrics and supporting design principles.

Performance management for LEAN SCM ensures above all that all planning processes are tightly synchronized and coupled along the entire supply chain. To this end, we offer KPIs and planning parameters that are tailored to the planning processes and objectives of LEAN SCM. By integrating them into the strategic and tactical renewal processes, your company can effectively pre-configure its supply chain to ensure that the best supply chain modes are always chosen. The LEAN SCM principles and metrics for performance management presented here will also help your company to establish a prospective view of supply chain behavior that helps to avoid firefighting due to unplanned events.

Overall, you can create considerable value for your company through effective performance management. Following performance management practices tailored to LEAN SCM objectives reduces variability along the supply chain and synchronizes operations very effectively. In the end, these practices will generate substantial cost savings and service improvements.

10

The Planning System Landscape for LEAN SCM

Readers who have read the previous nine chapters of this book have demonstrated their interest and confidence in LEAN SCM and its benefits. The question now is: How can your company realize these LEAN SCM benefits? Chapters 8 and 9 explained in detail the changes in your SCM organization that are involved in implementing LEAN SCM. We hope you now understand and are prepared to make a commitment to the transformation processes that must be considered to ensure the successful implementation. Now, however, you may be asking yourself about the impact of LEAN SCM on your company's IT systems. This is absolutely a valid question.

Today, supply chain and production planning are typically embedded in transactional ERP systems, which nowadays are often supported by so-called APS. Do these system architectures sufficiently support LEAN SCM concepts or is further IT support needed? To help you answer this question, we list the key IT requirements below (Figure 10.1) that are based on the core characteristics of LEAN SCM concepts.

These requirements reveal that ERP/APS system architectures do not sufficiently support LEAN SCM. In particular, the pre-configuration of tactical LEAN SCM parameters and the support of production planning on an operational level are not effectively supported by conventional ERP/APS system architectures. This raises the question of how best to acquire the necessary IT support for LEAN SCM. Can organizations secure and leverage huge ERP and APS investments they have made in the past? Can current IT system architectures be transformed and enhanced to support LEAN SCM principles? For those who are curious about the answers to these questions, the following chapter provides the relevant information.

In Section 10.1, we offer insights into the progress that has been made in recent decades in IT system support for planning processes. In this context,

Key IT requirements	Reasoning	ERP	ERP+ APS
Global end-to-end supply chain transparency	Timely and accurate information about customer demand, inventory levels and capacity availability is a core pre-condition for implementing a pull-based SCM operating model and optimal pre-configuration of LEAN SCM parameters.	✗	✔
Integrated end-to-end planning	LEAN Planning requires an integrated view of the whole supply chain while considering interdependencies between local and global planning, such as the impact of cycle times on required supply chain inventories.	✗	✔
Simultaneous planning	Rhythm Wheel-managed production requires simultaneous planning of materials, capacities and sequences due to extensive interdependencies that are characteristic of process industries with their dynamic changeover times.	✗	✔
Support of LEAN Planning on a tactical level	The pre-configuration of tactical LEAN SCM parameters, such as cycle times and inventory replenishment levels, is data-intensive and requires algorithm-rich calculations across all sites and inventory-keeping locations.	✗	✗
Support of LEAN Planning on an operational level	LEAN SCM requires that IT systems deal appropriately with process and demand variability, e.g., by triggering short-term adjustments of Rhythm Wheels or structured monitoring of relevant performance indicators.	✗	✗

FIGURE 10.1
IT requirements for LEAN SCM.

we emphasize that today's planning solutions should be extended or customized by adding the capability of LEAN Planning support. Section 10.2 introduces today's conventional IT system architectures with their planning components. We show how your company can leverage its existing IT solutions to create a LEAN-capable IT architecture. Chapter 10 concludes by highlighting complementary LEAN SCM add-ons that should be implemented to enable the required IT support for LEAN Planning (Section 10.3).

10.1 THE EVOLUTION OF IT PLANNING SYSTEMS

From the beginning of the industrial revolution until the late 1950s, IT played almost no role in supporting the processes and activities that are nowadays grouped under the SCM umbrella. Hence, supply chain processes such as ordering materials, production planning and scheduling, managing shipments, and so on were almost always performed manually. This certainly changed in the 1960s, when early software systems made it possible to realize the first process automation concerning replenishment control and later MRP. However, supply chain planning was still not significantly supported. Since that time, however, the evolution of modern IT has enabled mature supply chains to be managed more efficiently.

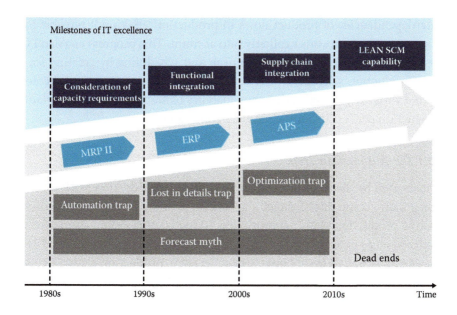

FIGURE 10.2
Evolution of IT planning systems.

Figure 10.2 depicts the most important milestones and corresponding dead ends that have occurred with respect to the evolution of IT support for SCM.

10.1.1 MRP II: Consideration of Capacity but Captured in the Automation Trap

With the advent of manufacturing resources planning systems (MRP II), the first major step in the evolution of IT support for SCM was taken. However, a phenomenon known as the automation trap clouded the brilliance of this progress.

10.1.1.1 Consideration of Capacity Requirements

In the 1970s, many companies used MRP systems that were capable of planning only for individual products independently from one another. Because this approach neglected capacity constraints, creating feasible production schedules remained a manual task.

In the 1980s, IT support for SCM made the first remarkable jump in its evolution. For the first time, systems were capable of considering the interdependencies of product flows in production, making it possible to take capacity constraints explicitly into account. This progress enabled IT systems to support production planning and scheduling by creating feasible production schedules. This technological advance is typically connected with MRP II, indicating the capability of managing the entire plant production process including capacity constraints.

10.1.1.2 Automation Trap

Encouraged by the technological enhancements embedded in MRP II systems, many companies formed a general aspiration for completely automating production planning processes. To realize this vision, efforts were undertaken to explicitly model specific production environments in MRP II systems, which were supposed to be reflected in production plans. Owing especially to technological constraints, it was soon recognized that these efforts did not pay off. Human intervention still remained essential to any effort to incorporate real-world conditions into the planning results.

10.1.2 ERP: Functional Integration but Lost in the Details

The advent of ERP systems meant the second milestone in the evolution of IT support for SCM. However, many companies found themselves lost in the details.

10.1.2.1 Functional Integration

Following the notable progress that MRP II systems brought about in IT support of production planning and scheduling, the next major advancement concerned integration across all functional units of an entire company. ERP systems aimed to support all activities within a company on a single software program. Hence, the software needed to serve the requirements of disparate departments such as finance, human resources, and marketing and sales as well as manufacturing planning and control. Through such extensive integration of functionality, it seemed undeniable that the efficiency of intra-company processes could be significantly increased.

10.1.2.2 The Lost-in-Details Trap

In the 1990s, driven by technological progress, companies attempted to establish decentralized production systems that were supposed to manage production processes that were tailored to the needs of local production environments. This idea seemed promising since such an approach was thought to overcome the shortcomings of centralized production systems. However, decentralized production systems did not fulfill initial expectations. This was due to two reasons in particular. First, the attempt to account for every detail in a production environment in planning systems did not bring notable benefits because the interdependencies of various factors proved to be too complex to be modeled accurately. Second, additional scale-related problems occurred, since detailed consideration of local production specifics required a vast amount of input data. To obtain and maintain this, detailed data was a tough challenge in practice and in most cases was far from being achieved. As a consequence, it was often impossible to avoid unsatisfying planning results. As a result of these challenges, decentralized production systems have not been broadly adopted in process industries.

10.1.3 APS: Supply Chain Integration but Caught in the Optimization Trap

With the rise of advanced planning technology, the third major milestone in the evolution of IT support for SCM was achieved. Nonetheless, its potential benefits could not be readily enjoyed, as companies found themselves caught in the optimization trap.

10.1.3.1 Supply Chain Integration

The implementation of ERP systems paved the way for remarkable progress in the functional integration of processes within a company. However, room for improvement remained with respect to the integration of supply chain stages. One aspect concerned production planning across separate sites. Particularly in cases with several ERP systems in place, cross-factory production planning lacked efficiency. Moreover, ERP systems typically do not support the organic integration of supply chain partners such as suppliers and customers. The remedy for this shortcoming was provided by modern APSs. Production planning across sites as well as the integration of suppliers and customers is typically supported by these systems. Hence, APS made it possible to realize a truly integrated supply chain, enabling end-to-end production planning, including the integration of external partners.

10.1.3.2 The Optimization Trap

With advanced planning technology, data from along the entire supply chain are potentially available. Based on such data and significantly increased data-processing capabilities, via cloud computing or in-memory technology, for example, modern optimization engines promise to optimize complex production plans and schedules within a very short timeframe. However, companies increasingly realize that optimization is not a silver bullet for solving all planning challenges. The absence of comprehensive planning results is a major obstacle in this context. Typically, responsible planners can understand the provided results only with difficulty and thus view optimization engines as black boxes. Lack of acceptance by production planners is often the consequence of such process opacity, implying that systems are not consistently used.

10.1.4 The Forecast Myth: An Overarching Obstacle

In addition to the abovementioned traps, which have impeded the evolution of IT planning systems in its core phases, the forecast myth must be seen as an additional and overarching obstacle. Over the entire course of the evolution of planning systems, forecasts have been used not only for long- and mid-term planning but also to trigger production in the short term. This approach has been treated as a fundamental axiom and has never been questioned. This is astonishing insofar as supply chain managers typically face severe challenges due to poor forecasting accuracy for most of the products in a company's portfolio. High inventory levels and poor customer service often result, because the wrong products are in stock. In other cases, rush orders are sent to production to fulfill customer demand quickly when forecasts turn out to be wrong. On the one hand, this stresses capacity utilization and the need for excess capacity, while on the other, production plans are constantly changed.

10.1.5 IT for LEAN Planning: How to Escape the Optimization Trap and the Forecast Myth

Your company can overcome the described pitfalls of conventional ERP/ APS system architectures by reflecting LEAN SCM principles in its IT systems. First, utilizing consumption pull for appropriate products in the portfolio will help to avoid the forecast myth. If production is triggered by consumption, products are manufactured only when really needed by customers, avoiding excess inventory for some products and low customer service levels for others. Furthermore, frequent re-scheduling can also be prevented, reducing nervousness along the entire supply chain. Second, pre-configuration of the supply chain significantly simplifies planning on an operational level, thus avoiding the optimization trap. In addition to low planning effort, comprehensive planning results reap another major benefit. Employees who understand the output of IT solutions typically show high confidence in and acceptance of a system that encourages them to use it consistently. By overcoming the abovementioned shortcomings, implementing LEAN SCM-capable IT solutions helps companies exploit the potential of modern IT more effectively (Figure 10.3).

However, today's ERP/APS system architectures have not broadly adapted to functionalities that accommodate the capability of supporting LEAN Planning processes. As outlined earlier, key requirements are therefore typically not fulfilled. As we show in the following sections, this gap should be closed by leveraging your existing IT system architecture and a few complementary LEAN IT add-ons.

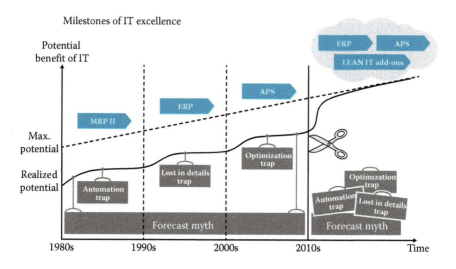

FIGURE 10.3
LEAN SCM capability helps to fully exploit the potential benefits of IT.

Summary

This section reviewed the evolution of IT support for all planning processes and activities that are grouped under the SCM umbrella. From minor support in the 1960s to the comprehensive services provided today, significant progress has been made. We have highlighted major advancements with the rise of MRP II, ERP, and APS solutions. Certainly the capability of considering capacity constraints with MRP II and the integration of several supply chain stages with APS must be considered fundamental milestones in the planning area. We have, however, discussed several dead ends, such as the optimization trap and the forecast myth, that have prevented IT from providing even better support. We have concluded that by building LEAN SCM principles into its IT systems your company can avoid these pitfalls of conventional ERP/APS system architectures. In the following sections, we explain how your company can achieve such results while leveraging your existing IT system architecture.

10.2 ENABLING LEAN PLANNING: HOW TO LEVERAGE PAST IT INVESTMENTS

As we have shown, today's conventional ERP/APS system architectures are not sufficient to implement LEAN Planning within an organization. Core requirements such as the pre-configuration of tactical LEAN SCM parameters and the production planning on an operational level are not effectively supported. So does your company need to build up a completely new system architecture to cover these requirements? Do past investments in IT system architecture mean wasted money? No, not at all! It is rather the opposite that is the case. The elements of today's conventional ERP/APS system architectures represent an important backbone on which to implement LEAN SCM. However, we show in this section that in some cases the functionalities of typical APS components in particular need to be complemented.

You probably recognize your company's IT architecture for supply chain planning in Figure 10.4. It is characterized by an ERP and MDM backbone as well as interfacing APS modules (market demand planning, supply

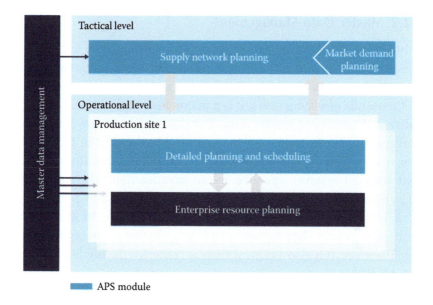

FIGURE 10.4
Common IT planning system architecture of companies in process industries.

network planning, detailed planning, and scheduling). In the following sections, we provide insights concerning these modules. In this context, we describe important functionalities that are already covered while also highlighting existing gaps that must be bridged to implement LEAN SCM in your planning systems.

10.2.1 Enterprise Resource Planning

ERP systems represent the backbone of today's IT system architectures in process industries. They support all business processes in an organization, from purchasing and finance to marketing and sales. However, ERP systems are not capable of effectively supporting supply chain planning for highly complex global supply chain networks and production environments. Therefore, companies in process industries typically complement their ERP systems with APS technology that supports supply chain planning on both the global and local level. In such ERP/APS system architectures, planning results from the APS system are seamlessly transferred to the ERP system, where the plans are executed. Owing to limited supply chain planning capabilities, ERP systems alone do not have the capability to incorporate LEAN SCM.

10.2.2 Master Data Management

Nothing impacts planning results more than the quality of master data such as lead times and minimum lot sizes. However, in global supply chains with dozens of production plants, separate local IT systems, and thousands of system users, it is a major challenge to maintain high quality and consistent master data. For that reason, many companies in process industries have implemented MDM solutions. In these solutions, master data are maintained globally and automatically synchronized with all systems operated by an organization. This capability makes it possible to maintain consistent master data at any point along the supply chain at any time. With end-to-end planning in scope, such harmonized master data represents a key facilitator of successful planning. Therefore, it is also a critical success factor for the implementation of LEAN SCM.

10.2.3 Market Demand Planning (APS Module)

Market demand planning is typically the starting point of the overall planning cycle. Its purpose is to create an unconstrained mid- and long-term demand plan. Such a demand plan is the basis for further planning and execution of supply chain activities, such as capacity planning and sourcing decisions. Today's APS solutions typically include a market demand planning module that supports the corresponding planning process. Figure 10.5 summarizes typical functionalities.

In contrast to traditional planning concepts, LEAN Planning uses forecasts on a more aggregated level to pre-configure a supply chain. However, a demand plan is still required. By providing excellent support in this

Key functionalities	Reasoning	LEAN enablement
Forecasting and manual enrichment	Forecasting, supported by various tools (e.g., data cleansing, statistical forecasting and advanced macrotechniques) and completed by manual enrichment to facilitate a consensus future demand plan.	✔
Aggregation and disaggregation	Aggregation and disaggregation of data enables flexible reviews, entries and changes on various levels while keeping them consistent at all levels.	✔
Measuring forecast accuracy	By monitoring key forecast-related metrics (e.g., forecast accuracy), conclusions can be drawn regarding how to reduce future forecast errors.	✔

FIGURE 10.5
Typical functionalities of APS market demand planning modules.

area, the market demand planning module represents a very helpful element of LEAN SCM.

10.2.4 Supply Network Planning (APS Module)

To create a truly integrated supply chain, IT needs to support planning across separate sites along the supply chain. The supply network planning module in modern APS solutions enables this capability and actively supports the planning of replenishment, production, supply, and distribution along the entire supply chain. The key in this context is the propagation of demand in the supply chain network. We summarize the core functionalities of supply network planning modules in modern APS solutions in Figure 10.6.

The described functionalities demonstrate that supply network planning modules in modern APS systems are essential to LEAN SCM. As mentioned earlier, they provide important global end-to-end transparency and the capability of end-to-end planning in this context. However, there is still a piece of the puzzle missing, as common supply network planning modules do not support the pre-configuration of LEAN SCM parameters.

10.2.5 Detailed Planning and Scheduling (APS Module)

A detailed planning and scheduling module is designed to support local production planners in finding feasible production schedules for all their assets. Especially in complex production environments with multiple production steps and parallel production lines, such an APS module can

Key functionalities	Reasoning	LEAN enablement
Global end-to-end supply chain transparency	Timely and accurate information about supply chain data such as demand, inventory levels, and capacity availability are the foundation of an optimized planning process that considers the entire supply chain.	✔
End-to-end planning	Integrated view of the whole supply chain enables the consideration of interdependencies in supply chain planning.	✔
Supply chain optimization	Optimization algorithms and heuristic approaches aim to optimize the supply chain *but do not support the pre-configuration of LEAN SCM parameters.*	⚡

FIGURE 10.6
Typical functionalities of APS supply network planning modules.

Key functionalities	Reasoning	LEAN enablement
Simultaneous planning	As opposed to typical ERP systems, simultaneous planning of materials, capacities, and sequences is provided, which enables the creation of feasible plans.	✔
Dynamic setup times	Capability of monitoring and considering sequence-dependent (dynamic) set-up times in planning, which is crucial, especially in process industries.	✔
Advanced scheduling heuristics	Support of automated planning and scheduling to reduce the workload for planners. *But common heuristics do not support rhythm wheel scheduling and monitoring.*	⚡

FIGURE 10.7
Typical functionalities of APS detailed planning and scheduling modules.

provide a significant benefit. Figure 10.7 summarizes the key function-alities of modern detailed planning and scheduling modules to support LEAN SCM.

As we have previously discussed, having the capability of simultane-ous planning while considering dynamic set-up times is especially cru-cial in this context. However, typical detailed planning and scheduling modules do not support rhythm scheduling and monitoring. This gap needs to be closed by appropriate IT add-ons, which we describe in Section 10.3.

Summary

In this section, we addressed the concern that past investments in IT systems might become obsolete when implementing LEAN SCM. As we have emphasized, the opposite is the case. The typical IT sys-tem architecture in process industries provides an excellent basis for LEAN SCM. Typical components such as an ERP system, an MDM solution, and core APS modules fulfill key requirements for imple-menting LEAN SCM within your company. Therefore, your company can readily leverage past IT investments in ERP and APS solutions. However, several gaps remain that prevent IT systems from achiev-ing LEAN SCM capability. Particularly with respect to the tactical and operational planning levels of LEAN SCM, additional support is required. As we show in detail in the next section, only a few comple-mentary LEAN SCM IT add-ons are necessary to enhance your com-pany's current IT system architecture.

10.3 LEAN PLANNING ADD-ONS TO COMPLETE THE IT SYSTEM

To address the abovementioned shortcomings of today's conventional ERP/APS system architectures, a few LEAN SCM IT add-ons must be implemented to completely cover the IT requirements of LEAN SCM. Especially on the tactical and operational planning levels, this additional support is highly required. Figure 10.8 depicts the corresponding LEAN SCM IT add-ons. It is important to note that these add-ons should be seamlessly integrated into existing IT system landscapes.

In the next sections, we offer further insights into the IT add-ons that are needed to effectively support LEAN Planning.

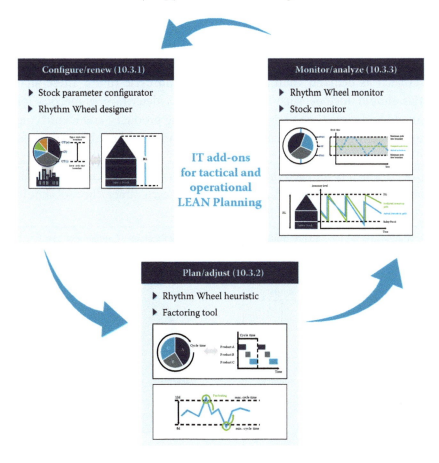

FIGURE 10.8
LEAN SCM add-ons to complement current IT planning systems.

10.3.1 Configuring and Renewing Tactical LEAN SCM Parameters

We explained in Chapter 6 that in order to pre-configure a supply chain, it is necessary to determine and periodically renew tactical LEAN SCM parameters (Figure 10.9). Effective IT support should be provided by two additional LEAN SCM IT add-ons, which we describe in the following sections.

10.3.1.1 Stock Parameter Configurator

ERP/APS architectures are not capable of supporting the tactical level of LEAN Planning, that is, the pre-configuration of LEAN SCM parameters. It is possible to close one part of this gap by implementing an IT add-on that configures LEAN replenishment (=stock) parameters such as the inventory replenishment level. The functionalities of such a Stock Parameter Configurator need to cover three dimensions.

The first dimension includes advanced analytics for analyzing input data such as demand volumes, variability, and cycle times. These analytics should include such functionalities as determining metrics for demand and cycle time variability, segmenting products, and identifying and correcting outliers. Second, such an IT add-on should be equipped with optimization algorithms that are capable of determining optimal stock parameters for each production asset in the supply chain. In addition to covering such factors as lead times and cycle times, the tool should

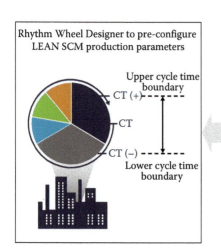
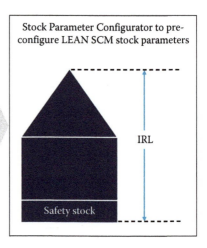

FIGURE 10.9

LEAN SCM IT add-ons to support the pre-configuration of supply chain parameters.

also consider variability in the supply chain. Third, the Stock Parameter Configurator ideally should be able to account for supply chain interdependencies. Such an IT add-on would help in optimizing inventory levels along the whole supply chain, for example, by leveraging pooling effects. Such multi-stage inventory optimization functionality would make it possible to reduce supply chain inventories while maintaining the targeted service levels, as described in Section 6.3.2.

Note that such an IT add-on should be designed as a decision support tool. Derived parameters should be used only as suggestions and should remain adjustable by planners to allow for the maximum degree of flexibility.

Figure 10.10 shows the Stock Parameter Optimizer devised by the CAMELOT Group. The Stock Parameter Optimizer configures the required stock parameters such as the IRL and safety stock levels based on inputs such as demand and supply variation. It functions as a seamlessly integrated add-on to the SAP APS solution and has already been integrated into several IT system architectures of leading companies in process industries.

10.3.1.2 Rhythm Wheel Designer

While the Stock Parameter Configurator focuses on stock parameters, another important IT add-on is required to close the gap in current IT architectures that prevents them from supporting the tactical layer of LEAN SCM. This additional add-on should address the pre-configuration of parameters for Rhythm Wheel-takted production. Because this pre-configuration determines the design of the Rhythm Wheel, we call the corresponding IT add-on the Rhythm Wheel Designer. Like the support provided by the Stock Parameter Configurator, the functionalities of the Rhythm Wheel Designer need to cover three dimensions.

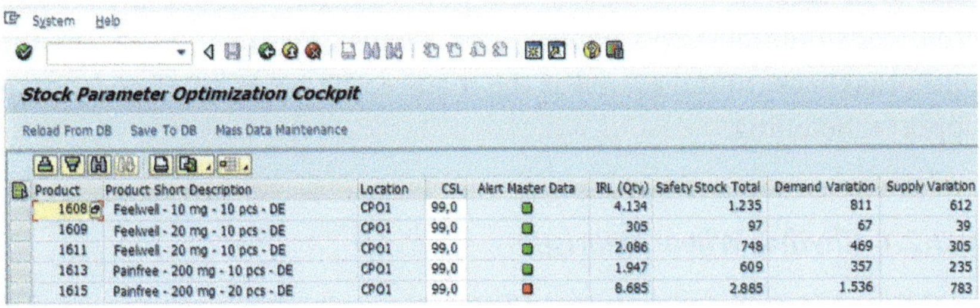

Stock Parameter Optimization Cockpit

Reload From DB   Save To DB   Mass Data Maintenance

Product	Product Short Description	Location	CSL	Alert Master Data	IRL (Qty)	Safety Stock Total	Demand Variation	Supply Variation
1608	Feelwell - 10 mg - 10 pcs - DE	CPO1	99,0	◻	4.134	1.235	811	612
1609	Feelwell - 20 mg - 10 pcs - DE	CPO1	99,0	◻	305	97	67	39
1611	Feelwell - 20 mg - 10 pcs - DE	CPO1	99,0	◻	2.086	748	469	305
1613	Painfree - 200 mg - 10 pcs - DE	CPO1	99,0	◻	1.947	609	357	235
1615	Painfree - 200 mg - 20 pcs - DE	CPO1	99,0	◻	8.685	2.885	1.536	783

FIGURE 10.10
CAMELOT's Stock Parameter Optimizer as an integrated SAP add-on.

First, the Rhythm Wheel Designer should support the determination of optimal production sequences. Managing the complexity that often arises with a large number of products and the sequence dependency of changeover times requires that optimization approaches are embedded in the solution. Genetic or ant algorithms should be considered in this context. Second, the Rhythm Wheel Designer should be capable of calculating the optimal Rhythm Wheel cycle time, minimum and maximum cycle boundaries, and the resulting production quantities for each product. It should also support the design of all Rhythm Wheel types, including the High-Mix Rhythm Wheel. It is therefore necessary to calculate optimal production rhythms, which requires the application of the latest findings from operations research as well as the corresponding translation into appropriate optimization engines. Third, the Rhythm Wheel Designer should be equipped with functionality for predicting the impact of the Rhythm Wheel design on relevant performance indicators such as overall equipment efficiency and the estimated changeover time per Rhythm Wheel cycle. This functionality provides important support for planners seeking the right pre-configuration of their rhythm-managed production assets.

Figure 10.11 depicts a screenshot of the Rhythm Wheel Designer when it is fully integrated into the SAP APS solution. State-of-the-art technology has been incorporated to allow for a maximum degree of user-friendliness. Such features as an advanced graphic interface and drag-and-drop functionality ensure effective and convenient use for planners.

10.3.2 Planning and Adjusting Production Based on Actual Consumption

In the operational LEAN SCM Planning process, a production schedule needs to be created based on pre-configured LEAN SCM parameters and actual consumption (see Chapter 7 for details). Figure 10.12 recaps the conceptual background as well as the required LEAN SCM IT add-ons to support scheduling.

10.3.2.1 Rhythm Wheel Heuristic

As outlined earlier, conventional ERP/APS architectures are not capable of supporting the operational level of LEAN Planning. In this context, there is a fundamental gap regarding the creation of LEAN production

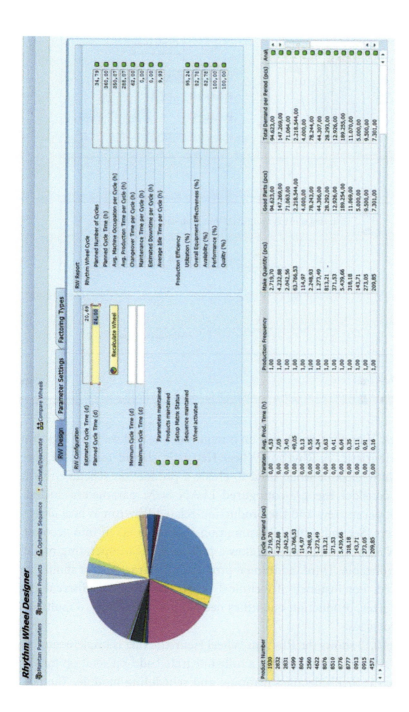

FIGURE 10.11

CAMELOT's Rhythm Wheel Designer as an integrated SAP add-on.

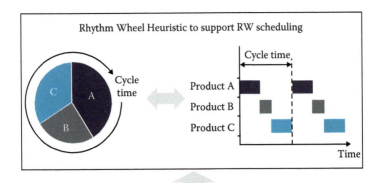

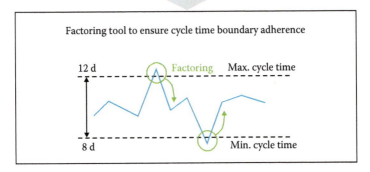

FIGURE 10.12
LEAN SCM IT add-ons to support the creation of LEAN production schedules.

schedules. Therefore, an IT add-on is required, one that we call a Rhythm Wheel Heuristic. A Rhythm Wheel Heuristic should take three core aspects into account when used to create a production schedule. First, it must consider the pre-configured LEAN SCM parameters, including stock parameters and the production-related Rhythm Wheel design parameters. Second, actual consumption should be taken into account, as LEAN SCM is designed to follow a pull-based SCM operating model where applicable. Third, additional production characteristics should be considered. For instance, if technical constraints require adherence to fixed lot sizes, production quantities need to be rounded up to multiples of such lot sizes.

Figure 10.13 shows the Rhythm Wheel Heuristic that is implemented in the SAP APS environment. It is a fully integrated add-on solution for PP/DS, which is SAP's detailed planning and scheduling module. The bottom part of the depicted planning board visualizes the Rhythm Wheel

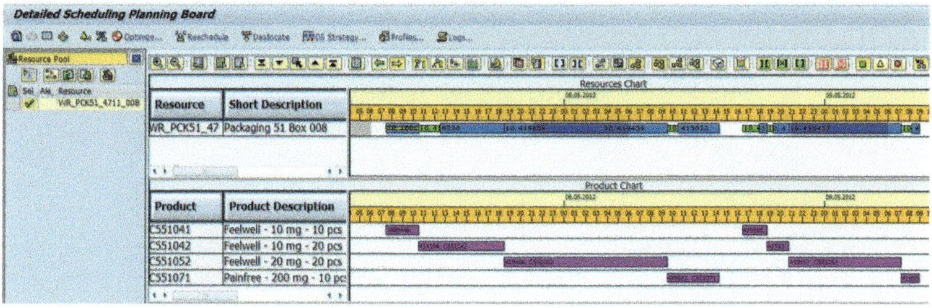

FIGURE 10.13
CAMELOT's Rhythm Wheel Heuristic as an integrated SAP add-on.

schedule. It is characterized by its repetitive production sequence. The production quantities vary slightly from cycle to cycle as they depend on actual consumption.

10.3.2.2 Factoring Tool

As we have outlined in previous chapters, we recommend defining minimum and maximum cycle time boundaries. On the one hand, such boundaries help to reduce variability and thus to maintain the takt along the supply chain. On the other, boundaries create the flexibility and agility that are needed to react to actual consumption and short-term requirements. To keep a given cycle time within these boundaries, production quantities should be reduced in some cycles, while other cycles might require the use of idle time. This calls for factoring. Especially when a given asset is responsible for a large number of products, factoring can be a very challenging task if a planner must perform it manually. Therefore, we recommend implementing a Factoring Tool. Note that such a Factoring Tool should make it possible to select from several potential factoring types (see Chapter 6 for more details). This makes it possible to select the optimal factoring type, depending on a company's particular situation.

Figure 10.14 shows the selection screen of CAMELOT's Factoring Tool. The selection of a suitable factoring type is a tactical decision. Therefore, the corresponding selection mask is integrated into the Rhythm Wheel Designer. The factoring itself, of course, is executed on an operational level where the Factoring Tool complements the Rhythm Wheel Heuristic by adjusting the production schedule according to short-term requirements.

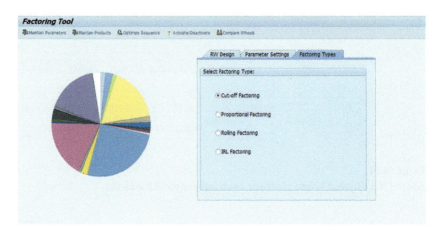

FIGURE 10.14
Selection screen of CAMELOT's Factoring Tool.

10.3.3 Performance Monitoring for the Renewal Process

The control phase of the tactical and operational LEAN Planning process entails evaluating the quality of the pre-configured LEAN SCM parameters. Figure 10.15 recaps the conceptual background as well as the required LEAN SCM IT add-ons to support the required performance monitoring in this context. These add-ons provide crucial input for the tactical renewal process.

10.3.3.1 Rhythm Wheel Monitor

The Rhythm Wheel Monitor addresses one of the remaining gaps in conventional ERP/APS architectures for supporting the operational level of LEAN Planning. The Rhythm Wheel Monitor is required to support the control phase, in which the quality of the pre-designed Rhythm Wheel is evaluated and used as an input for the next design phase. This ensures continuous improvement of Rhythm Wheel-based production planning.

To measure the quality of the Rhythm Wheel design, the Rhythm Wheel Monitor should compare designed Rhythm Wheels with the Rhythm Wheels that are ultimately scheduled. Based on such a comparison, key performance indicators such as cycle time attainment, cycle time variation, and run-to-target (see Chapter 9 for details) must be automatically calculated.

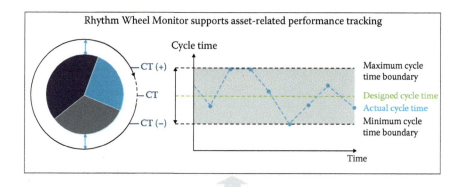

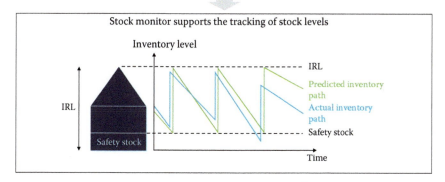

FIGURE 10.15

LEAN SCM IT add-ons to support LEAN-specific performance monitoring.

To provide the best possible support for the quality assessment of the Rhythm Wheel design, the Rhythm Wheel Monitor should provide a cycle-related view on the one hand and an item-related view on the other. A cycle-related view allows for direct comparison between the designed Rhythm Wheel and the Rhythm Wheel that is ultimately scheduled. Additionally, the Rhythm Wheel Monitor should deliver cycle-specific metrics and provide an overview of all cycles. An item-related view shows item-specific metrics over various cycles and enables the evaluation of a specific item. The results gained from both views make root cause analysis possible and are required as input for the next Rhythm Wheel design phase to improve overall performance.

In Figure 10.16, a screen shot of CAMELOT's Rhythm Wheel Monitor is depicted. As a fully integrated SAP add-on, it gathers all relevant input data automatically, runs sophisticated analysis, and plots the corresponding results on an advanced graphical interface.

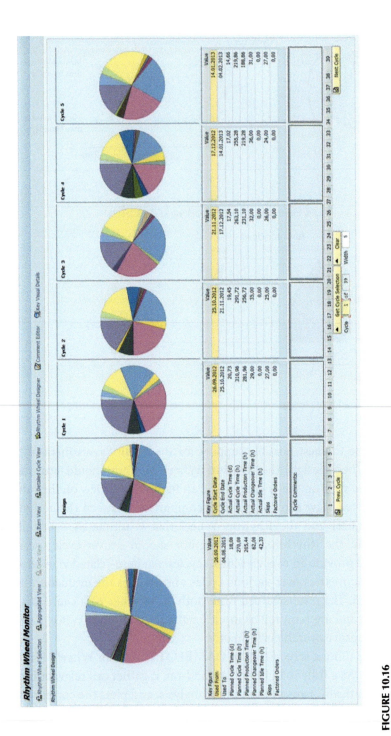

FIGURE 10.16

Cycle view in CAMELOT's SAP-integrated Rhythm Wheel Monitor.

10.3.3.2 Stock Monitor

The remaining gap that prevents conventional ERP/APS architectures from supporting LEAN SCM can be closed by the Stock Monitor. Analogous to the asset-related Rhythm Wheel Monitor, the Stock Monitor helps to evaluate the suitability of pre-configured parameters on the replenishment side. Corresponding results serve as input for the next renewal of supply chain stock parameters.

The evaluation of pre-configured stock parameters entails comparing the predicted inventory path against the actual stock development. A large deviation in this path indicates that stock parameters might not have been pre-configured appropriately. If a root cause analysis confirms this presumption, we strongly recommend aligning stock parameters in the next tactical renewal process.

Figure 10.17 contains a screenshot of CAMELOT's Stock Monitor, which is an integrated SAP add-on. By visualizing deviations of actual inventory developments (red line) from predicted developments (blue line), planners are effectively supported in the tactical renewal process. The depicted situation reveals that actual inventory developments have been predicted with considerable accuracy for the first two products. On the other hand, the predicted inventory path for the third product deviates widely from the one that was finally observed. This is a clear indication that stock parameters for this product should be refined in the next tactical renewal process.

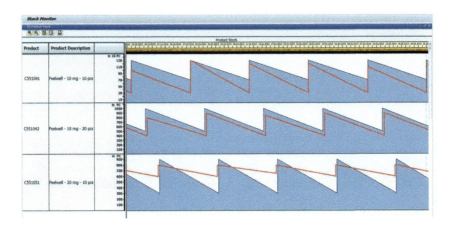

FIGURE 10.17
Inventory analysis in the Stock Monitor.

BOX 10.1 INTEGRATION OF LEAN SCM IT ADD-ONS INTO THE SYSTEM ARCHITECTURE OF A LEADING PHARMACEUTICAL COMPANY

Following its decision to implement LEAN SCM, a leading pharmaceutical company discovered that one of the key challenges in this effort was ensuring sufficient IT support for its re-designed planning processes.

The overall guideline was the objective of seamlessly integrating the required LEAN SCM IT add-ons into the existing SAP system environment, which consisted of SAP ECC (the ERP solution), SAP APO (the APS solution), and SAP MDM (the Master Data Management solution). Figure 10.18 shows how the integration of CAMELOT's LEAN Suite add-ons into the existing IT system architecture finally looked.

Figure 10.18 shows how seamless integration of SAP DP/SNP and tactical LEAN IT add-ons was achieved, enabling automatic transfer of demand data, which is crucial for pre-configuring LEAN SCM

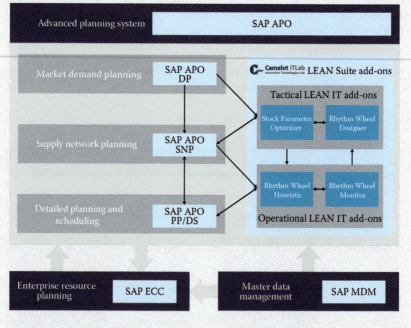

FIGURE 10.18
Integration of LEAN Suite add-ons into the SAP system architecture of a leading pharmaceutical company.

parameters. Through an interface between the Rhythm Wheel Designer and the Stock Parameter Optimizer, it is possible to consider the interdependencies between stock parameters and asset-related parameters. The Rhythm Wheel Heuristic automatically receives these pre-configured parameters from the tactical LEAN IT add-ons as well as the actual consumption data from SAP SNP. Based on this input, the Rhythm Wheel Heuristic creates corresponding production schedules in SAP PP/DS. Through seamless interfaces to PP/DS and the Rhythm Wheel Designer, the Rhythm Wheel Monitor finally gathers information needed to compare the pre-designed production schedule with the actual one. Derived performance indicators are then made available so that the pre-configuration of LEAN SCM parameters can be appropriately aligned in the next renewal process.

The existing SAP MDM solution is an important facilitator for the integration of LEAN IT add-ons into existing system architectures, ensuring the consistency of master data along the entire supply chain and thus facilitating reliable LEAN end-to-end Planning.

With the successful integration of LEAN IT add-ons into its existing SAP system architecture, the pharmaceutical company has taken an important step toward a truly LEAN supply chain. And to stay ahead of competition, the next level of LEAN SCM Planning development is already envisioned in this innovation driven organization. It is based on the next generation of the SAP HANA computer platform, with a new type of database, just built on in-memory technology. Here a large amount of granular data (bid data volumes) such as customer orders, shipping and production data, and so on are all stored and compressed in-memory, without the need of any data aggregations. This breakthrough new technology brings the analytics layer and the application layers, side by side, on a single in-memory platform together. It joins, up-to-date unreached, new integrated planning and predictive analytics capabilities. Therefore it is already envisioned to build up all of the tactical renewal cockpit functionalities as a pivotal element in the new sales & operations planning (S&OP) processes. With the data access at all granulation levels and the real real-time transformation capabilities, the company will be able to uncover trends and anticipate behavior of the supply chain during the S&OP meetings online. Predictive analytics for the cycle time oscillation and trends of the inventory replenishment levels will turn them into real actionable measures for the end-to-end supply chain synchronization. Furthermore, enhanced with real-time simulation capabilities and what–if scenario possibilities, this new LEAN Planning maturity level promises a true supply chain decision support in future.

Summary

In this section, we introduced LEAN SCM IT add-ons that are required to close gaps in conventional IT system architectures that prevent them from effectively supporting LEAN Planning. These IT add-ons support the tactical and operational levels of LEAN Planning along three dimensions: configuring/renewing, planning/adjusting, and monitoring.

In this context, we first explained that the pre-configuration and renewal of LEAN SCM parameters should be supported by the asset-related Rhythm Wheel Designer and the Stock Parameter Configurator for the replenishment side. Second, we highlighted the functionality of the Rhythm Wheel Heuristic and the Factoring Tool, which support the creation of production schedules as well as adjustments when demand occurrence deviates from expectations. Third, we introduced two IT add-ons that support monitoring within LEAN Planning. The Rhythm Wheel Monitor supports the evaluation of the quality of pre-configured production parameters, while the Stock Monitor supports the evaluation of defined stock parameters. Both add-ons provide highly important input for the renewal of supply chain parameters during the tactical renewal process.

Finally, we provided insights into how a leading pharmaceutical manufacturer successfully integrated LEAN IT add-ons into its existing SAP IT system architecture.

CHAPTER SUMMARY

In addition to organizational changes, integrating LEAN SCM into a company's IT system architecture is a crucial success factor. As we have argued in this chapter, key requirements should be fulfilled to solve this challenge.

In Section 10.1, we traced the significant progress in IT support for supply chain planning that has occurred since MRP was developed in the 1960s. The subsequent advent of MRP II and ERP systems as well as APS solutions represented remarkable milestones in this evolution. However, several dead ends, such as the so-called optimization trap and the forecast myth, have prevented IT from providing even better support. We have found that, by incorporating LEAN SCM principles into IT systems, it is possible to avoid these pitfalls of conventional ERP/APS system architectures.

Although today's conventional IT system architectures in process industries are not able to sufficiently support LEAN SCM, we showed that there is no basis for concern that past investments in IT systems will become obsolete with the implementation of LEAN SCM (Section 10.2). Indeed, the opposite is the case, as conventional IT system architectures provide an excellent basis for LEAN SCM. We showed, however, that in order to completely fulfill all IT requirements of LEAN SCM, some additional IT add-ons are necessary.

We completed the chapter by providing insights into complementary IT add-ons that are required to provide sufficient IT support for LEAN Planning (Section 10.3). In this context, we explained tactical and operational LEAN IT add-ons and their functionalities in detail. Finally, we introduced an industry case in which a leading pharmaceutical manufacturer successfully integrated LEAN IT add-ons into its existing SAP IT system architecture, illustrating what effective IT support of LEAN SCM looks like.

11

The LEAN SCM Journey

Early on in this book, we noted that process industries face huge challenges in a business environment marked by VUCA (variability, uncertainty, complexity, ambiguity). To meet those challenges top management must draw up and systematically attend to a daunting agenda that include maintaining reliable supply for consistent delivery to customers. This in turn means reacting robustly to supply disruptions while at the same time adapting to rapidly changing market environments. LEAN SCM provides the organizational structure with which your company can overcome the challenges of the VUCA world. After your company successfully adapts its supply chain to the LEAN SCM paradigm, it will reap the benefits of reliability and agility.

However, transforming your company's supply chain in accordance with the LEAN SCM paradigm is a journey that requires considerable effort, depending on the starting point. Nothing less than a major shift in the organizational mindset is required, because the major principles of SCM change dramatically. LEAN SCM will require changing your company's business processes, functional roles and responsibilities, and IT systems. Do not expect the journey to be an easy one. Nevertheless, if your company adheres to the following three guidelines, it will successfully complete the LEAN SCM journey and fully capture its benefits:

- Build strong commitment to and leadership for LEAN SCM—for organizational alignment.
- Create a holistic LEAN SCM architecture—for process and systems alignment.
- Establish LEAN SCM program management—for active governance.

11.1 BUILDING STRONG COMMITMENT AND LEADERSHIP FOR LEAN SCM

Every transformation initiative needs management buy-in and leadership commitment to be successful. However, a supply chain transformation on the scale of a transition to LEAN SCM from conventional SCM practices requires even stronger commitment. LEAN SCM is a conceptual paradigm change in supply chain and operations planning that changes everyday processes for many roles in the supply chain organization (see Figure 11.1). Although every change encounters resistance, the paradigm changes associated with LEAN SCM will very likely mean that significant barriers in your company will complicate the journey.

Consider, for example, the role of supply chain planner. Under LEAN SCM concepts and processes, he or she will be responsible for assuring up-to-date inventory target settings that are synchronized along the entire supply chain. He or she will therefore need to participate in monthly tactical renewal processes and devote considerable time to analyzing demand and supply trends that impact the supply chain. Traditionally, this role and the person carrying it out was not involved in tactical or operational planning tasks at all, working more on setting objectives (for others). Supply chain planners are therefore unlikely to jump into the new LEAN SCM Planning environment with unbridled enthusiasm.

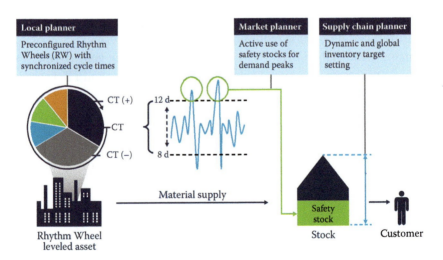

FIGURE 11.1
Changes in the supply chain under the LEAN SCM paradigm.

Next, consider the local production planner. He or she typically works hard to fulfill incoming orders and has likely acquired individual techniques with which to "optimally" schedule and reschedule production to satisfy incoming demand. Continuous actions and reactions by shifting orders give the appearance of adopting a customer orientation but more often than not this approach merely make a lot of noise at production sites, generating additional "waves" in the upstream supply chain. This is not the LEAN SCM way of planning. Under LEAN SCM the planner receives a simple, predefined production schedule based on a preconfigured Rhythm Wheel. A planner accustomed to the traditional business-as-usual approach will not like the idea of losing the freedom to individually reschedule production and might even fear becoming replaceable by some sort of standardized production planning concept. Moreover, the traditional supply chain planner will need to accept the impact of LEAN SCM scheduling on upstream supply chain stages as well, and be convinced that the Rhythm Wheel approach stabilizes both local asset utilization and more importantly all productive activity along the entire supply chain.

Finally, consider the market planner. Market planners typically concentrate on building significant safety stock to address demand and supply variability. Safety stock is by definition never touched by planning runs in existing ERP or APS systems. However, in the course of daily execution of supply chain processes, the use of physical safety stock often causes at least mild panic as planners overreact and place emergency orders to production sites. Guided by LEAN SCM concepts, however, productions assets are managed and inventories are replenished according to preconfigured and takted Rhythm Wheel schedules. This means supplying products based on average Rhythm Wheel cycle times and configured cycle time variations that absorb the major part of demand variability (Figure 11.1). Nevertheless, some demand signal peaks will be completely cut off during a LEAN SCM Planning runs and therefore have to be filled from planned safety stock. Thus, using the new LEAN Planning approach, safety stock will be used actively for some demand peaks—a paradigm change in supply chain planning. It will remain a challenging task, however, to convince the market planner and the supply chain organization not to panic, but to stay calm and trust that the tactical parameter settings will cover demand volatility. This will be neither easy for nor self-evident to the market planner.

The supply chain and production planning organization is generally wedded to its old way of working, and members of the organization are very skeptical of change. It should come as no surprise that individuals carrying out business processes that have followed the traditional forecast-based input myth and push-based planning system approach for more than 50 years will resist the transition to a new paradigm. To overcome such organizational concerns, your company will need active leadership and strong commitment from key stakeholders. And it will need such commitment not only for the moment, but also for the long term to accomplish a complete transformation to the LEAN SCM paradigm. Only when leaders show commitment and lead the way will others follow.

To support such a paradigm change in supply chain planning and bring the supply chain organization on board, we mention two methods of proving the benefits of LEAN SCM that have worked successfully in the past. First, top management must be convinced with a realistic and well-researched business case. A simulation case study that compares the status quo with a LEAN SCM scenario can be of invaluable help. With a dynamic simulation model of your company's supply chain, you can objectively estimate the impact of LEAN SCM concepts on resulting supply chain performance. Second, the IT organization also has to be convinced to adapt the existing ERP/APS-based system landscape and to enhance its functionalities to provide application support for the LEAN SCM Planning paradigm. The best way to do so is to construct a small-scale IT prototype. Such a prototype can achieve proof-of-concept that demonstrates how LEAN SCM concepts and processes come to life in existing IT systems. Such a simulation will also show the IT organization that its investment in APS software was not in vain, as LEAN SCM functionality requires only complementary additions that will allow the IT organization to continue leveraging its existing enterprise software.

Thus, when introducing and implementing LEAN SCM in your company, you must win the commitment of top management, the business, and the IT organization from the very beginning. And be sure to sustain this commitment for the entire journey, because there will be new barriers and new resistance with every step you travel.

11.2 CREATING A HOLISTIC LEAN SCM ARCHITECTURE

In addition to pledging leadership and commitment to LEAN SCM, your company needs to adopt an architecture and roadmap as a platform from which to successfully launch its LEAN SCM journey. It is not easy to build a house without knowing what it is supposed to look like architecturally before beginning construction. The same holds true for a supply chain transformation: implementation cannot begin without knowing what the "targeted ideal state" is supposed to look like.

Effective LEAN SCM architecture is holistic. This means considering and integrating all processes, organizational roles, performance management activities, and IT systems. Only by coordinating and linking these dimensions with each other can your company achieve a properly functioning and harmonized process along the entire supply chain. We illustrate such an architecture (much simplified, of course) in Figure 11.2.

Consider LEAN SCM demand and supply planning processes. They must be set up in a new way to support agile, consumption-based pull replenishment. Prior to adopting LEAN SCM, your company's planners were likely accustomed to traditional push replenishment, relying on the quality of the demand forecast and then netting it against target stock levels and actual available stock before propagating it to the next node of the

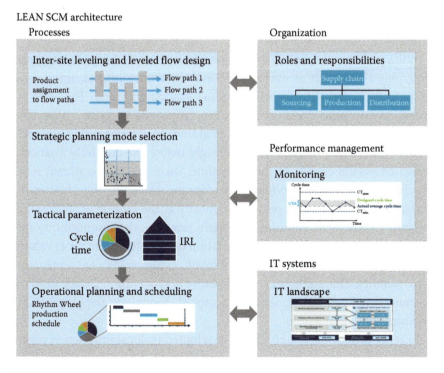

FIGURE 11.2
Holistic and integrated LEAN SCM architecture.

supply network and eventually down to the production sites. The production sites were then in charge of creating their own production schedules. Yet how long can production sites really follow such plans? Not long, since everyone understands that all forecasts are more or less wrong from the start, and that all such plans and the corresponding schedules have to be changed frequently. This has always meant constant firefighting along the supply chain as well as on the shop floor in everyday operations. When your company has embraced LEAN SCM, it can escape the forecast trap by completing the transition to real consumption-based pull replenishment and relinquishing its dependency on forecast accuracy, at least in the course of operational supply chain planning processes. This will entail establishing new tactical parameter renewal processes for configuring Rhythm Wheels and setting and adjusting inventory target levels. This is perhaps a monumental change in your company's planning processes—separating the tactical supply chain configuration from operational supply chain planning.

As we have seen, LEAN SCM heavily impacts the roles and responsibilities of the (global or end-to-end) supply chain planner and the (local) production planner. The two need to be included in the LEAN SCM journey carefully.

In the past, supply chain planners were accustomed to translating strategic objectives, such as overall corporate inventory targets, into tactical guidelines and KPIs for each product group and brand. The work of operational inventory planning was typically done by others. Planners only monitored and reviewed the tactical inventory target guidelines and the ex-post results—and usually only right before the quarterly financial reporting period ended. At the same time, the work of operational supply chain replenishment and network planning, including the adaption of planning parameters, was performed mostly at the local level. Following the new LEAN SCM Planning approach, however, the responsibility for the tactical setting and dynamic adaptation of inventory parameter values for each product is to be managed by supply chain planners as a primary responsibility. This additional responsibility and operational work need to be understood by management and accepted throughout the global SCM organization.

Another unique perspective must be adopted by local production planners. Their former planning and scheduling work will be divided into two parts. The first part covers the asset-related tactical preconfiguration of the production order sequence based on the Rhythm Wheel logic. This is typically done and reviewed quarterly. The second, operational, step taken by local production planners simply follows the designed sequence, resulting in reduced scheduling complexity and fewer nervous rescheduling actions. So the new LEAN SCM Planning concepts require and promote fewer "rescheduling heroes," those who unintentionally create undesirable variability along the supply chain when coping with very sporadic demand signals from sales organizations. This makes it obvious, then, that here also some organizational implementation and transformation barriers have to be overcome.

The same holds true for performance management. Traditional KPIs such as customer service level and inventory turnover remain valid and provide insight into the strategic performance of the supply chain. They do not, however, reflect some key LEAN SCM Planning objectives: reducing variability in the network and improving asset performance through the takt-leveled and synchronized supply chain. Therefore, the performance management system will have to incorporate metrics such as cycle time attainment to designed Rhythm Wheels, which indicates the extent to which the takt and synchronization along the supply chain are being maintained. In this way, LEAN SCM KPIs facilitate direct "temperature checks" of supply chain configuration stability as designed, adaptability to consumption surprises, and the resilience of "out-of-takt asset operations." Tactical LEAN SCM performance metrics are key inputs for monthly parameter renewal. Therefore, LEAN SCM performance indicators provide more of an ex-ante view of trends in supply chain behavior. They provide early signals of the need for adaptation and resilience along the supply chain and are therefore the central control parameters of a LEAN supply chain organization.

Last but not least, your company's IT systems will need to be adapted to the new requirements of LEAN SCM as well. The planning processes that are currently mapped in existing ERP and APS systems need to be enriched and customized in accordance with the newly envisioned LEAN Planning approach. In addition, implementing some LEAN SCM add-on applications such as a Rhythm Wheel design tool and a dynamic stock parameter configurator is necessary for supporting LEAN Planning processes with full organizational acceptance.

We hope you now understand the interactions between processes, organizational roles, performance management, and IT systems that are required to implement LEAN SCM in your company. Can you imagine what would happen if you attempted to change only one of the above-mentioned dimensions while leaving the others untouched? Correct: LEAN SCM concepts could not be successfully transferred into your organization. LEAN SCM will not work at all unless all dimensions of the LEAN SCM framework are addressed. Your company's LEAN SCM initiative will succeed only by creating a holistic LEAN SCM architecture and a transformation roadmap that ensures that all relevant dimensions are aligned and coordinated with each other.

11.3 ESTABLISHING LEAN SCM PROGRAM MANAGEMENT

As we have shown, adopting the LEAN SCM paradigm might entail a considerable amount of change in your company. In your experience, you have probably witnessed or participated in many projects that did not deliver the expected outcomes because they were not managed properly or were sliced into smaller project steps and managed as "one-off initiatives." We hope by now that you do not perceive a LEAN SCM initiative as a quick one-off project designed to achieve only short-term improvement. When adopting the LEAN SCM paradigm, your company takes on a mid- to long-term transformational journey that requires more than a mere project: it requires a *program*. And, to avoid seeing your company's LEAN SCM transformation end like so many single "one-off" projects, you need to establish tight LEAN SCM program governance (see Figure 11.3).

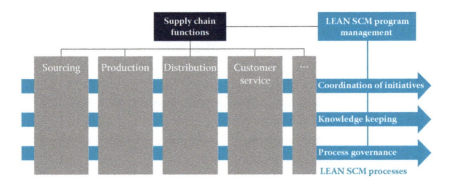

FIGURE 11.3
Functions involved in LEAN SCM program management.

Companies often spend considerable effort launching well-intentioned lean manufacturing, supply chain planning, or supply chain visibility initiatives. All too often they run out of steam on the way to completing a single one of those initiatives. One obvious reason for false starts is that such individually well-intended and meaningful initiatives interact with each other. In the worst-case scenario, they counteract each other. Therefore, it is especially important that your company practices LEAN SCM program management that coordinates single initiatives into an integrated whole. This ensures that all initiatives are aligned with each other and are effectively integrated.

During your company's LEAN SCM journey, it will gather knowledge from many new experiences while learning LEAN SCM concepts. It is important not to lose this knowledge because some program members might decide to leave your company and take their experience-based knowledge with them. To maintain such valuable knowledge, your company's LEAN SCM program management team should form a dedicated staff to manage knowledge related to LEAN SCM. When internal adjustments or external trends crop up during the LEAN journey, such knowledge keepers can evaluate and integrate them into the ongoing LEAN SCM concepts. Once integrated into the existing concepts, the knowledge can then be spread across the company to benefit the entire organization.

Now, imagine that your company has successfully implemented LEAN SCM Planning architecture. Do not be fooled by the illusion that everyone involved will adhere naturally to the same production and replenishment planning processes for the next 5 years. It is difficult to maintain standardized processes in a global corporation without effective process governance. Process governance ensures the long-term sustainability of implemented LEAN SCM processes, ensuring that such processes are not changed without alignment and permission of the LEAN SCM governance organization to guarantee standardization and continued high quality.

To summarize the most important points of this chapter, be sure to treat your company's LEAN SCM journey as a program, not a project. Effective LEAN SCM program governance will ensure proper coordination and integration of the various initiatives involved in the overall process, retain valuable knowledge of LEAN SCM in your company, and ensure long-term process sustainability.

CHAPTER SUMMARY

The VUCA world poses new challenges for companies in process industries. LEAN SCM has become the most effective way to meet those challenges. To implement LEAN SCM successfully, however, you must not underestimate the effort involved in changing the mindset of the organization. This should come as no surprise, as transitioning to the LEAN SCM paradigm is not a simple process improvement initiative. It involves a paradigm change in supply chain and production planning, targeting a quantum leap in supply chain performance.

Therefore, your company will need strong commitment from its leadership team to convince people throughout the organization to follow the new SCM paradigm. Such commitment is the only way to persuade people to leave their old ways of working behind and to stop resisting change.

Ultimately, the transformation to LEAN SCM must be regarded from a holistic perspective. Since processes, roles and responsibilities, performance management, and IT systems will change dramatically, it is important to ensure that these dimensions are compatible with each other. A holistic LEAN SCM architecture will bring about the necessary integration.

Last but not least, it is essential not to regard a LEAN SCM transformation as a "one-off" project, but instead to see it as a mid- to long-term program. To accompany the transformation, your company should establish a LEAN SCM program management team that oversees the coordination of interacting initiatives, retention of valuable knowledge, and the process governance that is necessary to achieve the long-term sustainability of LEAN SCM.

After all, the LEAN SCM journey is ongoing. There is always room for improvement, even though your company might already be a supply chain champion. The idea of continuous improvement is embedded in many LEAN SCM processes (e.g., in the renewal processes) and represents a major part of the LEAN philosophy. If you are willing to accept this commitment, welcome aboard and have a safe LEAN journey!

Part IV

How Your Industry Peers Gained Benefits by LEAN SCM

12

Read How Top-Industry Players Share Their Experiences with LEAN SCM

Thought-leader companies of the process industries are always seeking to improve their supply chain performance by the means of innovative concepts. They have the courage and endurance to walk the sometimes tough way of change—against external and internal barriers.

These leading companies have already made big steps into the direction of LEAN SCM. Some companies have adopted part of the tools and concepts; others have already started a holistic LEAN journey.

12.1 MOTIVATION AND APPROACHES TO LEAN SCM

In this final chapter, you can read how your business peers adapted their supply chain paradigm, what obstacles they faced, and which benefits they enjoyed by LEAN SCM:

- AstraZeneca's Lean SCM Journey
- Eli Lilly's Synchronized Lean Production
- Novartis' Buffer Management Concept
- Leveled Flow Design of a Major Scandinavian PharmaCo
- AstraZeneca's Rhythm Wheel Excellence Site
- The Lean Production Initiative at PCI—a BASF Company

All these pioneering companies have enjoyed significant benefits of their transformation toward LEAN SCM—and we are sure that the same holds true for the rest of the process industry companies as well.

12.1.1 AstraZeneca's Lean SCM Journey

Profile of Case Contributor

Andy Evans, head of Global Supply Chain Planning, has worked in AZ for 25 years (through ICI, Zeneca and then AZ) joining originally as a chemical engineer and has worked across all parts of the supply chain. In his early career, Andy Evans held a variety of engineering and production management roles which included working in France. More recent roles have including global supply chain manager, engineering manager for an API site undergoing major expansion, establishing a new logistics department on one of AZ's largest manufacturing sites and then subsequently leading it plus the packaging manufacturing during a major Lean transformation. His current role is the head of Global Supply Chain Planning.

Before joining AZ, he graduated with a first class honors degree in biochemical engineering from Swansea University in 1987.

12.1.1.1 Company Profile

"Our Lean Supply Chain Visibility initiative reshaped our operations and supply chain management operating model with fundamental new global capabilities to realize end-to-end lean and agile supply chains and a step change in information systems."—Andy Evans

- One of the world's leading pharmaceutical companies with 2011 sales of $33 billion.
- A 70-year track record of innovation that includes the introduction of many world-leading medicines.

- Active in over 100 countries; corporate office in the United Kingdom; strong presence in key markets; growing presence in important emerging markets.
- Over 62,000 employees: 47% in Europe, 31% in the Americas and 22% in Asia, Africa, and Australasia.
- Over $4 billion in R&D each year with 11,000 people in R&D organization.

12.1.1.2 Executive Summary

AZ started its lean journey to respond to emerging cost pressure and higher market dynamics in managing global supply chains in the pharmaceutical industry. Following a clear lean vision for our entire supply chain organization, we established a demand-driven supply chain approach based upon lean planning methods, high end-to-end transparency, and strong process governance. The higher efficiency and responsiveness of our supply chain organization improved key figures such as working capital, lead times, and asset utilization considerably. Furthermore, our lean supply chain strengthened our capability to provide highest customer service despite growing market uncertainties.

12.1.1.3 Company's General Situation

AZ must ensure that our medicines are provided to patients wherever they are based right at the time they need them. Highest customer service is the most prominent objective we have. However besides global supply reliability at the same time operations and SCM have to compensate prize pressure and patent expiries with excellent operational efficiency to maintain operating margin and create shareholder value. Right after the merger between Astra of Sweden and UK-based Zeneca in 1999, SCM and operations had a prominent role to create a unified operating model and to manage the transformation of the company. Historically, manufacturing and supply chain were not seen as a source of competitive advantage in the pharmaceutical industry. However, the contribution of supply chain to the company's success is substantial: Our experience shows that a 1% reduction in cost of goods sold yields $150–200 million savings while three additional stock turns result in $500 million savings.

Our journey toward lean SCM aimed to create and continuously improve a supply chain organization that is best in class and copes with the substantial Pharma market challenges.

12.1.1.4 Lean Challenge

AZ realizes double digit growth in emerging markets which for SCM creates the necessity to manage substantial SKU proliferation and introduces continuously higher demand variability not at least coming from an increasing amount of tender-based business. In parallel, the global manufacturing footprint is permanently adapted and optimized by selective outsourcing to contract manufacturers (CMOs), product transfers, and investment in new sites in critical emerging markets.

These demand uncertainties and the growing complexity have to be managed in AZ's product supply network which consists of complex global supply chains. Manufacturing steps especially upstream at API/drug substance manufacturing or the formulation of the drug product can take months to finish and are spread across different sites or CMOs often across continents.

Historically, Pharma SC planning approaches had a strong local or regional site focus with limited consideration of the overall end-to-end network and a low responsiveness to the market due to the applied silo-focused MRP planning logic. Those challenges were further amplified by the fact that market stocks were not owned and market distribution not managed by operations and SCM. Insufficient customer service levels, excess inventories, unnecessary long lead times, and lack of globally aligned process improvement were issues also for AZ in those days.

In addition to shop floor lean manufacturing programs in 2009, we consequently extended our lean journey to include a lean end-to-end SCM operating model to further improve supply reliability, cost efficiency, and agility of product supply. This initiative was called "Lean Supply Chain Visibility" (LSCV) and aspired to:

- Implement an innovative lean end-to-end SCM paradigm change leveling capacity utilization in a demand-driven mode
- Enable complete visibility of end-to-end supply chain information and informed decision making
- Realize a step change in global end-to-end SCM collaboration

12.1.1.5 Approach

LSCV was a 3-year full global SCM transformation covering 23 sites, 100+ marketing companies, and more than 350 directly involved operations/ SCM people.

The core of AZ's lean end-to-end SCM operating model was the introduction of methods that support our shift away from traditional forecast-based planning approaches to responsive supply chain processes. To enable consumption-based planning in the supply chain, our replenishment processes follow now mainly a pull-logic using, for example, predefined IRLs (inventory replenishment levels). VMI approaches further support demand-driven supply chain operations and reduce the bullwhip effect in our global supply chains. In our manufacturing sites, we introduced the concept of Rhythm Wheels for improved scheduling of products and more efficient synchronization across the supply chain. Fundamental to realizing consumption driven and leveled end-to-end SCM was the segmentation of our product portfolio considering demand variability and assigning the appropriate supply methods to the steps in the end-to-end product supply chains.

The second key element of our new SCM operating model is our fundamentally redesigned new AZ SCM and planning process and organization model.

Critical to successfully realizing, our new SCM operating model was a strong supporting leadership and continuous stakeholder engagement across the globe as well as the consequent approach to holistically redesign and implement new processes, a new SCM organization and to enable the new processes with new IT systems to provide the necessary information visibility (see Figure 12.1). Last but by far not least, a comprehensive change management supported the necessary transition to the new ways of working and the required culture and behavior changes.

Building on the cornerstone of the SCM and planning operating model, we detailed new standardized global and local processes for the different planning time horizons which deliver the necessary capability to enable end-to-end SCM decision making, we assigned accountabilities and responsibilities to new harmonized global, regional and local roles, and finally integrated the new processes and roles into a comprehensive globally and functionally integrated SC planning cycle and calendar (see Figure 12.2).

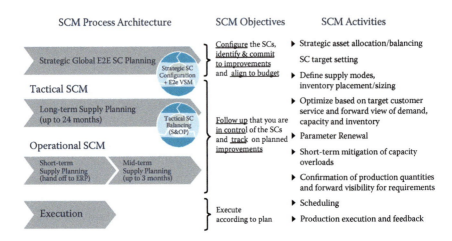

FIGURE 12.1
AZ's lean end-to-end supply chain management & planning operating model.

Resulting from the process design and in line with the lean end-to-end SC planning approach, we defined and implemented three new key roles:

- Global supply planner: Manages the end-to-end brand supply chains and leads S&OP, defines supply methods, monitors end-to-end targeted improvements, ensures balance between agreed customer service, inventory, and conversion costs.
- Asset planner: Site representative who commits the asset to yield, lead time, supply variation, and rhythm cycles for the asset.
- Marketing company account manager: Represents the voice of the customer and agrees customer service levels with the commercial side.

FIGURE 12.2
AZ's lean SCM approach integrated the end-to-end SCM processes with a global planning cycle.

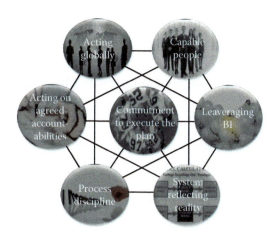

FIGURE 12.3

AZ's collaborative lean end-to-end SCM behaviors.

On the information system side, we designed and implemented a new global supply network planning solution and fully linked local scheduling that support the new end-to-end LEAN SCM Planning processes as well as the information requirements during the planning cycle.

Our regionally phased and global product supply chain specific go live of the new operating model, processes and IT support was accompanied by substantial capability development and training to implement the globally orchestrated lean end-to-end SCM concept. Central to embedding, the collaborative new ways of working were well-defined required behaviors that are continuously supported by global and local planning process governance (see Figure 12.3).

12.1.1.6 Results/Benefits

AZ benchmarks are very good on inventory against its peers. We have been able to secure customer service while leveling capacity usage and reducing planning effort.

The adoption of lean SCM principles across the various layers of AZ's supply chain helped us to remove nonvalue-adding activities and let the customer be the key trigger for our operations. The higher efficiency and responsiveness of our supply chain organization improved key figures such as working capital, lead times, and asset utilization considerably.

Has this lean SCM journey therefore now come to an end? Clearly not. We continue to apply lean supply methods to our product supply chains, we continuously globally enhance the maturity in our planning cycle and decision making, and we continue to implement innovative lean SCM tools. The groundwork is successfully laid but continuous improvement is key to lean.

12.1.2 Eli Lilly's Synchronized Lean Production

Profile of Case Contributor

Ronald W. Bohl, senior director, Supply Chain, has 32 years' experience in manufacturing and manufacturing consulting. For the last 10 years, he has been responsible for all supply chain business processes operating at Eli Lilly Pharmaceutical Company. Mr Bohl has made presentations at many professional society meetings and conferences, company management meetings, and software users' group meetings. He is the editor of the chapter "Capacity Planning" in the current edition of *The Production and Inventory Control Handbook*. He has written many articles for publication, including "Ten Commandments of Planning and Scheduling," "Give Your Master Scheduler a Chance to Succeed," and "Advance Planning and Scheduling Issues and Answers."

12.1.2.1 Company Profile and Case Summary

"It is not about improving the accuracy of the forecast and reducing the amount of uncertainty in the future, it is about eliminating the need for certainty."—Ronald W. Bohl

- As a leading global pharmaceutical company with a strong 130-year heritage, Eli Lilly delivers healthcare products to customers in 125 countries around the world.

- Eli Lilly operates. a global supply chain with research and development facilities located in eight countries and manufacturing plants located in 13 countries.
- The headquarters are situated in Indianapolis, the USA. Eli Lilly has more than 38,000 employees worldwide who generate more than US$ 23 billion in revenues.

12.1.2.2 Executive Summary

Eli Lilly adopted a supply chain excellence initiative to reduce costs and meet changing market requirements. To reduce the need for certainty in supply chain planning and efficiently synchronize production and distribution, Eli Lilly introduced four LEAN SCM concepts to its supply chains: aggregate forecasts, pull-control using VMI, postponement strategies, and Rhythm Wheels for production planning and scheduling. These innovations enabled us to substantially reduce lead and cycle times as well as inventories along the supply chain. Furthermore, such supply chain optimization reduced the need to invest in costly new resources.

12.1.2.3 Company's General Situation

To compete successfully in its market, Eli Lilly needs to respond to the typical challenges of the global pharmaceutical business. It faces pressure from healthcare reform as well as changes in the global marketplace and consumption patterns that create uncertainty for our whole business. Like other companies in the pharmaceutical industry, Eli Lilly's key issues are strengthening the product pipeline and mastering the phase-out of key patents. Furthermore, Eli Lilly wants to make significant investments in emerging markets and new products.

Cost reduction programs were launched at Eli Lilly in response *to* these issues. A major achievement along this road was the adoption by Eli Lilly of a lean production philosophy to reduce complexity and improve supply chain flexibility. As highlighted in Figure 12.4, our lean supply chain program aims to find the sweet spot between often conflicting supply chain objectives. Within Eli Lilly, it is the global supply chain organization that introduces, designs, implements, and assures effective use of lean demand and supply processes that allow us to resolve our issues and our uncertainties and add value to our customers.

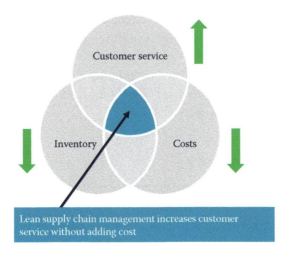

FIGURE 12.4
Lean principles in the supply chain optimize both cost and service.

12.1.2.4 Lean Challenge

At Eli Lilly, our SCM directors faced two challenges in particular: eliminating the need for certainty about future market developments and adopting measures to satisfy differentiated customer needs without major investment in new machinery or personnel.

The future is uncertain, and the further ahead you look the more uncertain it is. Ask yourself for instance what business trips will you be taking or what weddings or celebrations you will be attending 12 months from now. You will have to admit that you know very little about the personal future of someone you ought to know the most about: yourself. Conventional supply chain planning approaches aim however to determine manufacturing decisions 11–12 months before the product is actually in the hands of a customer. So we went to our sales and marketing colleagues and asked them for forecasts. We asked for forecasts at the SKU level, at the strength level, and 12 months out—and we needed them to be as accurate as possible. Based on this information, we let our planning systems determine manufacturing quantities and production sequences to make a specific drug at a specific dosage. Figure 12.5 illustrates the issues associated with forecasting for supply chains.

Now that means that we had to make manufacturing and supply chain decisions long before we knew what customers would really demand. It

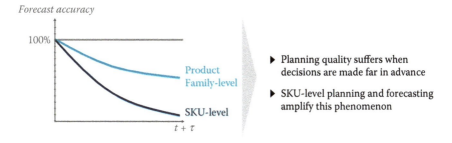

FIGURE 12.5
Resolving dependency on forecasts is a key challenge for supply chain planning.

is hard enough to predict your personal future 12 months out; how can we expect to know what the future holds for our products in the marketplace at this high level of granularity? So the real issue here was not about improving the accuracy of forecasting and reducing uncertainty about the future, it was about eliminating the very need for certainty. This is where lean SCM comes in.

A second challenge was managing more finely differentiated market demand along the supply chain. We do not only have to master more orders from our customers, but we must simultaneously compete in markets that are asking for smaller quantities. Take for example the packing orders at one of our manufacturing sites: Last year, we had about 4,300 packaging orders moving through a manufacturing site that operated more than 12 distinct packaging lines. And we were looking at a significant increase to as much as or more than 7,700 packaging orders coming into that particular facility over the next couple of years (see Figure 12.6).

To solve this challenge, a typical approach would be to add more packaging lines and more people. However, this approach would have added substantial cost to our supply chain operations. Therefore, we were looking for better ways to solve this puzzle. For Eli Lilly, the most promising approach was to adopt lean production processes, which helped us avoid the need for costly investments in new production resources.

12.1.2.5 Approach

Eli Lilly has established a program called Operating Standards for Supply Chain Excellence that helps us maintain excellence in our supply chain processes within manufacturing sites, in marketing, in affiliates,

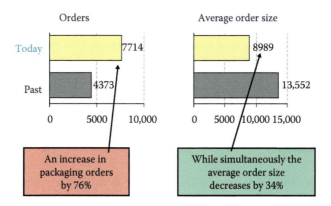

FIGURE 12.6

Manufacturing has to manage more orders and smaller average order sizes today.

and at distribution centers (DCs). The objective is to adopt some basic concepts that define how to create production plans, ensure that the plans are valid, and execute those plans on a day-to-day basis. To reduce complexity while improving flexibility along our supply chain, and to establish true lean and demand-driven SCM, we rely on four fundamental concepts:

- Simplified forecasting with central planning
- Replacing Demand Requirements Planning (DRP) with Vendor Management Inventory (VMI)
- Postponement strategies
- Synchronized lean production using Rhythm Wheels

The first concept applies to our demand statement. Under the old way of forecasting at Eli Lilly, affiliates provided SKU-level forecasts with a 3-year horizon. That forecast drove our production plan all the way along the supply chain. Not only did this create a lot of volatility, but even more importantly, even though we measured the accuracy of that forecast over the next 3–4 months, but it was also nothing more than a statistical continuation of that which we already knew. So, we still did not really see in that forecast all of the business potential that could in fact be realized. So the issue here was to make the uncertainty of the forecast irrelevant.

Our new approach complements SKU-level information with more aggregate product family forecasts provided by our business areas (see Figure 12.7). Because aggregate forecasts are more stable and accurate

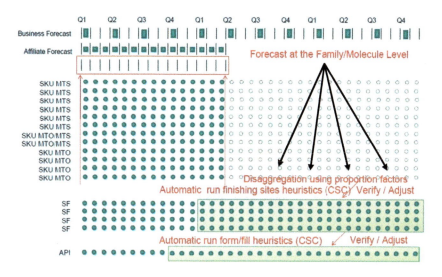

FIGURE 12.7
Simplified forecasting through aggregation and central planning.

than SKU-level forecasts, especially if the forecast horizon is extended, this helped us to simplify and improve our forecasting and supply chain planning. Furthermore, these product family forecasts provided a better basis for making strategic business decisions about buying equipment or sourcing changes.

The second measure involved replacing our DRP approach with a VMI approach. Here, we switched from traditional "push" control to a "pull" control approach to manufacturing. Under our previous DRP logic, manufacturing had no information about demand and forecast signals or available inventory at the DCs, but was driven by due date requirements for shipment quantities derived from our forecasts.

Our VMI system follows a buffer management approach and thus demand-driven pull logic (see Figure 12.8). For every product at every DC, minimum and maximum inventory boundaries are defined based on guidelines developed by the global supply chain organization. These boundaries account for the optimal shipment and production order quantities as well as for the safety stocks required to buffer demand and supply variability to meet the agreed-to service levels. The new objective for manufacturing is now simple: Keep inventory at DCs between the minimum and maximum inventory boundary. This ties production to customer demand and makes detailed due date quotations obsolete.

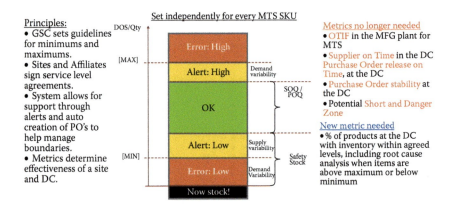

FIGURE 12.8
Replacing DRP with VMI.

Third, we have established a postponement strategy that allows us to respond to actual sales more efficiently and reduces our dependency on forecasting accuracy. Consider, for example, the manufacturing and distribution process at Elanco, the animal health division of Eli Lilly. Our old approach was to manufacture the bulk product, finalize the product by labeling it, and then ship it to regional DCs. Owing to considerable transportation lead times along the supply chain, these manufacturing decisions had to be made far ahead, before the actual product was sold. As a consequence, we required a substantial amount of inventory to buffer against demand variability and forecasting errors.

Our new strategy is to take away demand variability at the SKU level by postponing product finalization and consolidating upstream demand along the supply chain, where it is more accurate (see Figure 12.9). Therefore, we moved our finishing sites into the regional hubs that finish orders and finalize products. By doing so, we established a push–pull boundary in our manufacturing process whereby finalization follows pull logic and products are finished to demand while bulk products are manufactured based on forecasts at the product family level. This reduced the need for certainty in the manufacturing process and increased our market responsiveness.

As we explained in the introduction, a key challenge for our manufacturing operations is managing higher-order numbers while simultaneously decreasing average order sizes. So how did we do this without adding personnel and lead time, and without making things less frequent?

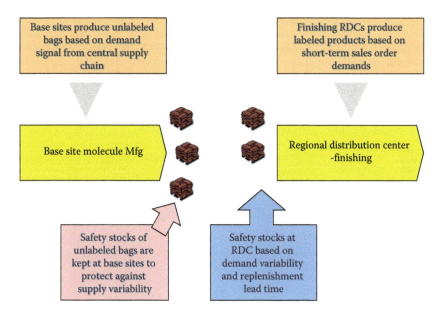

FIGURE 12.9
Postponement strategies in Eli Lilly's supply chain.

To solve this challenge, we adopted a Rhythm Wheel concept across our packaging sites whereby we schedule products in a fixed sequence, as shown in Figure 12.10. In accordance with the High-Mix Rhythm Wheel concept, we produce, for instance, high-volume A-products 10 times a year, while assigning B- and C-products to a lower production frequency. By optimizing our schedules and production sequences according to the Rhythm Wheel, we were able to reduce changeover times. Overall, the Rhythm Wheel helped us to reduce lead and cycle times along our supply chain and increase market responsiveness at our production assets, which also supports the smooth operation of our VMI processes.

Finally, we synchronized our production processes with transport to DCs. Using fixed sequences across the packaging lines allows us to efficiently consolidate shipments to each destination and avoid sub-optimization.

12.1.2.6 Results/Benefits

In summary, we are developing processes and solutions that shift focus from dependency on forecasting while simultaneously helping manufacturing

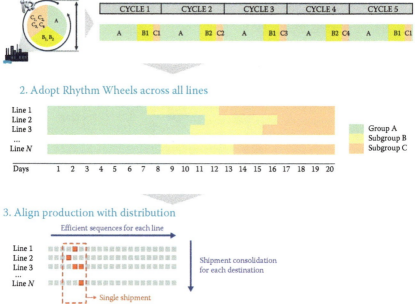

FIGURE 12.10

Synchronization of lean production and distribution with Rhythm Wheels.

become more flexible and reliable through synchronized lean production using Rhythm Wheels. All these measures together help us to bring value to our customers.

A key benefit of our lean SCM approach is the significant reduction in lead and cycle times along the supply chain. Using lean principles, we have reduced lead times from 34 to 10 days and cycle times from 10 to 3 days. Furthermore, shortening lead times and cycle times reduces inventories and increases our responsiveness to market demand. We have found that we do not need a separate inventory program: we just focus on cycle and lead times along the supply chain.

Lean principles help us run our supply chain more efficiently and avoid unnecessary investment in production resources. At packing sites, for example, we can now manage higher-order numbers and lower average order sizes without adding new packaging lines or hiring new people. Overall, we are able to avoid a 14% cost increase per pack through lean SCM.

No matter what challenges we will face at upcoming implementation stages of the synchronized lean concept, it is the future of the pharmaceutical industry and the future of Eli Lilly.

12.1.3 Buffer Management at Novartis

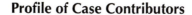

Profile of Case Contributors

Christophe Vidonne, head of Supply Chain Management, joined Novartis Animal Health in 2000. Since his more than 20 years of professional experience in the retail and pharmaceutical industry, he exposed a variety of roles such as Finance Business Planning & Analysis, Production, Distribution & Supply Chain Management and Information & Technology. At Novartis Animal Health, he first held the position as the head of Supply Chain & Business Planning & Analysis in France before becoming the head of Supply Chain Management at the Basel headquarters.

Before starting his professional career, he successfully completed his master's in information & technology at the Instituts Nationaux des Sciences Appliquées and in biochemistry at the Université Claude.

Dr. Ralph Billo, has been head of Global Supply Chain Management at Novartis Animal Health since 2003. Prior to this role, he was member of Country Operations and assumed different Region Management roles from 1991 to beginning of 2003. In addition, he was also in charge for the Customer Service organization in the Basel supply hub from 1997. He joined the Farm and Companion Animal Health Division of Ciba-Geigy in May 1986 as a product manager and was in charge of various products in the farm animal business. In the geographical areas under his responsibility, he integrated various acquired businesses and implemented FAR

(Fast Action for Result) and Profit Walkthrough projects; following the merger of Ciba-Geigy and Sandoz to Novartis, he managed the integration activities in geographical areas under his responsibility. In 1991, he moved from Product Management to Region Management and in 2003 to Supply Chain in Technical Operations.

He graduated in biology and holds a PhD from the University of Basel, Switzerland. Key studies during his PhD thesis were realized in Berkeley, California, as well as in Heidelberg, Germany.

12.1.3.1 Company Profile

"Controlling and monitoring of inventory has long been fully integrated into supply chain functions. Transportation and warehousing cost management are also key tasks for supply chain managers. From an end-to-end perspective, what really counts is what remains visible and measurable for our customers. It is not only about key performance indicators and delivery service but also the ability to properly communicate accurate and meaningful supply chain information in a multi-cultural and cross-functional environment across geographies."—Christophe Vidonne and Ralph Billo

NOVARTIS

- Industry: Pharmaceuticals
- Products: Parasite control, vaccines, anti-infectives, and other medicines for animals and farmed fish
- Employees 2010: Approx. 2,700
- Revenue 2010: $1.208 billion
- Headquarter: Basel, Switzerland

12.1.3.2 Executive Summary

Novartis Animal Health (NAH) is a leading global provider of solutions for the prevention and treatment of various widespread animal diseases and parasite infestations. To maintain and even strengthen our market position, we are eager to continuously improve our supply chain performance as one of the key drivers of the company's success. Therefore, a few years ago, we triggered a state-of-the-art supply chain planning initiative with a special focus on redesigning our replenishment framework. The main change concerned a transformation from push-to-pull replenishment for selected products within our portfolio. The successful realization of the benefits of pull replenishment was enabled by the implementation of buffer management as a LEAN replenishment mode. This enabled us not only to reduce supply chain inventories but also to increase our planning accuracy while reducing required planning efforts. These positive impacts underline the benefits of our supply chain initiative and prove that we have come a long way toward a truly integrated supply chain.

12.1.3.3 Company's General Situation

NAH is a core business of the Novartis Group, a leading pharmaceutical company. NAH is a leader in developing new and better ways to prevent and treat diseases in pets, farm animals, and farmed fish. As a pharmaceutical company, NAH faces the same challenges as other companies in the industry—specifically in terms of continuous volume growth, increased complexity, and ongoing focus on quality standards, all of which require high management attention.

It is our belief that a functioning supply chain is a critical factor in successfully addressing these challenges and achieving our long-term business objectives. Therefore, a few years ago, we triggered a state-of-the-art supply chain planning initiative to "move to an integrated supply chain." By enabling efficient and effective management of global information and material flows, the initiative aimed to contribute to the overall supply chain target of reducing working capital without jeopardizing service.

12.1.3.4 Lean Challenge

One of the main areas of focus within the global supply chain initiative was replenishment planning. Critical observations about our supply chains indicated that we would benefit from an appropriate re-design of our replenishment framework. Three major observations are worth mentioning in this context. First, our planning process was based solely on forecasted demands, which we refer to as "push replenishment." As forecasts finally always turn out to be wrong, low planning accuracy resulted in frequent replanning activities and caused uncertainty across the entire supply chain. Second, our planning process was complex and

FIGURE 12.11
Key challenges of our former replenishment planning process.

characterized by many manual process steps. In combination with the large number of replanning activities we had to undertake, the required planning effort was enormous. Third, we observed high levels of inventories along our supply chains. This related primarily to the already mentioned lack of planning accuracy, together with long lead times due to the time-consuming planning process, lack of sufficient flexibility to react to actual demand, and large replenishment quantities (see Figure 12.11).

Owing to the negative implications of these factors for our supply chain performance, we were encouraged to actively address these challenges by redesigning our replenishment framework.

12.1.3.5 Approach

One of the major measures we adopted in redesigning our replenishment framework included the realization of "pull" via the implementation of buffer management as a LEAN replenishment mode. Three dimensions appeared to be very important in this context:

- Evolving from "push" to "pull"
- Operationalization of buffer management
- IT support as key enabler of buffer management

12.1.3.6 Evolving from "Push" to "Pull"

A cornerstone of our aligned replenishment framework is the transformation from "push" to "pull" replenishment. Instead of using forecasts (push), replenishment is triggered by actual consumption (pull). With pull replenishment in place, we expected to reduce required planning efforts, shorten lead times, improve the allocation of available production capacity, and decrease overall inventory levels due to newfound flexibility for reacting to actual consumption.

However, fully implementing pull replenishment entailed further strategic decisions:

- Selection of products suited for pull replenishment
- Alignment of the push/pull boundary

12.1.3.7 Selection of Products Being Suited for Pull Replenishment

As one size does not fit all, we did not intend to apply pull replenishment to all products in our complex portfolio. Therefore, we initiated an analysis of our products' demand and supply characteristics as well as our supply chain constraints. In this context, we considered factors such as demand variability, volume, and profitability. Based on a corresponding segmentation of our product portfolio, we identified the products that were best suited for pull replenishment. As our business environment is very dynamic, we will review our segmentation results at regular intervals to ensure the ongoing suitability of applied replenishment modes.

12.1.3.8 Alignment of the Push/Pull Boundary

The next step was to align the push/pull boundary for products that were selected to move from push to pull replenishment. With forecast-based replenishment, the push/pull boundary is located near the customer interface, whereas consumption-based replenishment allowed us to move the push/pull boundary upstream to bulk production (see Figure 12.12). This alignment yielded two major benefits. First, we could now react more flexibly to fluctuations in market demand for pull-managed products. Second, we could now lower the required forecast granularity. Instead of detailed forecasts at the SKU level, our pull managed products require demand forecasts only at the bulk level. As demand for bulk materials is less volatile, it can be predicted much more accurately. As a consequence, we were able to reduce our forecast errors and thereby improve overall planning accuracy.

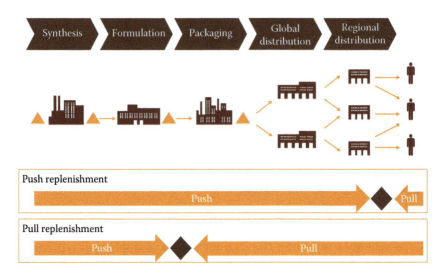

FIGURE 12.12
Alignment of the push/pull boundary.

12.1.3.9 Operationalization of Buffer Management

To realize pull replenishment, we decided to implement buffer management because it appeared to be best suited for our requirements. Prior to actual implementation, of course, the conceptual backbone of buffer management needed to be established and corresponding parameters required a pre-configuration. The following major steps were essential in this context:

- Buffer size determination
- Assignment of replenishment intervals
- Assignment of minimum order quantities

12.1.3.10 Buffer Size Determination

Within buffer management, the reorder point represents one of the key replenishment parameters. When inventory levels fall below this point, a replenishment signal is triggered indicating that replenishment is required. We refer to the reorder point as buffer size since the corresponding inventory level functions as a buffer once replenishment is triggered. We followed a structured approach to determine appropriate buffer sizes. On the basis of the definition of buffer components and corresponding

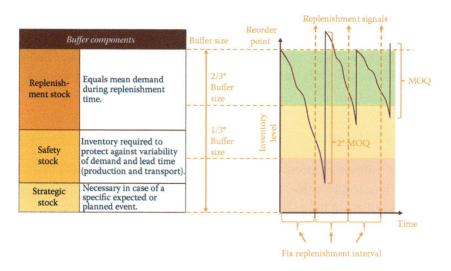

FIGURE 12.13
Operationalization of buffer management.

impacting factors, the resulting buffer size was calculated by the following formula using a safety factor of 1,3:

$$\text{Buffer size} = (\text{production lead time} + \text{transportation lead time} + \text{order lead time}) \times \text{safety factor}$$

Furthermore, we divided the inventory buffer into three zones: green, yellow, and red. Each of the zones represents one-third of the total buffer size (see Figure 12.13). On the basis of the defined zones, we have built up our monitoring and alerting system. If, for instance, the inventory level of a given product is in the red zone for a certain number of days, the planner is expected to evaluate the necessity of a buffer size increase.

12.1.3.11 Assignment of Replenishment Intervals

Prior to implementing buffer management, order lead time (= replenishment interval) for all products equaled 3 months, that is, replenishment was triggered at this frequency. As required inventory levels increase with longer order lead times, we decided to reduce this lead time for selected products when we implemented buffer management. Within buffer management, replenishment signals are typically triggered every time a given product's inventory is reviewed. Therefore, we had to be sure to review the inventory status for selected products more frequently than every 3 months. We made the final assignment of review cycles on the basis of a classification according to product-specific characteristics such as volume and value. Finally, replenishment intervals were assigned ranging from 2 weeks to 3 months.

12.1.3.12 Assignment of Minimum Order Quantities

To prevent replenishment ordering in very small quantities, we also assigned minimum order quantities (MOQ) to each product. Hence, replenishment quantities were allowed only to either equal the MOQ or to be a multiple of it. In this way, we also made it possible to synchronize replenishment orders with production and transportation lot sizes, allowing for smoother materials flows and thus lower supply chain inventories.

12.1.3.13 IT Support as Key Enabler for Buffer Management

Buffer management is certainly a very LEAN concept which can be easily applied in practice and does not require much human intervention. However, considering our complex global supply chains and the large number of differentiated products in our portfolio, appropriate IT support was critical to enabling efficient replenishment planning. It is particularly important to reduce manual efforts while enhancing global supply chain visibility of product-specific stock levels, both of which are supported through partnership with IT. Prior to investing in an IT solution, we formulated a number of requirements to ensure sufficient support for the application of buffer management. Figure 12.14 lists some of the key requirements that we believe are very important in this context. In particular, having

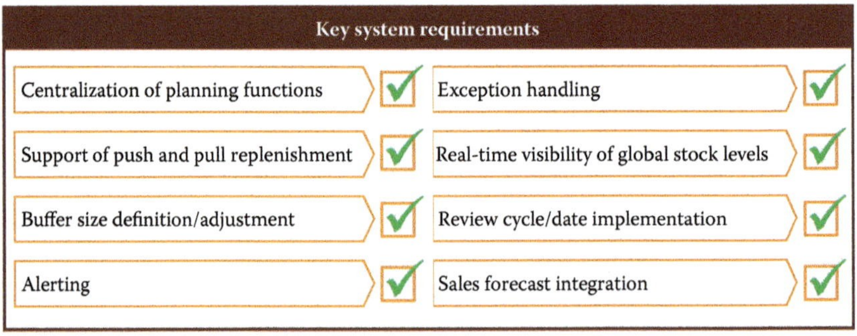

Key system requirements	
Centralization of planning functions ✓	Exception handling ✓
Support of push and pull replenishment ✓	Real-time visibility of global stock levels ✓
Buffer size definition/adjustment ✓	Review cycle/date implementation ✓
Alerting ✓	Sales forecast integration ✓

FIGURE 12.14
Key system requirements.

alert functionalities is important as it makes it possible to establish an early warning system. With the help of such a system, planners are able to react quickly and appropriately when there is a need to intervene.

12.1.3.14 Results/Benefits

Following implementation of pull replenishment via buffer management, NAH has increased the competitiveness of our supply chain and taken an important step toward realizing the benefits of a fully integrated supply chain. It is worth highlighting three major benefits, since they address the key challenges that we had previously faced.

First, we increased our planning accuracy. The move from push to pull can be seen as the main reason behind this development. Now that pull replenishment has enabled us to move the push/pull boundary upstream to bulk production, we are able to use forecasts with lower granularity, which means lower forecast errors and thus higher planning accuracy. Second, we have notably diminished our planning effort. Beyond the pre-configuration of a few replenishment parameters, buffer management enables us to manage replenishment with minimum levels of resources. Third, we have reduced our inventories along the supply chain. This is due mainly to newfound flexibility to react to real demand, shorter replenishment intervals and thus smaller replenishment quantities, and more effective allocation of production capacity. It is important to note that we were able to reduce our inventories without sacrificing service.

In the face of this positive development, we are confident that we will continue to maintain the competitiveness of our supply chains and be well prepared for future challenges.

12.1.4 Leveled Flow Design to Enable LEAN Planning

12.1.4.1 Company Profile

We will now look at the LEAN SCM experiences of a global pharmaceutical company that wishes to remain anonymous. Its senior vice president claims, "Lean planning concepts lead not only to the efficiency increase in our supply chain, but also made it more agile for the future growth!"

12.1.4.2 Key Facts

- The company is an innovative European R&D-based pharmaceuticals and diagnostics company with an emphasis on developing medicinal treatments and diagnostic tests for global markets. It develops, manufactures, and markets human and veterinary pharmaceuticals, APIs as well as diagnostic tests.

- The company develops products that are sold worldwide and marketed in over a hundred countries. Human pharmaceuticals sold through its own network or by marketing partners account for about 80% of its net sales.
- The company's own sales network covers almost all key European markets. Outside Europe, it operates through marketing partners. Sales through partners account for just under one-third of the company's net sales.

12.1.4.3 Executive Summary

Unbalanced planned utilization of different production assets leads to bottlenecks and underutilized capacities at the same time. The workaround to master this challenge was continuous rescheduling at the shop-floor. This was not only a pure fire working, but also created a suboptimal solution, due to the short-term perspective of the responsible planner. To solve this by a takted planning and sequencing approach like Rhythm Wheel, three preparation steps were undertaken: product portfolio segmentation, resource portfolio segmentation, and product-asset reallocation. As a result, leveled capacity utilization between assets was achieved and bottlenecks were eliminated.

12.1.4.4 Company's General Situation

The senior vice president remembers: "When I started working in the pharmaceutical industry, frequent innovations and drug patents were the bases for a blockbuster strategy with high margins, which could be realized even during economic downturns. In recent years, however, the expiration of blockbuster patents and a dwindling innovation pipeline have increased competition with generic drugs." As a consequence, the company was facing cost reduction pressure along with most other pharmaceutical companies. They had begun several traditional shop floor initiatives to increase operational efficiency. Among other things, they worked on changeover reduction, efficient problem solving, and total productive maintenance. In other words, they applied traditional elements of lean manufacturing to improve and stabilize our production processes.

Considering increased efficiency to be vital to sustainable competitiveness, they looked for further levers of efficiency in order to optimize productivity and reduce costs. As a consequence, they began supporting its lean manufacturing activities on the shop floor with lean SCM concepts, with both elements striving for the same target. We recognized that there is potential in setting up production planning in line with their shop floor capabilities, better utilizing their strengths while mitigating the weaknesses in available resources. Of course, the company did not and do not want to replace traditional operational excellence initiatives. Rather, they want to complement them from an SCM perspective, optimizing not only their manufacturing but also their planning processes.

12.1.4.5 Lean Challenge

Over the past few years, the company had experienced high capacity utilization on some lines in the packaging area. Thanks to frequent manual reallocation of production orders to alternative lines, they avoided huge overloads—luckily their planners knew what they were doing. If the company had allocated the portfolio according to the originally defined primary resources, we would have had resource utilizations as illustrated in Figure 12.15—clearly not an acceptable situation.

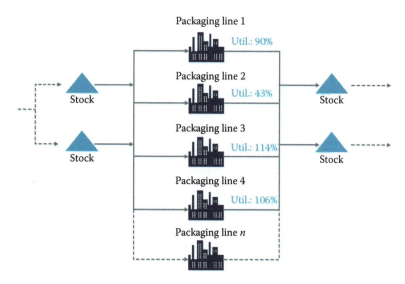

FIGURE 12.15
Average line utilization according to designed resource allocation.

The daily rescheduling of production orders smoothed the capacity utilization, but the company still had to run additional shifts on their bottleneck line to satisfy overall demand. Although they made it through the last year thanks to the planner's manual interventions, they had neither predictability nor transparency in production planning. It was normal that in 1 week, for example, a given product was produced on line 2 and during the next week it was scheduled on line 4. With a portfolio of about 1000 SKUs and a corresponding number of production batches, rescheduling activities required great manual effort.

The company also knew that capacity requirements would increase in upcoming years. A variety of new products were about to be launched and all of these new products would consume additional packaging department capacity. It was unclear whether they could cope with the product launches using the existing resource pool or if investments in capacity were inevitable. However, the company had the feeling that improved product allocation could be a potential lever for optimizing production efficiency. They hoped that they could further reduce production and changeover times by matching product characteristics and machine capabilities in order to improve their capacity situation in packaging.

12.1.4.6 Approach

The company applied a three-step approach during the strategic reallocation of our product portfolio: (1) analysis of the product portfolio, (2) analysis of the resource portfolio, and (3) product allocation. Figure 12.16 provides a brief overview of these steps.

As a first step toward strategic reallocation, the company analyzed their product portfolio for the packaging lines in scope and segmented it according to demand volume and demand variability. Figure 12.17 illustrates the results of the portfolio analysis. The company had to manage a highly heterogeneous mix including a high number of small-volume products with relatively high demand variability. This heterogeneity applied not only to the complete portfolio but could also be found within individual product family clusters.

Following product portfolio analysis, they focused on the resource portfolio and had a closer look at the key characteristics of the four packaging lines in scope. Here, they focused on two dimensions: line speed and flexibility in terms of product changeovers. Figure 12.18 depicts the outcome of the resource analysis: The packaging lines differed significantly—mainly

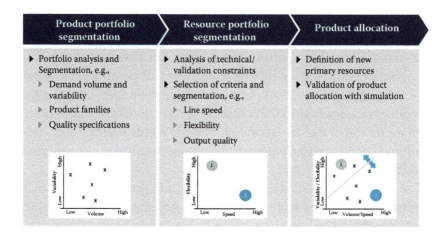

FIGURE 12.16
Three-step approach to strategic resource allocation.

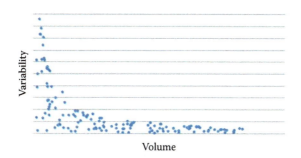

FIGURE 12.17
Overview product portfolio.

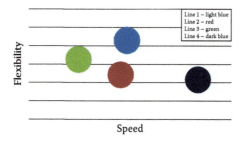

FIGURE 12.18
Overview resource portfolio.

in speed but also in flexibility. While this illustration shows average values to create transparency and shared understanding regarding the available pool of resources, they used product-specific run rates and changeover times for the exact analysis later on. Clearly, at that point the company's hope was that an improved allocation rule might help them to utilize the strength of each resource in the best possible way: The differences between the lines indicated that the allocation of high volumes on fast lines and small volumes on flexible lines might offer some potential to improve their production efficiency.

In addition to the product and line specifics, additional constraints such as technical restrictions had to be taken into account: Not every product could be produced on every line, so there were certain limits to the reallocation of products. For example, while plastic bottles could be run on one line, glass bottles had to be packed on another one. However, depending on a complex set of rules, the company could realize quite a bit of flexibility by conducting rotating product changes. This practice was sufficient to enable them to optimize the distribution of production volumes over all four lines—at least to a certain extent.

After the analyses, the products were reallocated. On the fastest line, the company implemented a repetitive 3-week production pattern. All high-volume products that fell into this pattern were assigned to this line. The remaining products with high volumes were assigned strictly to the faster lines, while products with small volumes but a high number of production batches (i.e., set-ups) went on flexible lines to reduce overall production and changeover time. This freed up capacity. Product families played a secondary role—it has to be admitted that this was counterintuitive to the company at first, since splitting product families meant higher changeover effort. Actually, there was a negative effect on changeover times if they separated product families. However, this (small) negative impact was more than compensated for by the capacity savings that could be realized by effectively leveraging the high production rates on the fast lines. After all, it made sense to have product families as a secondary criterion only. The company tried to keep them together if appropriate, but the main focus was on speed and volumes.

To analyze various future scenarios and validate the results, the company used discrete event simulation. They tested a range of scenarios and compared them with their benchmark solution, the original allocation of production orders. Not only did simulation prove quite powerful in making results visible and reliable, but it was also needed due to high process complexity. Sequence-dependent set-up times and product-specific run rates had to be modeled in detail to capture the capacity requirements as precisely as possible.

To ensure that the simulation model itself was valid, test runs for the baseline scenario were conducted. Since the simulation model replicated the baseline scenario with very high accuracy, the company could be sure that the results of the simulation runs would deliver feasible results. Figure 12.19 provides a snapshot of the simulation model.

In various simulation scenarios, they then tested several allocation strategies. As a first step, they tried to find the optimal allocation for the current situation. All exogenous factors were the same as in the baseline scenario. This first scenario delivered both an optimized product allocation for the near future as well as some generic proof of the improvement potential of strategic reallocation. In the next step, additional allocation strategies were customized to fit distinct exogenous influences, for example expected future demand volumes, new product launches, and adjusted line capacities. These future scenarios enabled them to sense what might happen in the medium-to-long-term horizon and which actions regarding strategic product allocation or operational excellence in general could be taken to maintain high production efficiency.

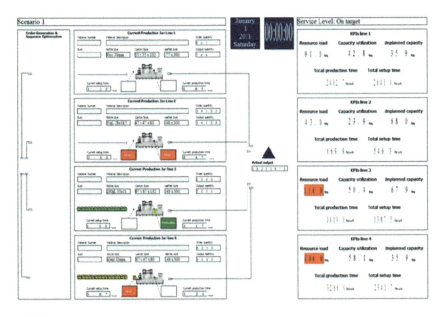

FIGURE 12.19
Simulation model for strategic product reallocation.

12.1.4.7 Results/Benefits

The simulation runs showed that strategic product reallocation has the potential to solve current and upcoming capacity issues and reduce the manual effort involved in planning and ad hoc reallocation of products.

In the baseline scenario, frequent rescheduling of orders was required to cope with capacity bottlenecks, but the optimized allocation utilized the resource pool more effectively. Instead of strictly keeping product families together, high volumes went on fast lines, low volumes on slow but more flexible lines. As a result, enough capacity was set free to cover most of the estimated demand for new products that will be launched during the upcoming 2 years. Figure 12.20 illustrates the improved capacity utilization.

The company has experienced increased stability in planning, since the majority of products are now assigned to their primary resources that are maintained in their ERP system's master data. Logically, rescheduling is still an option in periods of exceptionally high demand, and the company is convinced that their experienced planners will be able to further optimize the overall situation. However, their production planning—and the question whether available production capacity is enough—no longer depends on a regular reallocation of products.

In addition to the direct impact on production efficiency and capacity bottlenecks, the various simulation scenarios helped to create transparency

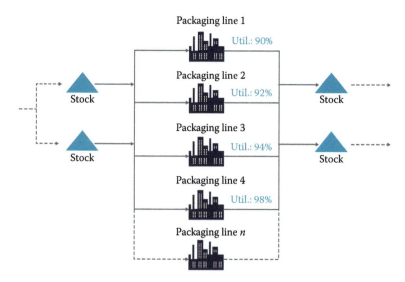

FIGURE 12.20
Average line utilization after strategic product reallocation.

regarding their product portfolio and production environment, revealing areas open to further improvement, a process that the company has already begun.

Overall, by additionally improving their production planning from a LEAN SCM perspective, the company was able to fully realize the benefits of their lean manufacturing activities.

12.1.5 AstraZeneca Excellence with Rhythm Wheel Takted Site

Profile of Case Contributor

André Wulff is former head of Supply Chain Planning at AstraZeneca (AZ), Germany. He joined AZ in 2003 and has held several positions as a supply chain planner, head of supply chain management, and project manager, Supply Chain–SAP APO. During his more than 14 years of working experience, he has gained profound Supply Chain Management knowledge as well as extensive experience in project management and the implementation of lean principles. Prior to beginning his working career, André Wulff studied supply chain management, logistics, and engineering after qualifying as a certified electronics installer at Siemens AG.

12.1.5.1 Company Profile and Case Summary

"There is no better planning concept than the Rhythm Wheel, I am convinced of this."—André Wulff, former Planning Head, AZ Germany.

- Industry: Pharmaceuticals
- Products: Cardiovascular, Gastrointestinal, Infection, Neuroscience, Oncology, and Respiratory

- Employees: 1,300 employees
- Sales (2011): $33 billion
- Headquarter: Wedel, Germany

12.1.5.2 Executive Summary

AZ started a global lean supply chain initiative under pressure imposed on product margins. The Rhythm Wheel was a promising concept in light of our effort to reduce inventory costs and improve planning and production at our production site in Wedel, Germany. We combined the specific Rhythm Wheel with a consumption-based pull replenishment mode. We implemented a Rhythm Wheel scheduling heuristic in our planning systems with explicit consideration of minimum make quantities for our low-volume products. As a result, we saw a significant reduction in changeover times using the optimized Rhythm Wheel sequence and a reduction in inventory levels by more than 20%. Furthermore, the Rhythm Wheel planning approach was widely and readily accepted by our employees due to its transparency and the reduced planning effort it made possible.

12.1.5.3 Company's General Situation

Like most other companies in the pharmaceutical industry, we were facing pressure due to more restrictive governmental regulations on introducing and setting prices for new medications. Since AZ depended on patenting its own research and development process, the pressure on margins increased as some of those patents were due to phase out in the near future. To ensure profitability, we responded with a global lean supply chain initiative. The goal was to create transparency along the supply chain and reduce working capital and production costs.

As part of that initiative, we sought to harmonize production planning processes across the company. A consistent planning approach would allow us to quickly allocate products to various production sites, and would make it possible to exchange valuable knowledge between planners. The Rhythm Wheel concept was identified as a promising solution with which to standardize planning. We decided that the packaging site at Wedel should start with a pilot of this new planning concept. Until that point in time, we had been using a complex planning concept at Wedel.

12.1.5.4 Lean Challenge

One of the main challenges posed by our former planning approach was that we did not have sufficient transparency in our production planning. Our planning approach did not deliver stable production patterns, but instead generated a completely new production sequence at a specific resource with every planning run. What made it even more complex was that almost all products could be scheduled for all of our 13 packaging lines, depending on the production version. This led to short-term reallocations of products to alternate packaging lines, which made it hard for us to achieve stable short-term production plans. Furthermore, the mix of our product groups on each line could result in long changeovers, since extensive cleaning was required between two production runs with different active ingredients. The left-hand side of Figure 12.21 shows a typical production schedule for two product groups.

Although the planning and scheduling were performed with the use of a highly sophisticated heuristic, this complex planning approach still did not return the desired results. Production plans were nontransparent and could change completely from one day to the next.

12.1.5.5 Approach

The solution to our complex production planning process was the introduction of the Rhythm Wheel concept with dedicated lines for each product group. First, we allocated the product groups to dedicated resources. In this way, we reduced planning complexity and benefited from learning effects. By grouping several products that include the same active ingredient, we could reduce changeover times due to less extensive cleaning efforts (see the right-hand side of Figure 12.21).

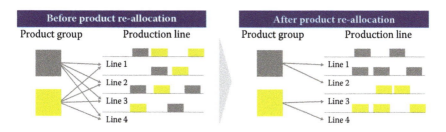

FIGURE 12.21

Production schedules were less complex after the reallocation of product groups to dedicated lines.

The next step was to implement the Rhythm Wheel concept on each of the lines. We opted for a High-Mix Rhythm Wheel and an IRL replenishment mode. The High-Mix Rhythm Wheel defined a minimum make quantity for low-volume SKUs. This minimum make quantity covered demand for several Rhythm Wheel cycles, such that we did not need to produce them in every cycle. Our high-volume "A" products on the other hand were scheduled in every cycle at flexible quantities. In this way, we could keep inventory levels low for our most valuable SKUs. We determined the optimal production sequence by sorting the products by dosage, and then by format in increasing order. Even MTO products were scheduled at exactly the right position in the cycle. In this way, we could reduce changeover times. Figure 12.22 shows a sequence-optimized Rhythm Wheel for one production line.

For all SKUs, pull was implemented by defining an IRL replenishment mode. Each Rhythm Wheel cycle, the inventory of a given SKU was checked and compared with the IRL. If the current inventory was below the IRL, the production quantity was determined as the difference between the IRL and the current inventory level for high-volume products. Low-volume products were produced with minimum make quantities as soon as their current inventory levels were below the IRL. If the current inventory was above the IRL, we forced a "skip" decision, which meant that the product was not going to be produced in that cycle. A "skip" could

FIGURE 12.22
Production sequence of one of the Rhythm Wheels at Wedel.

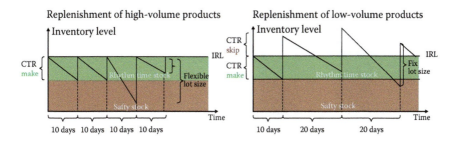

FIGURE 12.23
Replenishment of high- and low-volume products.

occur only for low-volume products, as their minimum make quantities lifted the current inventory level above the IRL (see Figure 12.23).

We implemented this replenishment logic with a so-called consumption trigger report (CTR). In the CTR, we listed the products in the optimized Rhythm Wheel sequence, their IRLs, and their current inventory levels. Based on this information, the make/skip decision and the production quantity were shown (see Figure 12.24). With the CTR, we were able to transparently show what had actually been consumed by our customers and what needed to be replenished. This means of visualization was of great help to planning and the shop floor.

The average planned cycle time was 10 days. But since skip decisions were possible, the Rhythm Wheel cycle time could be much shorter than the planned 10 days. If demand was high, cycles could be lengthened beyond 10 days due to flexible lot sizes. To prevent a given cycle from becoming either too short or too long, we introduced cycle time boundaries: The cycle time was allowed to react flexibly to demand between a minimum of 8 and a maximum of 12 days. This stabilized our capacity utilization,

Produkt	Produktbezeichnung	IRL	Inventory	Decision	Menge	Min. Qty.	Days Su
45013		MTO	4.002	Skip	-	MTO	99
45356		2129	7.933	Skip	-	8000	7
421218		4105	9.224	Skip	-	8000	7
421170		990	888	Make	3000	3000	4
423470		1033	1.634	Skip	-	4000	6
421081		76903	47.025	Make	29878	1	2
45489		46067	47.846	Skip	-	1	1
421219		13400	7.245	Make	8000	8000	1

Designed RW sequence

FIGURE 12.24
The consumption trigger report.

FIGURE 12.25
Integrated planning landscape in SAP APO.

although we had skip decisions for our low-volume products and could react flexibly to demand.

To obtain an integrated planning solution, we implemented the Rhythm Wheel planning logic and the CTR with APO PP/DS functionality (see Figure 12.25 for the integration of the Rhythm Wheel concept into the planning landscape). This gave us the benefit of real-time integration with our ERP software.

The Rhythm Wheel schedule was updated in a daily planning run. In this way the latest data, such as current inventory levels, were considered directly, and the results of a planning run were immediately transferred to our ERP system. Furthermore, we could also easily conduct mid- and long-term planning with the PP/DS Rhythm Wheel heuristic. Although production was scheduled based purely on real consumption, the following Rhythm Wheel cycles were simulated based on a projection of future inventory levels (see Figure 12.26). In this way, we could deliver important input to other business processes such as sales and operations planning and inventory planning for financial statements. For our planners, the heuristic delivered a capacity check, which showed how capacities would be used in the future.

One aspect of the Rhythm Wheel heuristic needed special attention: We wanted to use the heuristic for planning but still follow a strict pull

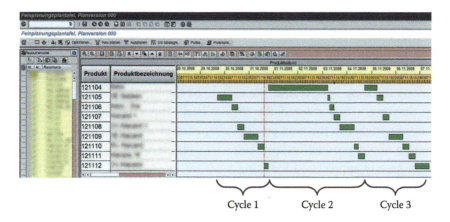

FIGURE 12.26
Rhythm Wheel scheduling in APO PP/DS.

principle. This at first sounded like a contradiction between two phi-losophies. However, we were able to resolve this contradiction and use the benefits of both philosophies at the same time by implementing a minor trick. We introduced an "X-line," which divides the time hori-zon into two parts. From any current day until the X-line, production was purely consumption driven. This meant that absolutely no forecasts were included in the actual production quantity. For the generation of the production schedule beyond the X-line, demand forecasts were used to simulate projected inventory levels and create the supply elements. In this way, we could capture the benefits of pull consumption but still maintain visibility over our production schedule and check projected capacity utilization.

12.1.5.6 Results/Benefits

By implementing the Rhythm Wheel concept, we were able to signifi-cantly increase the performance of our packaging site. By dedicating our packaging lines to certain types of active ingredients, and by producing in the optimized Rhythm Wheel sequence, we were able to reduce change-over times considerably. Furthermore, we achieved more than 20% in inventory savings by the introduction of pull principles through the IRL replenishment mode, because we produced only what had actually been consumed by our customers. Using minimum and maximum cycle time boundaries, we leveled production over time and achieved stable capacity utilization.

In addition to the inventory and capacity benefits, transparency in the production sequence was especially effective, generating great acceptance among the shop floor teams. We visualized the Rhythm Wheel sequence by drawing a big, colored Rhythm Wheel. The shop floor team then moved a sticker to the next product after each production run. The information about production quantities was visible through the CTR. The team greatly appreciated knowing what was going to be produced next, which enabled them to make preparations for the next production run. The increased visualization involved in the process made shop floor teams feel more engaged in the entire process, which enabled a smooth handover from planning to production.

With the integration of the Rhythm Wheel planning heuristic into our IT systems, we laid the cornerstone for a companywide harmonization of production planning. The results of the Rhythm Wheel at Wedel were so convincing that it was used as blueprint for a global roll-out within AZ.

12.1.6 The LEAN Production Initiative at PCI: A Company of BASF

Profile of Case Contributor

Thomas Semlinger, head of Production Construction Chemicals Europe (E-EBE), PCI Europe, joined PCI in 1998. Prior to assuming his current position as Head of Production at PCI's Augsburg site, he served as project engineer, project manager for projecting, and head of operating technology. Prior to joining PCI, Mr Semlinger studied at the Augsburg University of Applied Sciences in machine engineering and worked for more than 5 years for a company in the plastic processing industry, developing injection molding and specialist equipment.

12.1.6.1 Company Profile and Case Summary

- PCI Augsburg GmbH has been a renowned manufacturer of construction chemicals for 60 years. For decades PCI has been a market leader in tile fixing systems and other flooring products in the German-speaking market.
- PCI is a subsidiary of the worldwide-leading chemical company BASF with more than 700 employees and an annual turnover of €202 million in 2010.
- The entire German-speaking area in Europe is supplied by three factory locations in Germany. Under the PCI brand, products are distributed on the international market. The roots of PCI go back to the Augsburg site where PCI's headquarters are located.

12.1.6.2 Executive Summary

Meeting a 24-hour delivery promise and tightening inventory targets with increasing product complexity are key challenges for PCI's SCM and production planning. Applying a Rhythm Wheel concept to one of our production units clearly yielded superior performance. Owing to higher responsiveness in production, we reduced stock-outs during peak season by 80% while simultaneously achieving tangible inventory reductions. Key elements of the concept are optimization of the production sequence and reduced campaign size. Replenishment is scheduled by pull signals based on predefined inventory levels. We will be rolling out this improved approach to production changeovers based on lean manufacturing at other production units as well.

12.1.6.3 Company's General Situation

PCI stands for innovation and quality. Offering a wide range of products tailored to specific customer needs and promising to deliver them within 24 hours after ordering are key competitive advantages for us. To maintain this competitive customer service level in an increasingly dynamic market and further improve profitability, we identified flexibility and responsiveness in production as key areas for improvement.

In light of these challenges, we started a lean supply chain initiative to establish a more flexible production model and reduce working capital. The Rhythm Wheel was identified as a promising planning approach with which to respond to our business challenges. We decided to implement the High-Mix Rhythm Wheel approach in a pilot production line in our plant at Augsburg. After the successful pilot phase, we are now planning to transfer this concept to other production units.

12.1.6.4 Lean Challenge

PCI promises their customers outstanding service with short lead times. Supply chain management is in charge of ensuring this competitive service level. With increasing focus on net working capital, this promise is generating a major challenge for inventory and production management. The task is becoming even more complicated due to the growing complexity of the product portfolio and the corresponding raw material mix. Finally, the planning environment has to manage significant demand seasonality over the year.

Our production in scope focuses on the mixing and filling of powder products. The main challenge of our former production planning approach was low responsiveness to customer demand and long production campaigns. Production schedules were based mainly on historical demand figures and were fixed for several weeks. Furthermore, the significant length of production campaigns caused long production cycles for individual products, which additionally reduced opportunities to respond to short-term changes in demand and required higher inventory levels. As a consequence, changes in customer demand required significant manual rescheduling efforts and caused long-term turbulence in the supply chain.

12.1.6.5 Approach

We wanted our solution to optimize production scheduling, finished goods inventory, and service levels. This was achieved by realizing an optimal production sequence and reduced campaign sizes through the operation of a demand-driven Rhythm Wheel approach for production scheduling. We saw that a Rhythm Wheel approach suitable for high-mix environments would be capable of adding the necessary flexibility

FIGURE 12.27
Significant reduction of internal complexity through Rhythm Wheels.

to our operations and improving our current planning approaches (see Figure 12.27). The primary advantage of the chosen Rhythm Wheel solution for us is that products do not all have to be scheduled in every production cycle, thus taking into account product-specific differences in demand profiles.

Two steps were essential as we introduced our new approach to production planning: Determining an optimal production sequence and reducing campaign sizes. First, we determined the optimal production sequence for all products. This helped us to reduce changeover costs and time and lay the groundwork for our efficient Rhythm Wheel design. Furthermore, defining an optimal changeover sequence was also important to improving production yield, ensuring stable quality, and increasing available production capacity.

Our next step was to reduce the overall production cycle to increase production flexibility and minimize rescheduling activities and the risk of stock-outs. For all of our products, we optimized campaign sizes and improved production frequencies to respond more flexibly to changing market demand. The campaign sizes needed to be optimized along several dimensions, taking into consideration production sequences, hardware constraints, and delivery lot sizes. Following the chosen Rhythm Wheel concept, we defined constant production patterns for our high-volume products while our slow movers are now produced at stable but lower production frequencies.

Furthermore, our production process now follows a pull-driven planning approach whereby customer demand is the ultimate trigger for our

Traditional planning (MRP II)	Fixed sequence – variable volume
▸ Less focus on inventory and changeover costs	▸ Fixed sequence to improve product changeovers
▸ Production sequence not fixed	▸ Reduction of cycle times
▸ Requires longer planning horizons	▸ Higher flexibility through variable volumes

Rhythm Wheels are used to reduce internal complexity

FIGURE 12.28
Selected measures to improve production planning and scheduling.

production decisions along the supply chain. The weekly production program is planned based on the deviation of actual stock levels in comparison to clearly defined trigger levels. If the trigger levels are not reached, production of individual products is skipped. This helps us to avoid building up inventories in situations without actual customer demand and allows us to use the freed-up capacity for other products.

Encouraged by the very promising early results of our lean initiative, we are now working on further opportunities to improve our supply chain performance and to fully seize the benefits of our Rhythm Wheel-based production planning (see Figure 12.28). We believe that our new production planning approach provides the right foundation for approaching these additional measures and integrating them effectively in our production concept.

To fully exploit the opportunities made available by our LEAN SCM Planning approach, we identified reduced changeover costs as an area for immediate action. We believe that applying lean manufacturing tools at the shop floor level—such as increased teamwork, continuous improvement programs, and SMED—will further strengthen the advantages of our new planning concept. We are also working on pull-driven logistic processes. We are convinced that harmonizing our logistic processes and aligning them with the modified replenishment requirements of the Rhythm Wheel approach will provide additional success factors for our demand-driven production logic. The integration of production planning and workforce planning (e.g., shift models) will enable us to sustainably achieve the greater flexibility required for shorter production campaigns. Finally, we believe that our efforts to enhance our sales and operations planning processes will also help us derive even better parameters for production planning.

12.1.6.6 Results/Benefits

By implementing the Rhythm Wheel approach, we were able to significantly improve the performance of our pilot production line. By introducing a fixed sequence for production and reducing campaign sizes, we were able to increase our flexibility in responding to short-term demand changes as well as to improve planning stability significantly. In particular, we reduced the need for manual short-term rescheduling of production to almost zero.

Overall, reduced production complexity and improved availability of production capacity and utilization will additionally lead to increased market sales. Our customers will benefit especially from the new planning concept as we will be able to reduce the percentage of stock-outs during peak seasons by 80%. The pull principle has also enabled us to reduce working capital through a clear reduction in inventories compared with our historical inventory figures.

Following this highly successful pilot phase, we are now planning to transfer this approach to all other production units at PCI.

12.2 WHY LEAN SCM: SUMMARY OF KEY BENEFITS

As shown by the industry cases in this chapter, leading companies from pharmaceutical and chemical industries are increasingly adopting LEAN principles and concepts in their supply chains. All these companies found that a more efficient management of variability results in substantial performance improvement. By reducing variability in the supply chains, both customer satisfaction and cost efficiency were increased to a large extent (see Figure 12.29).

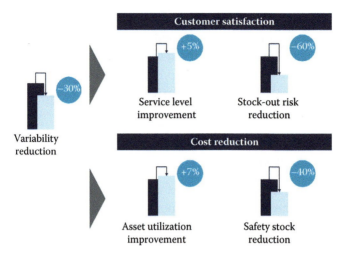

FIGURE 12.29

LEAN SCM leads to a reduction in supply chain variability and has therefore a positive impact on both working capital and customer service.

To seize the same benefits in your organization and master the challenges of today's VUCA world, you can use the LEAN SCM concepts, tools, and principles introduced in this book as a guideline for starting your LEAN journey successfully. You can be ensured that LEAN Supply Chain Planning will substantially improve the competitiveness of your supply chain and operations.

Bibliography

Amey, J. and M. Jarosch. 2012. Supply chain management organisation. Study findings of the University of Warwick and Camelot Management Consultants AG.

Bowersox, D. J.; Closs, D. J. and M. B. Cooper. 2002. *Supply Chain Logistics Management.* Boston: McGraw-Hill Irwin.

Carter, W. L. 2008. Think lean to escape the planning loop trap. http://www.techmankanata.com//a67/Planning-Loop-Trap.pdf.

Cecere, L. 2013. A practioner's guide to demand planning. *Supply Chain Management Review* 17(2):40–46.

Cohen, S. and J. Roussel. 2005. *Strategic Supply Chain Management: The Five Disciplines for Top Performance.* New York: McGraw-Hill.

Cook, A. D. 2012. *Forecasting for the Pharmaceutical Industry: Models for New Product and In-Market Forecasting and How to Use Them.* Farnham: Gower Publishing Limited.

Floyd, R. C. 2010. *Liquid Lean: Developing Lean Culture in the Process Industries.* New York: Productivity Press. 1st Edition.

Glenday, I. F. 2006. *Breaking Through to Flow.* 1st Edition. Cambridge: Lean Enterprise Academy Ltd.

Heinmann, M. and M. Klein. 2011. Strategic multi-stage inventory allocation in the process industry. Focus Topic Paper, Camelot Management Consultants AG.

Hines, D.; Found, P.; Griffiths, G. and R. Harisson. 2011. *Staying Lean: Thriving, Not Just Surviving.* 2nd Edition. Boca Raton: CRC Press.

Hopp, J. H. and M. L. Spearman. 2004. To pull or not to pull: What is the question? *Manufacturing & Service Operations Management* 6(2):133–148.

Hopp, W. J. and M. L. Spearman. 2009. *Factory Physics.* New York: McGraw-Hill.

King, P. L. 2009. *Lean for the Process Industries: Dealing with Complexity.* 1st Edition. New York: Productivity Press.

King, P. L and J. S. King. 2013. *The Product Wheel Handbook: Creating Balanced Flow in High-Mix Process Operations.* New York: Productivity Press.

Knein, E. L. and P. Streuber. 2010. Das Rhythm Wheel—Simulationsbasierte Entwicklung eines innovativen Konzepts zur Produktionsplanung und-steuerung in der pharmazeutischen Industrie. [The Rhythm Wheel—simulation-based development of an innovative concept for production planning and control in the pharmaceutical industry]. Diploma thesis at the University of Cologne.

Lee, H. L. and J. Amaral. 2002. Continuous and sustainable improvement through supply chain performance management. Stanford Global Supply Chain Management Forum, Stanford University, Palo Alto, California.

Machado, V. C. and S. Duarte. 2010. Tradeoffs among paradigms in supply chain management. *Proceedings of the 2010 International Conference on Industrial Engineering and Operations Management*, Dhaka, Bangladesh.

Myerson, P. 2012. *Lean Supply Chain: Logistics Management.* New York: McGraw-Hill.

Ohno, T. and N. Bodek. 1988. *Toyota Production System: Beyond Large-Scale Production.* New York: Productivity Press.

Packowski, J.; Knein, E. and P. Streuber. 2010. Ein innovativer Lean Ansatz zur Produktionsplanung und -steuerung—Das Multi-Echelon Rhythm Wheel Konzept am Beispiel einer pharmazeutischen Supply Chain. In *Dimensionen der Logistik Funktionen, Institutionen und Handlungsebenen*, eds. R. Schönberger, and R. Elbert, 101–115. Wiesbaden: Gabler Verlag.

Pool, A.; Wijngaard, J. and D.-J. Van der Zee. 2011. Lean planning in the semi-process industry: A case study. *International Journal of Production Economics* 131(1):194–203.

Riezebos, J.; Klingenberg, W. and C. Hicks. 2009. Lean production and information technology: Connection or contradiction? *Computers in Industry* 60:237–247.

Robinson, M. 2008. ERP/DRP and lean manufacturing are not compatible. White paper, Institute of Operations Management.

Romano, E.; Santillo, L. C. and P. Zoppoli. 2008. The change from push to pull production: Effects analysis and simulation. Selected Papers from the WSEAS Conferences in Spain.

Sabri, E. H. and S. N. Shaikh. 2010. *Lean and Agile Value Chain Management: A Guide to the Next Level of Improvement*. Plantation: J. Ross Publishing.

Schey, V. and R. Roesgen. 2012. Mastering complexity. Focus Topic Paper, Camelot Management Consultants AG.

Schönberger, R. and R. Elbert. 2010. *Dimensionen der Logistik: Funktionen, Institutionen und Handlungsebenen*. Wiesbaden: Gabler Verlag.

Shingo, S. 1996. *Quick Changeover for Operators: The SMED System*. New York: Productivity Press.

Simchi-Levi, D. 2010. *Operation Rules: Delivering Customer Value through Flexible Operations*. 1st Edition. Massachusetts: MIT Press.

Simchi-Levi, D.; Kaminsky, P. and E. Simchi-Levi. 2003. *Managing the Supply Chain: The Definitive Guide for the Business Professional*. 1st Edition. New York: McGraw-Hill.

Womack, J. P. and D. T. Jones. 2002. *Seeing the Whole: Mapping the Extended Value Stream*. 1st Edition. Cambridge: Lean Enterprise Institute.

Womack, J. P. and D. T. Jones. 2003. *Lean Thinking: Banish Waste and Create Wealth in Your Corporation*. 2nd Edition. Revised and Updated. New York: Productivity Press.

Womack, J. P. and D. T. Jones. 2005. *Lean Solutions: How Companies and Customers Can Create Value and Wealth Together*. 1st Edition. New York: Productivity Press.

Index